Biophysics

FOR

DUMMIES

A Wiley Brand

by Ken Vos, PhD

Professor at University of Lethbridge,
Lethbridge, Alberta, Canada

FOR

DUMMIES

A Wiley Brand

Biophysics For Dummies®

Published by: **John Wiley & Sons, Inc.,** 111 River Street, Hoboken, NJ 07030-5774, www.wiley.com

Copyright © 2013 by John Wiley & Sons, Inc., Hoboken, New Jersey

Published simultaneously in Canada

For general information on our other products and services, please contact our Customer Care Department within the U.S. at 877-762-2974, outside the U.S. at 317-572-3993, or fax 317-572-4002. For technical support, please visit www.wiley.com/techsupport.

Wiley publishes in a variety of print and electronic formats and by print-on-demand. Some material included with standard print versions of this book may not be included in e-books or in print-on-demand. If this book refers to media such as a CD or DVD that is not included in the version you purchased, you may download this material at http://booksupport.wiley.com. For more information about Wiley products, visit www.wiley.com.

ISBN: 978-1-118-51350-7 (pbk); ISBN: 978-1-118-51352-1 (ebk); ISBN: 978-1-118-51353-8 (ebk); ISBN: 978-1-118-51354-5 (ebk)

10 9 8 7 6 5 4 3 2 1

Contents at a Glance

Table of Contents

Introduction

· ·

*W*elcome to *Biophysics For Dummies*. Biophysics is a fascinating field of science that combines the study of the laws of physics with the study of systems involving living organisms (biology). The combination of these two fields makes biophysics interdisciplinary, which means biophysicists work side by side with people from many different backgrounds. Biophysics is a very diverse and interesting field; even if you spend your entire life studying biophysics, you can still discover new and interesting pieces of information.

About This Book

Biophysics For Dummies lays down the foundations for the fields of biophysics, including neurophysics, medical physics, health physics, and related fields that overlap with biophysics, presented in an easy-to-access manner. This reference book presents biophysics in plain English, so you can easily find what you're looking for. When you're reading, you don't have to begin at the beginning. You can go directly to the chapter or section that interests you and start reading. Of course, I prefer that you read it from cover to cover, but then again, I am a bit biased. If you're strapped for time and only want to read what you need to know, even when you're reading the chapter or section of interest to you, you can skip the sidebars and the paragraphs marked with the Technical Stuff icon without losing any of the essential info.

This book is unique in that the majority of the material is at the introductory level, but the material presented is at an advanced enough level that you can use the book as a stepping stone in your biophysics studies. This book also lays out in a clear step-by-step procedure how to apply concepts in physics to problems in biophysics and the life sciences. The book introduces topics in the five fundamental areas of physics: mechanics, fluids, thermodynamics, electromagnetism, and nuclear physics.

You may notice while reading the book that I have done a few things that I hope make your reading and search of information easier:

- I avoid using URLs. These URLs can change over time, so I have placed only the more important ones that probably won't change on the online Cheat Sheet. You can find all the important links in a single place for easy access with a single click at `www.dummies.com/cheatsheet/biophysics`.

- I italicize all the variables used in mathematical formulas, so you can easily identify them. I also italicize words when I define them. Many words in biophysics have special meanings, and understanding the terminology is an important step toward comprehending the subject.

- I use certain symbols differently than do some other biophysics books. The symbols are as follows:

  - N for the torque instead of τ (*tau*), which is used in many introductory books. (Many engineering books use M.) Some more advanced physics books use N for torque and in addition, τ looks very similar to t (time), T (period), and T (half-life). I would have too many physical quantities using similar symbols.

  - $P^{(a)}$ for absolute pressure, $P^{(g)}$ for gauge pressure, and P for power. I have too many sections where I use power and pressure at the same time, so I distinguish them this way.

  - E represents energy and F represents force. I distinguish between the different energies and forces by using subscripts. Some books use T or K for the kinetic energy and some use U or V for the potential energy. I use E_K and E_P instead for kinetic energy and potential energy.

Foolish Assumptions

As I write this book, I assume you, my dear reader, fall into at least one of the following groups:

- You're in college and taking an introductory biophysics course.

- You're interested in studying biophysics or some related field where knowledge of biophysics is useful.

- You're involved with the sciences and want to expand your knowledge base in biophysics.

- You have already taken algebra, geometry, and a science course in either biology, chemistry, or physics.

Icons Used in This Book

I use a few icons as markers in the margins. These markers are useful for helping you locate material or skip over material, depending on what you're searching for. I use them to indicate what I think is important for you to notice. These icons can help you navigate through the material.

When I present helpful information that can make your life a bit easier when studying biophysics, I use this icon.

This icon highlights important pieces of information that I suggest you store away because you'll probably use them on a regular basis.

This icon highlights common mistakes or errors that I see time after time from people who are taking a biophysics course.

This icon indicates in-depth examples. Try solving the problem and continue reading to see how to solve the problem.

This icon requires nonessential information, usually at least at a calculus background level. If you have a math phobia, then you may want to avoid reading these paragraphs. If you enjoy biophysics and mathematics, then I encourage you to read these paragraphs.

Beyond This Book

In addition to the material in *Biophysics For Dummies,* I also provide a free Cheat Sheet online at www.dummies.com/cheatsheet/biophysics. The Cheat Sheet adds a few extra tidbits that you will find interesting, such as solving biomechanical problems. You can also find other interesting bits of additional information online at www.dummies.com/extras/biophysics.

After reading the Cheat Sheet and online information, you may decide to pursue biophysics more in-depth, so I include URLs to the biophysical society, the association of medical physicists, and the health physics society. These links are a great starting point in search of answers to your biophysical questions.

Where to Go from Here

Science is about being curious and exploring, which is what attracted me to biophysics. As you read this book, feel free to jump around and start with the chapters and sections that interest you the most. If you need a particular section for your science course, such as kinematics or biomechanics, you can go straight there. You can also look in the index or the table of contents to find a topic that interests you. No matter what you decide to read, enjoy your adventure into the world of biophysics.

Part I
Getting Started with Biophysics

Go to www.dummies.com/cheatsheet/biophysics to discover some more informative Dummies content online about biophysics.

In this part . . .

✔ Get a thorough overview of what biophysics is, including its diverse fields, such as biomechanics, fluids, waves and sound, the electromagnetic force, and medical physics, so you can fully appreciate how it affects your daily life.

✔ Discover where you can find biophysics. You may be surprised to know who biophysicists are and where biophysics is used.

✔ Tackle mathematics, most of which should be a review for you if you've already taken a chemistry, physics, or calculus class. Biophysics does use mathematics, so having a decent grasp of the basic formulas and equations is important when you study biophysics.

✔ Comprehend some of the basics of biophysics, such as notation and terminology, that aren't used in everyday life and clear up a few common myths.

✔ Make the distinction between experimental and theoretical biophysics. Biophysics isn't mathematics, but mathematics is a tool used by both experimental and theoretical biophysicists.

Chapter 1

Welcoming You to the World of Biophysics

*B*iophysics is the study of biology and all sciences connected to the biological sciences using the principles and laws of physics. It's the ultimate interdisciplinary science combining biology, chemistry, and physics. If you love science, then biophysics is for you. The field touches on all aspects of all the natural sciences.

This chapter gives you the bird's-eye view of biophysics and what you'll find in this book. In this chapter, I explain the general features of biomechanics, the motion of fluids, waves and sounds, and electromagnetic force as well as radiation and radioactivity.

Getting the Lowdown on What Biophysics Really Is

No matter if you're stuck taking a biophysics course to meet your science course requirements or you're taking your first of many biophysics courses, you need to make sure you understand what you're studying. Just break down the word *biophysics. Bio* means life and *physics* means nature, so biophysics is the study of living matter, its motion, and its interaction with the natural universe. Chapter 2 expands on the explanation of what biophysics is, and Chapter 3 covers some of the basic terminology used in biophysics.

The following clarifies what biophysics really means:

- ✔ Biophysics uses techniques and methods from physics, mathematics, biology, and chemistry to study living organisms.

- ✔ Biophysicists design experiments or do computational calculations in order to understand biological processes. A few examples of these biological processes are

  - Photosynthesis

  - The on-off switching of genes

  - Memory and brain processes

  - Muscle control

- ✔ Biophysicists study how the senses work.

- ✔ Biophysicists try to understand why things behave the way they do in sports and improve the performance of athletes.

- ✔ Biophysicists study how molecules enter cells and how they interact.

- ✔ Biophysicists study how cells move, divide, and respond to the environment.

As you can see, biophysics is all of this and everything that deals with living organisms. Biophysics plays an essential role in medicine, sports, engineering, physics, biology, biochemistry, and environmental science to mention a few areas. Whenever you're considering something that involves a living organism and its interaction with its surroundings, you're using biophysics.

Grasping the Mechanics of Biomechanics

Biomechanics is an important part of biophysics. *Bio* means life, and *mechanics is* the study of the interaction of a physical object with its surroundings. Therefore, *biomechanics* is the study of a living object's interaction with its surroundings, which also includes the study of how living organisms move and the causes of this motion.

These sections explain a bit more about what biomechanics is. I discuss rules because biophysicists love rules, explain what happens when forces try to change an object's motion, and look at the motion of an object.

Surveying the rules

Biomechanics has many rules because things don't happen randomly or by chance. Things happen because of actions, and these rules tell you what the consequences of an action are. These rules are usually called *laws*, which can't be broken.

Some important laws in biomechanics are

- **Newton's first law of motion, the law of inertia:** This law tells you objects are lazy, and you have to force them to change their motion.

- **Newton's second law of motion, the law of acceleration:** If you force an object to change its motion, then this law tells you how the motion will change.

- **Newton's third law of motion, the law of action and reaction:** This law states that if one object applies a force to a second object, then the second object will apply the opposite force back on the first object.

- **(Law of) conservation of momentum:** This law tells you that the total momentum of an isolated system doesn't change even if the objects within the system are bouncing off each other.

- **(Law of) conservation of energy:** The law tells you that you can't create or destroy energy; you can only change it from one form to another.

- **The work-energy theorem:** If you want to change an object's kinetic energy, then you must do work on the object.

Chapter 4 introduces these rules of physics that are applicable to biomechanics. This chapter also explains what a force is and what energy is as well as the connection between forces and energy.

Focusing on statics

Statics, the situation when a biological system isn't moving, even if under the influence of forces, is another important part of biophysics. The physics of biological systems that aren't moving can be very complex. Chapter 5 lays out the procedure for solving problems in translational equilibrium, then problems in rotational equilibrium. Finally, it combines the two, which is called *static equilibrium.*

Meanwhile, Chapter 6 includes the following:

- ✔ Calculating the center of mass of a biological system
- ✔ Determining the effective weight of a biological organism
- ✔ Viewing biological organisms as machines and levers
- ✔ Examining different ways that biological organisms can be deformed
- ✔ Eyeing different properties of the organism when it's enlarged or shrunk.

Going the dynamic route

Biomechanics looks at the motion of biological organisms and the forces that act on them. Chapter 7 identifies what causes the forces that generate the motion. Two main types of motion are as follows:

- ✔ **Linear motion:** This type includes situations where the net force is one-dimensional such as in skydiving. You can study this type of motion by using forces or looking at the energy of the system.
- ✔ **Circular motion:** This type includes torques and rotational energy. It's useful in situations, such as in diving competitions or certain gymnastics events where the athlete is spinning and twisting.

Moving around with kinematics

Kinematics is the study of how biological organisms move without worrying about why. All you need to know is the acceleration, velocity, and position to describe an object's motion or a system of objects' motions. Chapter 7 is the "why" objects move, and Chapter 8 is the "how" objects move. Chapter 8 starts with describing the linear motion of objects and then switches to circular motion.

Eyeing the Physics of Fluids

Fluids are a collection of objects (usually molecules) that stick together as a group, but the objects move about randomly relative to each other, unlike solids where molecules will be fixed and not travel from one side to another side. Fluids play a key role in biophysics, such as blood transporting oxygen to the cells or the motion of sap in a plant.

These sections examine how fluids influence the world around them. I begin with the rules and forces in fluids, discuss the flow of different types of fluids, and finish with discussing how material enters and leaves our bodies.

Understanding fluid's mechanics and cohesive forces

Fluids obey rules and this section goes over some of the foundational rules. Some of these ideals are

- **Pascal's principle:** A (incompressible) fluid at rest will transmit a change in pressure to all points in the fluid equally. For example, fill a balloon up with water and then squeeze the top of the balloon. The water in the balloon will increase in pressure everywhere within the balloon.

- **Archimedes's principle:** Any object wholly or partially immersed in a fluid (or gas) has a force exerted on it by the fluid (or gas) called the *buoyant force,* which is equal to the weight of the fluid displaced by the object.

- **Conservation of mass:** The total mass of the fluid doesn't change unless you add or remove fluid from the system.

- **Bernoulli's equation:** The equation shows how the speed of the fluid will change from forces acting on the fluid. For example, if you pour a fluid out of your glass, it will pick up speed as it flows toward the floor.

- **Cohesive force:** It's the attractive force between molecules. This force keeps a water drop together and gives rise to surface tension. The force is called *adhesion* when it's between molecules that are different, say the fluid and the container.

Chapter 9 expands on these ideas and concepts related to fluids.

Tackling fluid dynamics

Fluid dynamics is the study of moving fluids. The properties of fluids are very important in many fields of biophysics. For example, you may be interested in how blood flows through restricted channels, how to throw a ball to maximize its curve, or how to optimize an irrigation system in environmental science.

Viscosity is a measure of a fluids resistance to change. For example, maple syrup is more viscous than water. Fluids can be split into two main groups:

- **Nonviscous fluids:** The first case corresponds to situations where the viscosity can be ignored

- **Viscous fluids:** In these fluids, the viscosity plays an important role and can't be ignored.

In the case of viscous fluids, you need to consider what type of fluid you have and the type of flow:

- **Newtonian fluids:** In a *Newtonian fluid,* the ratio of the stress to the strain is a constant, which is the viscosity.
- **Non-Newtonian fluids:** If a fluid is not Newtonian, then it's non-Newtonian. Water is Newtonian, whereas ketchup is non-Newtonian.
- **Laminar flow:** A viscous fluid flowing at low speeds will form layers with different speeds and little mixing between the layers. The layer closest to a boundary will try to match the boundary's speed.
- **Turbulent flow:** A viscous fluid flowing in an unpredictable manner with rapidly changing properties. The smoke rising from a campfire is a turbulent flow (except the smoke closest to the flame, which is laminar flow).

Chapter 10 looks more closely at the dynamics of fluids.

Moving through membranes and porous materials

Porous materials allow fluids to flow through them, such as water flowing through sand. *Membranes* are boundaries within biological organisms that separate two fluids. Membranes are usually very thin and play different roles in a biological system. For example, the eardrum *(tympanic membrane)* has air on both sides and vibrates when sound waves hit it, whereas the membrane within the lungs is semi-permeable, allowing oxygen molecules to go from the air into the blood and carbon dioxide to move from the blood into the air. These materials play a very important role in biological organisms and are an important area of biophysics.

You have probably noticed that perfume lingers in the air for a long time after it has been sprayed into the air. It takes the perfume a long time to dissipate unless you turn on a fan. This concept, called *diffusion,* is important in understanding how materials within a fluid are transported and how the material moves through a membrane. Chapter 11 starts with diffusion. Chapter 11 then discusses more about membranes and porous materials, including human *metabolism,* the conversion of food into energy, and the elimination of molecules from the human body.

Comprehending Waves and Sound

Waves are a means by which energy is transferred from one region of space to another region. As the wave propagates through space, it's usually associated with the temporary disruption of the material in that region. (You can think of the crest of a water wave as it moves across the surface of the water.) *Sound* is a pressure wave that causes the molecules in the gas, liquid, or solid to temporarily vibrate. They're important to the study of biophysics because biological systems need energy to do work. Music and communication between animals are very important.

The following sections break it up a little more. These sections mention how the wave disrupts the material as the wave propagates through the material, explains how sound is made, followed by how the ear hears those sounds, and discusses some applications of sound waves.

Disturbing the material

A *wave* propagating through a material will usually cause the material to be disturbed from its rest position. After the wave has passed, then everything usually returns to normal. In some cases, the energy in the wave will cause irreparable damage to the material, and it can't return back to its original state. Think of a sonic boom shattering a window.

Related to this is *harmonic motion,* where the material bounces back and forth or up and down. Water waves at the beach cause the water to go up and down in a repeating pattern. In many situations, the harmonic motion obeys *Hooke's law,* which states that the farther the material is distorted from its rest position, the stronger the force to restore the material back to its normal position. Many applications of waves and harmonic motion exist in biophysics. For example, you can use harmonic motion (Hooke's law) to find the weight of a virus.

The different types of waves include the following:

- ✔ **Longitudinal waves:** These types of waves have the material vibrate back and forth in the direction parallel to the wave's motion.

- ✔ **Transverse waves:** These types of waves have the material vibrate back and forth in a perpendicular direction to the wave's motion.

- ✔ **Electromagnetic radiation:** These are transverse waves, which are unique in that they do not need a medium to propagate through.

- ✔ **Sound waves:** These are longitudinal pressure waves.

- ✔ **Water waves:** Water waves can be of different types, but the ones that people are the most familiar with are the surface water waves that propagate toward the shore.

Chapter 12 takes a closer look at these types of waves and how waves interact with other waves of the same kind and how the waves interact with their surroundings.

Knowing how animals and instruments make sound waves

Sound is pressure waves that are created by the vibration of an object, such as the vocal folds in a human or the skin on a drum. The resonance of air within a cavity, such as a flute, can also create sound. A few properties of sounds include the following:

✔ Sound needs the vibration of matter for the sound to propagate. Unfortunately, science fiction movies show sound waves propagating through space, which is wrong.

✔ Sound waves are longitudinal pressure waves in gasses, but they can be longitudinal and transverse in a solid.

✔ Sound travels at approximately 1,130 feet per second (344 meters per second) in air near sea level. The speed of sound depends on many factors including the temperature and density of the air.

✔ The speed of sound is equal to the *wavelength* (the distance from one crest to the next) times the *frequency* (the number of crests that pass by per second).

✔ **Interference:** Sound waves interacting with other sound waves interact either *constructively* (with enhanced amplitude) or *destructively* (with decreased amplitude).

✔ **Resonance:** Sound waves trapped between boundaries interact with their echo. At specific frequencies they will have constructive interference, which is called *resonance*. For example, blowing across the opening of an empty bottle makes a loud noise.

Chapter 13 discusses these properties in greater depth and looks at similarities and differences between a guitar and the human voice, as well as other instruments such as the clarinet and flute.

Hearing sound waves

Hearing is a very complex phenomenon and an important subject in biophysics. In addition, comprehending how hearing works can give an understanding of how biological systems work and how information is sent to the brain and processed. When sound waves hit the human body, the majority of the sound bounces off the body and travels elsewhere. You wouldn't be able to hear the sound except for the fact you have ears.

The ear is a clever device that takes the sound wave in air and converts it to an electrical signal that the brain can understand. The outer ear channels the sound wave to the eardrum, which vibrates with the frequency of the sound. The motion of the eardrum causes the *ossicles* (the three small bones in the middle ear) to vibrate, which in turn cause the *oval window* (which is a membrane between the middle ear and the inner ear) to vibrate. The vibration of the oval window causes the fluid in the inner ear to vibrate. The motion of the fluid is detected by hair cells, which are the ends of the nerves that transmit the signal to the brain.

Check out Chapter 14 for more information about how humans hear and how sound waves traveling through the air are changed into electrical signals that are sent to the brain and why sound waves have a limited range.

Applying sound waves

Waves are a method of transmitting energy, and so sound waves allow animals to interact with their surroundings. Three different applications of that transmission of energy are

- ✔ **Doppler effect:** When the source of a sound wave and the listener are moving relative to each other, then the frequency according to the listener is different than what the source emitted the sound at.
- ✔ **Echolocation:** Some nocturnal animals use sound to find their way around in the dark by emitting a sound and listening to the echo dubbed *echolocation*.
- ✔ **Ultrasound imaging:** Imaging that uses pressure waves with very high frequencies. The waves' speed varies depending on the density of the material. The changes in speeds can be used to detect the boundaries between different materials and produce an image. Ultrasound imaging is one of the safest imaging methods used in medicine today.

Refer to Chapter 15 for more about these three and some of their applications and limitations.

Forcing Biophysics onto the World

Force is a method to quantify the interaction between objects. If there were no forces, then objects in the universe wouldn't interact, hence meaning no life. Through forces you know that the universe exists around you.

The following sections discuss the electromagnetic force, introduce radio-activity and radiation, which occurs within the nucleus of an atom when it's unstable, look at applications of radiation, and examine medical physics as an application of biophysics in medicine.

Binding with the electromagnetic force

The *electromagnetic force* is the force between charged particles. The proton and electron have charge, which is a fundamental physical property of these particles. Charges produce electric fields, and moving charges create magnetic fields. If the charged particles are accelerating, they create electromagnetic radiation.

Another way electromagnetic radiation is created is by the annihilation of a particle with its antimatter counterpart. The electromagnetic force is the most important force in biophysics, chemistry, and society. The electromagnetic force is what keeps molecules together, causes electrical pulses to travel down the nerves, allows you to see, produces friction between your feet and the ground, and a lot more.

A few important laws related to the electromagnetic force are

- ✔ **Gauss's law:** *Gauss's law* states that charges produce electric fields. The electric fields start at positive charges and end at negative charges. A version of this law does exist for magnetic fields. It states that no magnetic charges exist and all magnetic phenomena are a consequence of moving electric charge.

- ✔ **Maxwell-Ampere law:** The *Maxwell-Ampere law* states that moving charges create magnetic fields and electric fields that are changing in time create magnetic fields.

- ✔ **Faraday's law:** This law states that magnetic fields that change in time produce an electric field. This law is the foundation of the electric generator, the electric guitar, and magnetic resonating imagers (MRIs) to name just three of a multitude of applications.

- ✔ **Lorentz force:** The electromagnetic force is the interaction between charged particles. These electric and magnetic fields produced by electric charges propagate through space and come into contact with other charged particles. The Lorentz force explains how these fields exert a force on the other charged particles.

Chapter 16 discusses these laws in greater depth and different electrical power sources, electrical circuits, energy, and the transformation of energy from one type to another.

Getting a hold on radiation and how it battles cancer

Radioactivity is when an atom changes into a new atom and emits radiation. Some of the different kinds of decay are

- **Alpha decay:** The atom ejects an alpha particle (helium nucleus), losing two protons and two neutrons.

- **Beta decay:** In *beta decay,* a proton changes into a neutron (positive beta decay) or a neutron changes into a proton (negative beta decay).

- **Electron capture:** An electron is captured by the nucleus, changing a proton into a neutron.

- **Fission decay:** The atom splits into two new atoms.

- **Proton decay:** The atom ejects a proton, becoming a new element with the same number of neutrons and one less proton.

- **Neutron decay:** The atom ejects a neutron, becoming a new isotope with the same number of protons and one less neutron.

Radiation is a means by which energy is emitted through space. Radiation comes in two forms: electromagnetic radiation and particles. A few of the different forms of radiation are

- **Non-ionizing electromagnetic radiation:** Most types of electromagnetic radiation fall in this category. It includes radio waves, microwaves, infrared radiation, light, and low-energy ultraviolet radiation. The low-energy ultraviolet radiations (UVA, UVB, and UVC) are more like ionizing radiation than non-ionizing radiation.

- **Ionizing electromagnetic radiation:** This electromagnetic radiation has sufficient energy to eject an electron from an atom or molecule. It includes high-energy ultraviolet radiation, X-rays, and gamma radiation.

- **Alpha particle:** This is the nucleus of a helium atom.

- **Beta-negative particle:** This is an electron, but it was ejected from the nucleus.

- **Beta-positive particle:** This is a positron ejected from the nucleus.

- **Cosmic rays:** These are actually charged particles entering the atmosphere from space. The majority of the particles are hydrogen nuclei, helium nuclei, and beta-negative particles.

- **Neutron radiation:** Free neutrons are unstable with a half-life of ten minutes, and when atoms absorb the neutrons, it makes the atom unstable.

Chapter 17 discusses radioactivity and radiation in more detail. It also highlights some of the benefits and applications of radiation.

Working with radiation

Radiation is bad because it causes damage to the cells in the body. At high radiation doses, the cells die quickly and the effects are immediate. A lot of damage happens to the cells at moderate radiation doses. The body can't keep up in repairing the cells, and some are repaired incorrectly. In time, the mutant cells become cancerous.

Radiation is everywhere, which is called the *natural background radiation.* The world's average natural background radiation is 2.4 millisieverts per year, whereas in the United States it is 3.1 millisieverts per year. The natural background radiation does change a lot from one location to the next. In addition, radiation exists from medical visits, such as X-rays at the dentist. The average amount of radiation from medical sources for most counties is very low, but in the United States, it's 3.0 millisieverts per year, so the average person in the United States receives 6.1 millisieverts per year of radiation.

Chapter 18 examines the biological effects of radiation in more detail. It also highlights a few misconceptions about radiation and one cancer that is mostly preventable.

Using biophysics in medicine

A large field related to biophysics is medical physics and health physics. One part of medical physics is using radiation in medicine, which has more benefits than drawbacks. Some of the ways that medicine uses radiation include the following:

- **Nuclear medicine:** In nuclear medicine, radionuclides are produced for placement within the body or to form part of a radiopharmaceutical drug. These radioactive compounds are then used for both diagnosis and treatment.

- **X-rays and computed tomography (CT) scans:** Dentists and doctors use X-rays to image the body. CT scans use multiple high-energy doses of X-rays to obtain detail images of soft tissue.

- **Positron emission tomography (PET) scans:** A radionuclide is placed inside the body, which decays by emitting positrons. The positrons annihilate with the electrons inside the body to produce gamma rays, which leave the body and are detected. The gamma rays allow for a three-dimensional image to be produced.

Chapter 19 delves deeper into these methods and how they work. The chapter also outlines the benefits that outweigh the dangers.

Chapter 2

Interrogating Biophysics: The Five Ws and One H

If you ask most people when they take their first college or university physics course what physics is, many don't know. The situation is far worse when you ask people to explain biophysics. The purpose of this chapter is to help you answer this question, and go beyond. This chapter opens your eyes a bit, so you see that biophysics is everywhere, and no matter what you do in life, biophysics will play a role.

This chapter answers the hard five Ws (what, where, why, when, and who) questions and the one H (how) question about biophysics. Here I explain what biophysics really is, where you can use biophysics (I usually tell people you can use it everywhere, but I ease up and give you an easier answer), why biophysics is important, when you may need biophysics in life, who needs this knowledge after an entry-level college course (you may be surprised) and how biophysics may pop up in a career path that interests you.

Figuring Out What Biophysics Is

Biophysics is a natural science and the study of living matter, its motion, and its interaction with the natural universe. *Bio* comes from the Greek word for "life," whereas *physics* comes from the Greek work for "natural" or "nature."

Therefore, biophysics involves the study and application of the laws of the physical universe when living organisms are involved. An understanding of these laws will indicate how and why living organisms behave the way they do.

Objects that have *self-sustaining processes* are considered alive, so the cell is considered the basic building block of *living organisms*. Living organisms respond to stimuli, reproduce, and maintain some type of homeostasis. *Homeostasis* is the ability to maintain a constant stable condition. For example, people maintain a constant internal body temperature of 98.6 degrees Fahrenheit (37 degrees Celsius) when healthy.

Biophysics deals with small things, such as understanding the interaction between molecules within cells, comprehending the interaction of molecules within cells with external sources of energy such as radiation, deciphering the metabolism of molecules, and explaining the diffusion of molecules across a membrane. Biophysics is applicable at all length scales from molecules to the influence of forces on populations or the mechanics involved within sports or the environment.

Locating Biophysics: The Where

Biophysicists ask the fundamental questions and build the foundations for many different disciplines. Any natural science involved in the study of biological systems is connected to biophysics. In other words, everywhere you have living organisms you have biophysics.

The interdisciplinary nature of biophysics means that it's usually hard to find a cluster or group of biophysicists in their own department. Instead you can find them working within other departments or in the private sector. You can basically find biophysicists everywhere.

You can find biophysicists in the following fields:

- ✔ **Biochemistry:** The fields of biochemistry and biophysics are so closely related that the boundary between the two is very blurred. Many biochemists use biophysics in their research, or their research can be considered biophysical. In many cases, a biochemist can easily be referred to as a biophysicist and vice versa.

- ✔ **Bioengineering and biomedical engineering:** Engineers use the concepts and ideals from the natural sciences to devise and build tools, structures, and processes for use in society. These two disciplines use concepts from the three sciences: biology, chemistry, and physics. Bioengineering and biomedical engineering are large and rapidly growing fields. Some aspects of these fields that closely connect them to biophysics are that they mimic biological systems to create products, create devices to control biological systems, and modify the genetics of organisms (such as foods) to enhance a trait within the organism (for example, make it resistant to disease).

✔ **Biology:** Biophysics explains how and why things work the way they do within biology. For example, physics has only five types of fundamental energy (not including dark energy). Energy can't be created or destroyed, only changed from one form to another. Living organisms consume, transform, and use energy. The forms of energy and how they're transformed are biophysical processes. The mechanisms behind homeostasis are biophysical. Biophysics is involved from the small, such as molecular biology, to the biomechanics of large animal motion.

✔ **Environmental science:** Environmental science is a multidisciplinary field with contributions from biophysics, physics, biology, chemistry, geology, and soil science. Environmental science deals with energy systems, pollution problems and solutions, climate changes, agriculture, and natural resources.

✔ **Kinesiology:** *Kinesiology* is the study of human (and animal) motion, which includes biomechanics, a part of biophysics (check out the chapters in Part II for more information). The study of biomechanics includes things such as understanding how the body moves, how the nerves send signals to the brain, how the brain sends electrical signals to the muscles so they twitch (contract), and other physiological functions.

✔ **Medicine:** In hospitals, clinics, and research labs you'll usually find medical physicists, health physicists, and biophysicists. The biophysicists are more involved with the basic research that has the potential for medical applications.

✔ **Neuroscience:** *Neurophysics* is a branch of biophysics that deals with the nervous system. It covers a large range of scales from interactions at the molecular scale to the brain's function. The biophysicists are usually part of the neuroscience group, which is the interdisciplinary study of the nervous system. The field of neuroscience consists of researchers from biophysics, biology, biochemistry, chemistry, medicine, and psychology to mention a few.

✔ **Pharmacology:** *Pharmacology* is the study of the interaction of drugs with living organisms. Biophysicists are involved with *pharmaceuticals*, which are drugs with medicinal properties, and *radiopharmaceuticals*, which are drugs containing a radioactive isotope. The field includes the study of natural drugs and the synthesis of artificial drugs, their composition and properties, and their interactions with the body. The study of the interactions with living tissue is usually split into two areas:

- **Pharmacokinetics:** The study of the body's ability to absorb, distribute, metabolize, and excrete the drug

- **Pharmacodynamics:** The study of how the drug causes changes to the cells and the drug's physiological effects

Understanding Why Biophysics Is Important

Biophysics deals with how the laws of the natural universe work when the laws are applied to systems involving living organisms. An understanding of these laws explains why and how biological organisms behave the way they do. Having knowledge of these laws and the understanding of their applications in the natural sciences and medicine is important for the advancement of society.

Even the everyday person has some knowledge of biophysics: Biophysics is everywhere. For example, by understanding these laws, a person knows not to stick her finger into an open light socket when the power is on. Another example is a person knows that eating plutonium instead of his vegetables is bad, only if he knew that plutonium is radioactive and not very nutritious, and he understands the dangers of radioactivity.

The laws of biophysics tell you how cells interact with radiation, how the nerves work, how molecules are metabolized, and how to minimize energy expenditure or increase performance in sports. The better you understand the laws, the better you can use them to achieve your goal. Scientists throughout all the natural sciences and the other life sciences and engineers use biophysics all the time to guide them in their research, focus, and development of devices and applications.

Determining When Biophysics Is Relevant

Biophysics is always relevant and important. You need biophysics to understand the laws of nature. You need biophysics to understand why things are behaving the way they are. You need biophysics to understand how biological systems will behave under certain conditions.

A person driving his car doesn't really care how things work as long as the car works the way it's supposed to, but a mechanic needs to understand how a car works and knows the answer why, when something doesn't work the way it's supposed to. The mechanic is trained to use tools that make it possible to fix the car. The same is true for biophysics. Biophysics gives scientists and engineers the tools and knowledge to do their job and understand the problem. Without it, society would be similar to the Dark Ages with many of the things that everyone takes for granted being nonexistent. For example, nuclear magnetic resonance (NMR) was developed during World War II and then used in physics. It was then applied in the fields of chemistry and biochemistry, and eventually in medicine where it's called magnetic resonance imaging (MRI).

Finding Out Who Are Biophysicists

Biophysics isn't a secret society, and the members are usually proud to talk about biophysics. Biophysical society exists in many nations, and the members consist of biophysicists, biologists, chemists, physicists, and engineers. The people come from all over, such as universities, colleges, government research labs, and medical institutes to name a few. In addition, some universities even have a student biophysics society or club.

Organizations and societies for biophysics have been in existence for more than half a century. According to the Biophysical Society, a person can join many subgroups within their society. Remember, there is a lot of overlap between biophysics, biochemistry, and biology, so many of these subgroups can also be associated with biochemistry and biology. A few of these groups are as follows:

- ✔ **Bioenergetics:** People involved with *bioenergetics* are interested in how energy is used by biological systems. They look at processes that lead to the production and utilization of energy at the molecular and cellular level. Chapters 4, 9, and 11 examine energy at a closer level.

- ✔ **Biological fluorescence:** People involved in *biological fluorescence* use the technique as a nondestructive method of analyzing molecules within biological organisms. *Fluorescence* is the ability of molecules to absorb electromagnetic radiation and then emit electromagnetic radiation (usually at a lower frequency). I discuss electromagnetic radiation and absorption in Chapters 16 and 17.

- ✔ **Membrane biophysics:** People involved in *membrane biophysics* are interested in the mechanisms of ion transport across biological membranes. Chapters 9 and 11 look more closely at membranes.

- ✔ **Membrane structure and assembly:** People involved in *membrane structure and assembly* are interested in the biophysical properties of lipids, lipid assemblies, membrane proteins, and lipid-protein interactions relevant to membranes. Check out Chapter 11 for more insight into membranes.

- ✔ **Nanoscale biophysics:** People involved in *nanoscale biophysics* are interested in the study and control of biological, bio-compatible, or bio-inspired matter on the scale of atoms and molecules. Refer to Chapters 4, 6, 9, and 11 for more information.

✔ **Permeation and transport:** People involved in *permeation and transport* are interested in the study of biophysical mechanisms of permeation and transport of small molecules and biopolymers through cell membranes. Refer to Chapters 9 and 11 for relevant information.

Answering the Hows of Biophysics

Many questions focused on the how exist in biophysics. For example, two are "How do I perform biophysics?" and "How do I become a biophysicist?" The first one comes with experience; you apply the methodologies and techniques of physics to answer questions involving biological organisms. All sciences apply the scientific method, which can be summarized by these four:

✔ **Observations:** You observe or become aware of some event or phenomenon.

✔ **Hypotheses:** You propose theories to explain your observations.

✔ **Predictions:** Your theories have consequences, which you can predict.

✔ **Experiments:** You perform experiments to test the observations, hypotheses, and predictions. If any are found in error, then you have to repeat them. The experiments can be empirical in the lab or theoretical calculations.

The second question is interesting. Several universities do offer different levels of degrees in biophysics, ranging from bachelor of science all the way up to doctorates. At the undergraduate level, you have to take several physics, biology, chemistry, and biochemistry courses. At the graduate level you'll do research in a specific area of biophysics, so make sure you do some background research and pick something you're really interested in.

In addition to the biophysics programs, you can also choose from medical physics and health physics programs. They usually require graduate school, so you can't stop after four years, but who would want to stop doing biophysics?

Chapter 3

Speaking Physics: The Basics for All Areas of Biophysics

*I*f you want to know biophysics, then you need to know that many words in everyday speech come from physics. In physics though, these words have specific meanings that are blurred in everyday speech. For example, people have a tendency to interchange the words *mass* and *weight* to mean the same thing, but they're very different things in physics. Physics also likes to use mathematics for a few reasons.

This chapter provides an overview to the language of biophysics, including how your everyday words fit in the language. This chapter also explains the shorthand notation (mathematics) of biophysics, dealing specifically with the physics concepts that apply to all areas of biophysics.

When working through problems, I suggest you also write your own dictionary of each mathematical symbol and what the symbol means. Doing so is especially helpful if you're reading from more than one biophysics book (such as this book and your biophysics textbook). If you have a math phobia, then think of the math as shorthand notation for long-winded sentences or paragraphs.

A detailed review of this math is beyond this book's scope, so if you need a refresher on that math, I suggest you check out *Physics I For Dummies, Physics II For Dummies,* or *Physics Workbook For Dummies* all by Steven Holzner (John Wiley & Sons, Inc.).

Stretching Out in All Physical Dimensions with Units

Physical entities in biophysics have dimensions in addition to a numerical value and possibly direction too. *Dimension* refers to a physical property. For example, suppose I write 3 + 4 = 7. This is a purely mathematical expression with no connection to anything physical. However, if I wrote 3 feet (0.91 meters) + 4 feet (1.22 meters) = 7 feet (2.13 meters), then the physical dimension of length is associated with this expression.

Physics has five fundamental dimensions. In addition, a set of standards, called *units,* is required to make quantifiable use of these dimensions. The two common sets of standards are the

- ✔ **Système International (SI):** The *SI system* is based on the metric system. It's a decimal system with a single standard for each physical dimension. SI units are the most common set of standards used in global commerce and in universities and government scientific labs today. Organizations such as the National Institute of Standards and Technology (NIST) are always refining the standards, making them more accurate. In the following sections, I mention the standard for each of the physical dimensions.

- ✔ **United States Customary Units:** The United States commerce and general public use a set of units similar to the old British imperial units, both of which are based on the old English unit system. The standards for these units are now defined in terms of the SI unit standards. United States units are used in everyday life and in most of industry within the United States, although government labs and universities in the United States use the SI units. One of the roles of the National Institute of Standards and Technology (NIST) is to help industry within the United States to voluntarily switch to the SI units so they're more competitive globally.

The five fundamental physical dimensions and their units are as follows:

- ✔ **Length:** *Length* is the perception of extension to the spatial universe. The standard is the *meter,* and it's defined as the length light will travel in a vacuum during a time interval of 1/(299792458) of a second. The conversion between the set of units mentioned is 1 meter = 3.281 feet or 1 foot = 0.3048 meters. Length helps you know the location of objects and events around you.

- ✔ **Time:** *Time* is the perception of a sequence to events. The standard unit in both systems is the *second.* The second is defined as 9,192,631,770 times the period of the electromagnetic radiation emitted from Caesium-133.

- **Mass:** *Mass* is an intrinsic property of the object to resist acceleration by forces. The standard unit is the *kilogram* in SI units and the *slug* in United States customary units, so 1 kilogram = 0.06854 slugs or 1 slug = 14.59 kilograms.

  Many people, even some "experts" are confused about mass and weight. In fact, I once saw a science book for 12- and 13-year-old children say mass was the weight of the object, which is wrong. Weight is the gravitational force between the object and a very massive object like the Earth. Chapter 4 discusses weight in more detail. (By the way, the standard unit for force and weight is the *newton* in SI units and the *pound* in US customary units, so 1 newton = 0.2248 pounds or 1 pound = 4.448 newtons.)

  Density is the mass of an object divided by the volume it occupies. This means that density has the dimension of mass divided by cubic length, and it's not a fundamental dimension. In biophysics, when working with fluids and gasses, it's more convenient to work with density rather than mass. The mathematical formula for density is as follows:

  $$\rho = \frac{m}{V}$$

- **Temperature:** *Temperature* is a measure of the *thermal* (heat) energy per particle (matter or photon) in an object. The standard unit is the *kelvin* in SI units. A temperature of zero kelvin means the object has no thermal energy. In most situations the *Celsius* scale is used, with the only difference being at what temperature the zero degree is set. Another common unit used in the United States, outside of science, is the *Fahrenheit*. The conversions are as follows:

  - *Z* kelvin = *Y* degrees Celsius + 273.15. For example, 30 degrees Celsius + 273.15 = 303 kelvin
  - *Y* degrees Celsius = *Z* kelvin – 273.15.
  - *X* degrees Fahrenheit = (*Y* degrees Celsius) $\frac{9}{5}$ + 32. For example, (30 degrees Celsius) $\frac{9}{5}$ + 32 = 86 degrees Fahrenheit.
  - *Y* degrees Celsius = (*X* degrees Fahrenheit) $\frac{5}{9}$ – 17.78. For example, (70 degrees Fahrenheit) $\frac{5}{9}$ – 17.78 = 21 degrees Celsius.

- **Charge:** *Charge* is a property of objects that allows the object to interact with other objects through the electromagnetic force. The standard unit is the *ampere* in SI units, which is actually the unit for electric current. The unit for charge is the *coulomb*, which is equal to one ampere times one second.

Looking back: The changing concept of mass

Mass seems to be an easy thing to understand, but the idea of what mass is has changed over the years. For example, Aristotle believed that heavier objects fell faster than light objects. Then about 2,000 years later, Sir Isaac Newton proposed two types of masses in his *Principia:*

✔ *Inertial mass* is a physical property of an object that allows the object to resist changes to its motion caused by external forces. For example, you would probably have no problem rolling a watermelon down your apartment hallway, but you would have trouble pushing a train car down the train tracks. The mass of the watermelon is much smaller than the train car's mass.

✔ *Gravitational mass* is a physical property of an object that allows the object to be attracted to other objects that have gravitational mass. For example, the attraction between the earth and the moon is from their gravitational mass.

In the early 20th century, Albert Einstein proposed that mass is a form of energy and the two masses were equivalent. This proposal has been verified experimentally since then, and led to the idea of *nuclear power* — changing mass energy into other forms of energy. A half a century later a particle was proposed and came to be known as the *Higgs boson*. This particle gave other elementary particles their mass and was experimentally verified to exist in 2012.

Standards make communicating easier. For example, suppose I made my own set of units and told you my home is 1,000 klumpens from my office and it takes me 20 humilings to walk home. You'd have no clue what I was talking about. You still don't know how far my home is from the office and how long it takes me to get home, but if I told you my home is 2 miles (3.2 kilometers) from the office, and it takes me 40 minutes to walk home, then you know exactly what I mean, which is why standards are needed.

Grasping Scalars, Vectors, and Their Properties

Scalars and vectors are very important in biophysics because most physical quantity in biophysics is either a scalar or a vector. The following explains what the two are

✔ **Scalar:** A *scalar* is a quantity with only magnitude, but has no direction. An example is the temperature in a room. The temperature has a value of 68 degrees Fahrenheit (20 degrees Celsius), which is the magnitude. No matter whether you're facing north or south or standing or lying down, the temperature is the same value.

✔ **Vector:** A *vector* is a quantity with both magnitude and direction. For vectors, the direction is as important as the magnitude, even though some books give the impression that the direction isn't as important. For example, you throw a paper airplane at a friend sitting across the room. The velocity of the paper airplane is 3 feet per second (0.9144 meters per second) southward, which is a vector. The 3 feet per second is the magnitude of the vector and southward is the direction.

Vectors and scalars need a map. A map is called a *frame of reference*, and it gives the location of objects and events relative to some fixed point called the *origin*. These maps have a few extra things compared to what you're used to seeing; the map must have a *coordinate system* that labels all points in space relative to the origin.

In other words, people use a frame of reference, such as global positioning satellites (GPS) compasses, Internet maps, GPS apps, paper road maps, and other means, when finding their location on the earth and figuring out how to get to their destination from their current location (origin). The coordinate system labels the location of everything relative to your current location, so you know the distance and direction to different locations.

Instead of using north, south, east, and west, I use the following mathematical shorthand for my directions called *unit vectors*:

$$+\hat{x} = \text{east}; \quad -\hat{x} = \text{west}; \quad +\hat{y} = \text{north}; \quad -\hat{y} = \text{south}; \quad +\hat{z} = \text{up}; \quad -\hat{z} = \text{down}$$

As long as you remember that the letters with hats are just directions and nothing else, then you'll be fine.

The following are some properties of scalars and vectors. Here I restrict my discussion to two-dimensional vectors for convenience (refer to Figure 3-1 for an example).

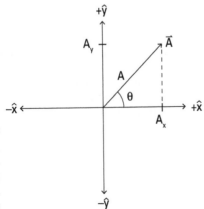

Figure 3-1: A two-dimensional vector.

✔ **Describing a two-dimensional vector:** You need two parameters to do so. A vector has both magnitude and direction, so one of the parameters is the vector's magnitude and the other parameter is the angle the vector makes relative to the +x-axis. Mathematically, these two parameters are represented (A, θ).

✔ **Writing two-dimensional vectors:** You can do so in Cartesian component form. *Cartesian component form* means any vector can be expressed as the sum of two vectors: one vector parallel to the x-axis and one vector parallel to the y-axis. Because the x-axis and y-axis give the directions, all you need are the magnitudes of the two vectors, which are represented by A_x and A_y.

✔ **Switching two-dimensional vectors between the two forms:** If you know the magnitude and direction, then you can find the Cartesian components. Mathematically, the two Cartesian components are related to the magnitude and direction (A, θ) by the formulas: $A_x = A\cos(\theta)$; $A_y = A\sin(\theta)$. If you know the Cartesian components of the vector, you can go the other way and find the magnitude and direction. Mathematically, the formulas are

$$A = \sqrt{A_x^2 + A_y^2}; \; \theta = \arctan\left[\frac{A_y}{A_x}\right]$$

WARNING!

If you're asked to calculate the angle when given the Cartesian components (A_x, A_y), then you must be careful. Your calculator can't tell which quadrant you're in and it may give you the wrong answer. In the arctan formula for θ, your calculator assumes A_x is positive. For example, your calculator gives the same angle for the vectors (2,4), θ = arctan (4/2) = 63.4 degrees, and (–2,–4), θ = arctan(–4/(–2)) = 63.4 degrees. The correct answer for the vector (–2,–4) is 180 + 63.4 = 243.4 degrees. You need to add an extra 180 degrees if A_x is negative.

✔ **Expressing two-dimensional vectors in the two forms:** You can write the vector as:

$$\vec{A} = \left(A_x, A_y\right) = \hat{x}A_x + \hat{y}A_y = \hat{x}A\cos(\theta) + \hat{y}A\sin(\theta)$$

which has both magnitude and direction.

✔ **Adding and subtracting vectors:** You can add or subtract only the Cartesian components of vectors that are parallel. For example, the up and down direction components of vectors add together, the north and south direction components of vectors add together, and the east and west direction components of vectors add together. Mathematically, it looks like:

$$\vec{A} + \vec{B} = \hat{x}\left[A_x + B_x\right] + \hat{y}\left[A_y + B_y\right]$$

✔ **Multiplying a scalar with a vector:** The scalar has to multiply each component of the vector. Mathematically:

$$c\,\vec{A} = \hat{x}\left[c\,A_x\right] + \hat{y}\left[c\,A_y\right] = \hat{x}\left[c\,A\cos(\theta)\right] + \hat{y}\left[c\,A\sin(\theta)\right]$$

This expression shows that a scalar can change only the magnitude, which changes from A to c A, and not the direction of a vector. **Note:** A negative constant will change the direction of the vector by 180 degrees.

✔ **Multiplying a vector with another vector (the dot product):** The multiplication of a vector with another vector produces a scalar. Consider the two vectors shown in Figure 3-2.

The dot product of the two vectors in Figure 3-2 can be calculated using these formulas

$$\vec{A} \bullet \vec{B} = A_x B_x + A_y B_y$$
$$= AB\cos(\theta_A - \theta_B) = AB\cos(\phi)$$
$$= A_{\|}B = AB_{\|}$$

The first line of this formula is useful if you know the Cartesian components of both vectors. The second line corresponds to the geometric method of calculating the dot product in Figure 3-2a. The third line corresponds to the geometric method of calculating the dot product illustrated in Figures 3-2b and 3-2c. This type of multiplication arises when calculating the *work done* on an object by external forces.

✔ **Multiplying a vector with another vector (the cross product):** The multiplication of a vector with another vector produces a vector. Consider the two vectors shown in Figure 3-2a. The cross product is

$$\vec{A} \times \vec{B} = \hat{z}\left[A_x B_y - A_y B_x\right]$$
$$= -\hat{z}AB\sin(\theta_A - \theta_B) = -\hat{z}AB\sin(\phi)$$
$$= -\hat{z}A_{\perp}B = -\hat{z}AB_{\perp}$$

The formula shows how to calculate the cross product.

Use the first line of this formula if you know the Cartesian components of both vectors.

The second line corresponds to a geometric method of calculating the cross product and corresponds to Figure 3-2a.

The third line corresponds to a geometric method of calculating the cross product and corresponds to Figures 3-2b and 3-2c.

The expressions in the formula give the magnitude and the direction. Notice that the direction is perpendicular to the plane formed by the two vectors. You can use the preceding formula to calculate the magnitude and determine the direction with the right-hand rule.

Figure 3-2:
The dot product of two vectors with the vectors in terms of their directional angle (a). The vectors with one vector expressed in components parallel and perpendicular to the second vector (b) and (c).

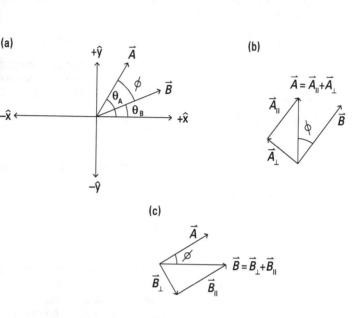

The *right-hand rule* gives the direction of the cross product (resultant vector) by following these three steps:

1. **Place the wrist of your right hand on the origin in Figure 3-2a.**

2. **Stretch your right hand and fingers in the direction of the first vector.**

3. **Curl your fingers toward the second vector.**

   If you're double jointed, then you want to curl your fingers toward the palm of your hand.

   Note which way your thumb is pointing, perpendicular to the plane formed by the two vectors, and that is the direction of the resultant vector.

   Try the two vectors in Figure 3-2a. My thumb is pointing into the page.

The cross product is the type of multiplication that arises when calculating the angular momentum of an object, the torque acting on an object by an external force, and the force acting on a charged particle moving through a magnetic field to mention a few applications.

Defining Physical Quantities

Several words in everyday use have very specific meanings in biophysics. For example, you can't use kilograms as a unit for weight or pound can't be used as a unit of mass.

These sections identify some important terms you need to know when studying biophysics and give the mathematical notation for each physical quantity. Describing scientific experiments or situations in biophysics can get long-winded if written out, so using mathematics as shorthand notation is helpful to keep things concise. Using mathematical shorthand also allows you to have a quantitative description and make predictions.

Plotting the position

The *position* is a vector with both magnitude and direction. It maps the location of all objects and events relative to some reference point. Creating a map is important to know where objects are located in space. For example, you're reading *Biophysics For Dummies* and suppose you're holding it 1 foot (30.48 centimeters) from your face. The magnitude of the position vector is 1 foot (30.48 centimeters), and in front of your face is the direction of the position vector. The mathematical description of the position is

$$\vec{s} = \hat{x}\,x + \hat{y}\,y + \hat{z}\,z$$

The position is a very important concept in biomechanics, where you need to know where things and objects are located. Refer to the chapters in Part II for more information.

Rotating to an angular position

The *angular position* is a vector with both magnitude and direction. The angular position vector gives the angle between the object (or event) relative to some reference direction (zero radians). It maps the location of all the objects and events lying in a circle. The angular position has units of *radians* instead of degrees. The direction of the angular position is perpendicular to the plane of the circle. The reason for this choice of the direction will make more sense later. (See the later section on angular velocity for more information.)

To change between radians and degrees: π radians equals 180 degrees. Calculators usually have a function key that allows you to switch between the two. Also, you can set your calculator to work in either radians or in degrees. The following sections are related to position and angular position:

Changing position: Displacement

Displacement is the change in position: final position minus the initial position. The displacement is a vector with both magnitude and direction. For example, the displacement of my head as I stand up from my chair is 2 feet (61 centimeters) upwards (unless I hit my head on the overhead light again).

The mathematical description of the displacement is

$$\Delta \vec{s} = \vec{s}_{final} - \vec{s}_{initial} = \hat{x}(x_{final} - x_{initial}) + \hat{y}(y_{final} - y_{initial}) + \hat{z}(z_{final} - z_{initial})$$

Figure 3-3 shows an example of the displacement of an object from its initial position to its final position. Notice the displacement vector doesn't care what path the object had taken, just the location of the initial and final positions.

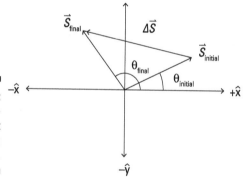

Figure 3-3:
The displacement of a moving object.

Altering the angle: Angular displacement

Angular displacement is the change in angle: the final angular position minus the initial angular position. The angular displacement is a vector with the direction perpendicular to the plane of the circular rotation. The angular displacement ($\Delta\theta$) is positive if the final angular position is counter-clockwise relative to the initial angular position, and it's negative if the final angular position is clockwise relative to the initial angular position.

The mathematical description of the angular displacement for the motion in Figure 3-3 is

$$\Delta\vec{\theta} = +\hat{z}\left(\theta_{final} - \theta_{initial}\right)$$

If the object is moving in a circle, then the *distance* traveled along the circle by the object is $\Delta s = r\,\Delta\theta$, where r is the radius of the circle.

Going the distance

Distance is the length of a route and is a scalar. This is what the odometer in a car measures. For example, I go to the car dealership and take a car for a test drive. When I get back from the test drive, I park in the exact same spot I started from. My displacement is zero because my final position equals my initial position, but the distance traveled by the car isn't zero and shows on the odometer. The distance doesn't equal the displacement.

The difference between distance and displacement is very important, so the two words can't be exchanged in casual conversation anymore.

Timing the change: Velocity

Velocity, also referred to as *instantaneous velocity,* is the time rate of change of the position, which means it measures how fast the position is changing at any given moment in time. Change of the position vector means a change in the magnitude of the position, a change in the direction of the position, or both.

The velocity is a vector because it's measuring how the position is changing, which is a vector. For example, you're driving a sports car 200 miles per hour (322 kilometers per hour), which is the magnitude, south on the freeway, which is the direction.

The mathematical shorthand notation for the velocity is

$$\vec{v} = \hat{x}v_x + \hat{y}v_y + \hat{z}v_z$$

Figure 3-4 describes the motion of a person bouncing on a trampoline. Figure 3-4a shows the position versus time graph of the person. The slope (which is the rise divided by the run) of the curve at the point p is equal to the velocity of the person at that time. If you calculate the slope of the curve at each point in Figure 3-4a and plot those values, you would obtain Figure 3-4b, which is the velocity versus time graph of the person.

You're probably wondering whether you can do the calculation in reverse. The answer is yes. If you calculate the area between the velocity versus time curve and the horizontal axis (time axis) from some initial time to some final time, this area is equal to the displacement during this elapsed time (final time minus initial time). The shaded area of Figure 3-4b illustrates the area. Remember that the area above the horizontal (time) axis is positive whereas the area below the horizontal (time) axis is negative. Figure 3-4b shows a negative area from time $t_{initial}$ to t_{final}.

Figure 3-4:
The position versus time graph (a). The velocity versus time curve (b).

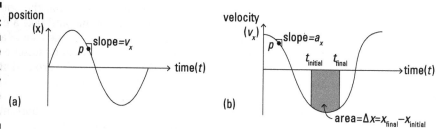

If you love calculus or your biophysics course is calculus-based, then this is for you. In calculus the relationships between velocity, position, and displacement are

$$\vec{v} = \frac{d\vec{s}}{dt}; \text{ and } \Delta\vec{s} = \int_{t_{initial}}^{t_{final}} dt\vec{v}$$

Scoping out speed

Speed is distance the object travels divided by the elapsed time it takes the object to move from the initial position to the final position. Speed depends on the path taken just like the distance. Another way to look at the speed: it is the magnitude of the instantaneous velocity vector.

Focusing on angular velocity

Angular velocity, also referred to as *instantaneous angular velocity,* is the time rate of change of the angular position, which means it measures how fast the angular position is changing at any given moment in time. Angular velocity is a vector with both magnitude and direction. The magnitude has units of radians per second where π radians = 180 degrees.

The mathematical description of angular velocity is

$$\vec{\omega} = \hat{x}\omega_x + \hat{y}\omega_y + \hat{z}\omega_z$$

In the case when the object is moving in a circle, the *speed* of the object is related to the magnitude of the angular velocity, $v = r\omega$, where r is the radius of the circle.

Figure 3-4 represents the relationships for linear motion. This figure is also applicable to angular motion. In Figure 3-4, replace the position, x, with the angular position, θ, and replace velocity, v_x, with angular velocity, ω.

Figure 3-4a becomes the angular position versus time graph. The slope of the curve at the point p is equal to the angular velocity. Figure 3-4b is the angular velocity versus time graph (see the shaded area in Figure 3-4b). The area between the angular velocity versus time curve and the horizontal axis (time axis) from some initial time to some final time, it's equal to the angular displacement during this elapsed time.

Angular velocity points perpendicular to the plane the object is rotating in. In many biophysical situations the angular velocity is one-dimensional. An example where that isn't true is in diving when a diver does a forward roll combined with a twist. The diver has two axes of rotation when he is spinning around as he falls from the platform to the pool.

For you calculus lovers, check out this formula. In calculus the relationships between angular velocity, angular position, and angular displacement are as follows:

$$\vec{\omega} = \frac{d\vec{\theta}}{dt}; \text{ and } \Delta\vec{\theta} = \int_{t_{\text{initial}}}^{t_{\text{final}}} dt\,\vec{\omega}$$

Examining the direction of angular variables

The direction of angular variables is perpendicular to the plane of rotation. Suppose you're sitting on a wooden horse on a carousel and it's traveling in a circle at 0.2 radians per second. (You have to make sacrifices when you're studying biophysics.) At one moment you're traveling toward the north, then toward the west, then toward the south, then toward the east, and on and on. So how do you pick a direction in the plane of rotation? The direction is constantly changing. The only direction not changing while you spin around is up (and down), which is the direction perpendicular to the plane. Therefore, you would say your angular velocity is 0.2 radians per second up.

The following sections are related to the velocities.

Figuring out average velocity

The *average velocity* is defined as the displacement of the object divided by the elapsed time it takes the object to move from the initial position to the final position.

Many people are more comfortable thinking in terms of miles per hour or kilometers per hour. In biophysics the use of meters per second or feet per second is conventional (and convenient in most situations). The conversions between the different units are

- ✔ 1 foot per second = 0.3048 meters per second = 0.6820 miles per hour = 1.097 kilometers per hour

- ✔ 1 meter per second = 3.281 feet per second = 2.237 miles per hour = 3.600 kilometers per hour

The mathematical description for average velocity is

$$\bar{\vec{v}} = \frac{\Delta \vec{s}}{\Delta t} = \frac{\vec{s}_{final} - \vec{s}_{initial}}{t_{final} - t_{initial}}$$

The bar above the arrow in this equation indicates this is the average velocity and not the (instantaneous) velocity.

Uncovering average angular velocity

The *average angular velocity* is the angular displacement of the object divided by the elapsed time. The mathematical description for the average angular velocity is

$$\bar{\bar{\omega}} = \frac{\Delta \vec{\theta}}{\Delta t} = \frac{\vec{\theta}_{final} - \vec{\theta}_{initial}}{t_{final} - t_{initial}}$$

Measuring acceleration

Acceleration, also referred to as *instantaneous acceleration*, is the time rate of change of the velocity and measures how fast the magnitude and or the direction of the velocity is changing at any given moment in time.

The mathematical description of acceleration is

$$\vec{a} = \hat{x}a_x + \hat{y}a_y + \hat{z}a_z$$

Figures 3-4 and 3-5 represent with the motion of a person bouncing on a trampoline. Figure 3-4b shows the velocity versus time graph of the person. The slope of the curve at the point P is equal to the acceleration of the person at that time. If you calculate the slope of the curve at each point in Figure 3-4b and plot those values, you'd obtain Figure 3-5, which is the acceleration versus time graph of the person.

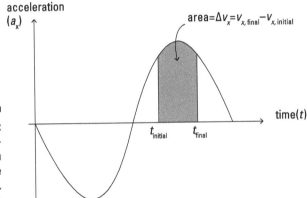

Figure 3-5:
The acceleration versus time graph.

In Figure 3-5, if you calculate the area between the acceleration versus time curve and the horizontal axis (time axis) from some initial time to some final time, it's equal to the change in the velocity. The shaded region in Figure 3-5 represents it.

For you calculus lovers, here is the relationship between acceleration and velocity:

$$\vec{a} = \frac{d\vec{v}}{dt}; \text{ and } \Delta\vec{v} = \int_{t_{initial}}^{t_{final}} dt\,\vec{a}$$

Comprehending angular acceleration

Angular acceleration, also called *instantaneous angular acceleration,* is the time rate of change of the angular velocity and measures how fast the angular velocity changes at any given moment in time. The angular acceleration is a vector with both magnitude and direction perpendicular to the plane of rotation. The units of the angular acceleration are radians per second squared. The mathematical description of angular acceleration is

$$\vec{\alpha} = \hat{x}\alpha_x + \hat{y}\alpha_y + \hat{z}\alpha_z$$

The graphical representation between the angular acceleration and the angular velocity is exactly the same as the graphical representation between acceleration and velocity. You can use Figures 3-4 and 3-5 as an example of how the angular variables are related.

In the case of an object moving in a circle's radius, you can split the acceleration into two parts:

✔ The *radial acceleration* is a measure of the change in the direction of the object's velocity. Its direction is toward the center of the circle.

✔ The *tangential acceleration* is a measure of the change in the magnitude of the object's velocity. Its direction is the same as the velocity (or opposite).

The mathematical descriptions of the magnitudes of the radial acceleration (a_r) and tangential acceleration (a_T) are

$$a_r = r\omega^2$$
$$a_T = r\alpha$$

Here is more calculus for the lovers of fine mathematics. The relationships between angular velocity and angular acceleration are as follows:

$$\vec{\alpha} = \frac{d\vec{\omega}}{dt}; \text{ and } \Delta\vec{\omega} = \int_{t_{initial}}^{t_{final}} dt\,\vec{\alpha}$$

Grasping average acceleration

The *average acceleration* is the change in an object's velocity divided by the elapsed time it takes the object's velocity to change from the initial velocity to the final velocity. The average acceleration is a vector with both magnitude and direction. The mathematical description for average acceleration is

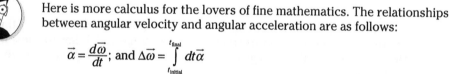

$$\bar{\vec{a}} = \frac{\Delta\vec{v}}{\Delta t} = \frac{\vec{v}_{final} - \vec{v}_{initial}}{t_{final} - t_{initial}}$$

For example, a very fast acceleration for any animal on rough ground is about 1 g (g is the acceleration due to gravity, which is 32.2 feet per second squared (9.81 meters per second squared)). To help give you a feel for what

1 g is, consider a car going from 0 to 100 miles per hour (161 kilometers per hour = 147 feet per second = 44.7 meters per second) in three seconds, the average acceleration is 48.9 feet per second squared (14.9 meters per second squared). This car has an average acceleration of 1.5 g's, which is fast.

Understanding average angular acceleration

The *average angular acceleration* is the change in the angular velocity of the object divided by the elapsed time. The average angular acceleration is a vector with both magnitude and the direction that is perpendicular to the plane of rotation.

The mathematical description for average angular acceleration is

$$\vec{\bar{\alpha}} = \frac{\Delta \vec{\omega}}{\Delta t} = \frac{\vec{\omega}_{final} - \vec{\omega}_{initial}}{t_{final} - t_{initial}}$$

Describing momentum

Momentum is the quantity that describes the motion of the object. Sir Isaac Newton determined that the momentum equals the mass times the velocity. The mathematical description for momentum is $\vec{p} = m\vec{v}$.

Interacting with others: Force

A *force* is how an object interacts with other objects. *Forces* allow you to quantify how the surroundings interact with an object and attempts to change the object's momentum. If an object didn't create forces on other objects, then the other objects wouldn't know it existed.

For example, consider a person singing (screaming) a song in the shower. Her voice produces a *sound wave* (noise) that propagates through space and hits your eardrum. The sound wave produces a force on the eardrum that causes it to vibrate and you hear the sound (noise).

The mathematical description for force is $\vec{F}_g$.

You can recognize specific forces as an uppercase *F* with an arrow and a subscript. The subscript tells you what type of force it is. For example, the subscript *g* here refers to the gravitational force and weight. Chapter 4 discusses the different forces.

Spreading force over an area: Pressure

Pressure is the force divided by the cross-sectional area. The *cross-sectional area* is the area over which the force is spread. For example, suppose you're going to push open the door. If you place your fingertip against the door and push, then the cross-sectional area is the area of your fingertip. But if you put your open hand against the door and push, then the cross-sectional area is the area of your hand. When the force is spread over an area, it's more convenient to talk about the pressure instead of the force.

The mathematical description for pressure is

$$P^{(a)} = \frac{\left|\vec{F}_A\right|}{A}$$

Notice the superscript on the symbol, *P,* for pressure. I use a superscript for two important reasons:

- ✔ It helps distinguish pressure from the power, which has the symbol P as well.
- ✔ There are two types of pressure
  - *Absolute pressure* measures the force per unit area.
  - *Gauge pressure* is equal to the absolute pressure minus the atmospheric pressure.

    I extensively use absolute and gauge pressures in Chapters 9 to 15.

Going 'round and 'round: Axis of rotation

The *axis of rotation* is the axis the object is rotating around. For example, the axis of rotation of the earth is the line from the South Pole to the North Pole. If you live on the equator, then you're making a big circle every 24 hours at an angular speed of 7.29×10^{-5} radians per second, whereas if you're standing at the South Pole, your angular speed is the same, but your speed is zero. This angular speed seems slow, but if you're at the equator, this corresponds to a speed of 1,040 miles per hour (1,674 kilometers per hour).

Distributing mass: Moment of inertia

The *moment of inertia* depends on how the mass of the object is distributed around the axis of rotation. It's proportional to the mass of the object and the radius of the object squared. In other words, it's a physical property of the object that allows the object to resist change to its angular motion caused by external torques. I discuss torque in the "Tackling torque" section later in this chapter.

For example, you would probably have no problem lifting a child and spinning the child in a circle around your body, but you would probably have trouble lifting a sumo wrestler and spinning the wrestler in a circle because of the large mass. However, if you place the sumo wrestler at the center of a carousel, you could probably spin the wrestler because most of the mass is at the axis of rotation.

The formal definition of the moment of inertia involves calculus. Assume my axis of rotation is parallel to the z-axis then:

$$I = \int \left(x^2 + y^2 \right) dm = \int \left(x^2 + y^2 \right) \rho \left(x, y, z \right) dx\, dy\, dz$$

Here *dm* is the amount of mass located at the point (*x*, *y*, *z*). ρ is the mass density (mass per unit volume).

Quantifying motion: Angular momentum

The *angular momentum* is the quantity that describes an object's rotational motion. The angular momentum equals the momentum of inertia times the angular velocity. The mathematical description for angular momentum is

$$\vec{L} = I\vec{\omega}$$

$$\left| \vec{L} \right| = \left| \vec{r} \times \vec{p} \right| = r\, p \sin(\theta) \left[\text{right-hand rule} \right]$$

The first expression for the angular momentum is for an object with a moment of inertia *I*, while the second expression is the magnitude for a particle (small object). The vector $\vec{r}$ is the shortest displacement vector from the axis of rotation to the particle and the vector $\vec{p}$ is the momentum of the particle. For example, the earth is very large with billions of people on it; however, compared to the size of the solar system, the earth is a speck of dust. You would use the first line to calculate the earth's angular momentum from spinning around its axis each day and the second line to calculate the angular momentum of the earth as it rotates around the sun.

Tackling torque

A *torque* is the rotational analogue of a force. It quantifies how the surroundings attempt to change the object's angular momentum. The torque is equal to the displacement from the axis of rotation to the object times the force acting on the object (cross product).

The mathematical description for the magnitude of the torque is

$$\left|\vec{N}_g\right| = \left|\vec{r} \times \vec{F}_g\right| = r\, F_g \sin(\theta)\left[\text{right-hand rule}\right]$$

Torque uses the symbol N. You're probably wondering why. Some books use the Greek letter tau (τ) because it looks like the Latin letter T and t for torque. However, t is used for time and T is used for half-life, period, and temperature. I use the letter t (for time) and T a lot in this book, so here I use N for torque. Refer to the section "Grasping Scalars, Vectors, and Their Properties" earlier in this chapter.

Working with work

Work is tied more closely to energy, but it's also related to force. You use energy to do work on an object. The *work done* on an object is related to the interaction (force) on the object.

Forces can possibly do *work* on moving objects by changing the object's motion. The work done on an object by a force is equal to the force times the displacement parallel to the force (dot product).

The mathematical description of the work done is

$$\Delta W = \vec{F} \bullet \Delta \vec{s} = \left|\vec{F}_g\right|\left|\Delta \vec{s}\right|\cos(\theta)$$

I have written the work as ΔW instead of the usual W because the amount of work done changes depending on the force and the displacement of the object.

The force may change over the path followed by the object (called the *trajectory*), in which case calculus must be used and the *work* is:

$$\Delta W = \int_{\vec{s}_{initial}}^{\vec{s}_{final}} dW = \int_{\vec{s}_{initial}}^{\vec{s}_{final}} \vec{F} \bullet d\vec{s}$$

Perusing power

Power is a measure of how fast a force is doing work on an object. Power is equal to the work divided by the elapsed time. The mathematical description is

$$P = \frac{\Delta W}{\Delta t} = \vec{F} \bullet \bar{\bar{v}} = |\vec{F}||\bar{\bar{v}}|\cos(\theta)$$

This expression for the power takes the total work done divided by the elapsed time, which is an average, so this expression for the *power* is sometimes referred to as the *average power*. Note that the expression uses the average velocity, which is the displacement divided by the elapsed time.

You can calculate the *instantaneous power*, using the calculus expression for the work done. The instantaneous power is

$$P = \frac{dW}{dt} = \vec{F} \bullet \vec{v}$$

Eyeing energy

Energy is a physical dimension that can't be directly observed; however, changes in the energy can be measured (or observed). Biophysics and other sciences discuss many different forms of energy, but only five fundamental forms of energy exist. The five fundamental energies and two other energies I discuss in this book are

- ✔ **Heat energy:** *Heat* is the energy associated with the random disorganized motion of atoms and molecules. Chapter 10 provides the mathematical formula for heat energy.

- ✔ **Kinetic energy:** The *kinetic energy* is the energy associated with an object's motion. The kinetic energy is equal to half the mass times the magnitude of the velocity squared. The mathematical description for kinetic energy is

$$E_K = \frac{mv^2}{2}$$

- ✔ **Mass energy:** An object's mass is a form of energy, which can be converted to other forms of energy. The mass energy equals the mass times the speed of light squared.

✔ **Potential energy:** *Potential energy* is associated with a conservative force. It's a measure of the potential of a force to do work on an object. I discuss several different potential energies in this book and explain each one in depth when I mention it.

✔ **Radiation energy:** *Radiation energy* is the energy within electromagnetic radiation such as light and X-rays. The energy of a photon is equal to the frequency of the electromagnetic radiation times Planck's constant.

✔ **Mechanical energy:** The *mechanical energy* of a system is the sum of all the kinetic energies and potential energies of all the objects contained within the biological system.

✔ **Total energy:** The *total energy* of the system is the sum of all the heat energy, kinetic energies, mass energies, potential energies, and radiation energy.

Part II

Calling the Mechanics to Fix Your Bio — Biomechanics

Pulleys with a Mechanical Advantage

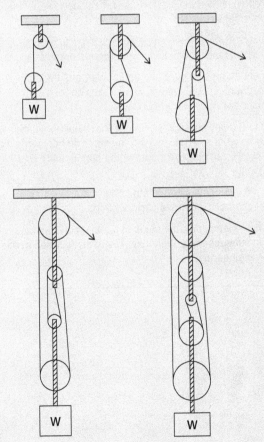

In this part . . .

✔ Understand the laws of physics relevant to biomechanics so you can use the ideals of forces and motion to study many biophysical systems.

✔ Familiarize yourself with rigid bodies, in particular combined with static equilibrium, and all the related principles.

✔ Find out about energy, work, and power, and the many biomechanical applications when these concepts are combined with conservation laws.

✔ Discover many physical principles that have applications in many fields of science, find out the limits on how large animals and plants can get, and visualize the human body as a machine.

✔ Grasp the dynamics and related physical properties of biomechanics to improve performances.

✔ See how *kinematics,* the study of motion, plays an important role in biomechanics, which applies to many areas from sports to bioengineering.

Chapter 4

Bullying Biomechanics with the Laws of Physics

*N*othing in the universe can break the laws of physics. In this chapter I introduce the fundamental laws of physics and explain how they're relevant to biological systems and biomechanics. *Biomechanics* is the study of forces acting on biological organisms. Biomechanics is relevant from the study of ions moving within an organism to the study of the motion of animals.

This chapter focuses on Newton's three laws of motion as well as the work-energy concept. These laws describe how objects move, based on their interactions with their surroundings. This chapter also describes conservative forces such as gravity. A conservative force will work against you in one direction, but going back in the opposite direction, the force will work with you. I also discuss nonconservative forces, such as friction, and explain work, power, and energy from a biophysics perspective.

Recognizing That the Force Is Always with You: Newton's Laws

The trick to Newton's laws is to remember and understand the words — free your words and your math will follow. The three Newton's laws are none other than a list of three commonsense sayings. (I use *commonsense* loosely in this case.) Unfortunately, this version of common sense isn't natural for many humans.

For example, it took thousands of years before people realized and accepted the fact that the earth wasn't flat or the earth moved around the sun and not the sun moving around the earth. Being able to appreciate these sayings takes time, so don't worry. I trust that you'll need less than thousands of years. In these sections I state the laws and explain what each law means.

Moving with inertia — Newton's first law of motion

Newton's first law of motion, the law of inertia, says that every object continues in a state of rest, or of uniform motion in a straight line, unless it's compelled to change that state by forces acting on it. Okay, that's not too bad; it's only one sentence after all. However, peel back the layers because many concepts and ideas are hidden in this one sentence:

- ✔ **Rest:** An object at *rest* stays at the same position with zero velocity and zero acceleration. (*Velocity* tells you how fast the position is changing, and the *acceleration* tells you how fast the velocity is changing.)

- ✔ **Uniform motion:** An object moving with *uniform motion* has no acceleration, which means the magnitude and direction of the velocity don't change. The object moves in a straight line at constant speed.

- ✔ **Forces:** *Forces* is how to describe objects interacting with other objects; this interaction causes accelerations. For example, I smash a tennis ball with the racket, so you can say the racket applied a force to the tennis ball, causing it to accelerate.

Newton's law of inertia tells you that an object likes to be at rest or in uniform motion. The only way to change an object's uniform motion (or rest) is through interactions with its surroundings (force). An object can't change its motion by itself; it needs to interact with its surroundings.

Stopping requires force — Newton's second law of motion

Newton's second law is a tad more complex, so I divide it into two sections. This law is especially important and used throughout biophysics, engineering, and physics. Become one with the law and embrace it. The first section is the original law of acceleration and the second subsection is Newton's law of acceleration adapted for rotating systems.

Going straight — the original law

Newton's second law of motion, the law of acceleration, means the time rate of change of the momentum of an object is equal to the net external force acting on the object from the surroundings.

Here are the concepts in this law, broken down in plain English:

- ✔ **External:** The *external* in Newton's law means only forces from sources that aren't the objects.

- ✔ **Net:** *Net* means you have to consider all the forces from all the external sources. Newton's law states that this net external force is how fast the momentum of the object will change.

  Suppose Corby is running down the field with the football. Licking his lips while running is an internal force and it won't change his momentum. However if Dean from the other team runs into Corby, then this collision is an external force on Corby, and it will cause his momentum to change.

- ✔ **Net external force:** The *net external force* is the vector sum of all the forces from the surroundings acting on the object. The mathematical description for net external force is as follows:

$$\vec{F}_{NET} = \hat{x}\, F_{x,NET} + \hat{y}\, F_{y,NET} + \hat{z}\, F_{z,NET}$$

$$F_{x,NET} = \sum F_x, \quad F_{y,NET} = \sum F_y, \quad F_{z,NET} = \sum F_z$$

Figure 4-1 demonstrates Newton's law of acceleration. Figure 4-1a shows a momentum versus time graph. The slope of the graph at a single point P is equal to the net external force acting on the object at that moment of time. If you calculate the slope of the curve between two points, the slope is equal to the average net external force acting on the object over that time interval.

Figure 4-1b shows a net force versus time graph. The area between the curve and the horizontal axis (the shaded region) is equal to the change in the momentum over that interval of time.

If you want to work some advanced math, the calculus definition of Newton's second law of motion, the law of acceleration, is as follows:

$$\vec{F}_{NET} = \frac{d\vec{p}}{dt}$$

In other words, the net force tells you how fast the momentum is changing, and the momentum of an object is equal to the object's mass times the object's velocity.

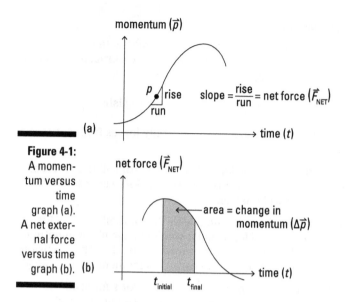

momentum $(\vec{p})$

p rise run

$slope = \dfrac{rise}{run} = $ net force $(\vec{F}_{NET})$

time (t)

(a)

net force $(\vec{F}_{NET})$

area = change in momentum $(\Delta\vec{p})$

time (t)

$t_{initial}$ t_{final}

(b)

Figure 4-1: A momentum versus time graph (a). A net external force versus time graph (b).

If the mass is a constant, then the time rate of change of the momentum is equal to the mass times the time rate of change of the velocity. *Acceleration* by definition is the time rate of change of the velocity. Therefore, the mass times the acceleration equals the net external force acting on the object when the mass is a constant. The mathematical description of Newton's second law of motion, the law of acceleration, is

$$\vec{F}_{NET} = m\,\vec{a}$$

Going around the bend — circular motion

The version of Newton's second law in the preceding section is the starting point, but it can be modified for when an object is going around in a circle. When an object is going in a circle, the torque is what's interesting. *Torque* is the rotational analogue of a force. It tries to make the object spin faster or slower.

Newton's second law of motion, the law of angular acceleration, states the time rate of change of the angular momentum of an object is equal to the net external torque acting on the object from the surroundings. In other words, the net external torque acting on an object is a measure of how fast the angular momentum will change. Take Figure 4-1 and replace momentum with angular momentum and net force with net torque to get an idea how this law looks.

$\overline{L}$ is the symbol for the angular momentum of the object, which is equal to the moment of inertia (I) times the angular velocity ($\overline{\omega}$). Don't forget that the angular momentum and the angular velocity are vectors. If the moment of inertia is a constant, then the moment of inertia (I) times the angular acceleration ($\overline{\alpha}$) equals the net external torque acting on the object. The mathematical description is

$$\overline{N}_{NET} = I\,\overline{\alpha}$$

Newton's second law gives a relationship between the angular acceleration of an object and the net external torque acting on the object. The *net external torque* is the vector sum of all the torques produced by all the external forces from the surroundings acting on the object. The mathematical description is

$$\overline{N}_{NET} = \hat{x}\,N_{x,NET} + \hat{y}\,N_{y,NET} + \hat{z}\,N_{z,NET}$$
$$N_{x,NET} = \sum N_x, \; N_{y,NET} = \sum N_y, \; N_{z,NET} = \sum N_z$$

If you're keen to do more math (who isn't?), the calculus definition of Newton's law of angular acceleration is as follows:

$$\overline{N}_{NET} = \frac{d\overline{L}}{dt}$$

Figure 4-2 provides a graphical interpretation of Newton's law of acceleration. Figure 4-2a shows the velocity versus time graph, demonstrating the connection to the net force through Newton's second law. The slope of the graph at a single point is equal to the acceleration, which is equal to the net external force acting on the object divided by the mass (m), which is a constant. Figure 4-2b shows the graph of angular velocity versus time graph for a rotational system with a constant moment of inertia, showing the connection to the net torque through Newton's second law. The slope of the graph at a single point is equal to the angular acceleration, which is equal to the net external torque acting on the object divided by the moment of inertia (I).

The unit for torque in SI units is *newton-meter* and not joule even though both are a kilogram-meter squared divided by second squared in fundamental units. The reason you use joule is because it's associated with work and energy, whereas newton is the unit of force and torque is the rotational analogue of a force, so newton-meter is used.

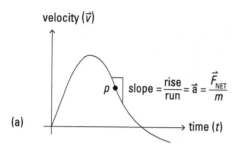

velocity ($\vec{v}$)

$$\text{slope} = \frac{\text{rise}}{\text{run}} = \vec{a} = \frac{\vec{F}_{NET}}{m}$$

(a)

time (t)

Figure 4-2:
A velocity versus time graph (a). An angular velocity versus time graph (b).

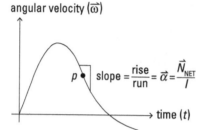

angular velocity ($\vec{\omega}$)

$$\text{slope} = \frac{\text{rise}}{\text{run}} = \vec{\alpha} = \frac{\vec{N}_{NET}}{I}$$

(b)

time (t)

You may be thinking how Newton's first two laws relate to biophysics and biomechanics in particular. For example, suppose you're hired to analyze the motion of Fred, a discus thrower, and the discus. To begin, according to Newton's first law of motion, the law of inertia, the discus will travel in a straight line unless Fred applies a force to the discus to keep it moving in a circle. Fred must also apply a torque to the discus so the discus spins faster according to Newton's second law of motion, the law of acceleration. After spinning around, Fred releases the discus, and according to Newton's first law of motion, the law of inertia, the discus now wants to travel in a straight line. The discus's angular momentum now becomes linear momentum as the discus flies off in a straight line. The discus won't travel in a straight line through the air because of gravity. This force causes the discus to have an acceleration toward the earth.

Interacting takes two — forces and Newton's third law of motion

This section examines Newton's third law of motion, which looks at how two objects interact. I introduce two forces that are common when two objects interact: the contact force and the normal force. (No abnormal force exists in nature.) These sections examine this law more closely.

Acting and reacting to Newton's third law of motion

Newton's third law of motion, the law of action and reaction, states if an object exerts a force on a second object, then the second object exerts a force on the first object that is equal in magnitude but in the opposite direction. This law of motion is different from Newton's first two laws. This law describes what happens when two objects interact with each other.

This law isn't obvious. Consider the following example. I ask Gemma to clasp hands with a sumo wrestler and ask her to push as hard as she can against the sumo wrestler's hands. Neither Gemma nor the sumo wrestler is moving, even though they're pushing against each other. Most people say the wrestler is pushing with more force, which is incorrect. Newton's third law, the law of action and reaction, states the magnitude of the force produced by the wrestler on Gemma's hands is equal to the magnitude of the force applied by Gemma on the wrestler's hands.

To help you see this, place a piece of paper between Gemma's hands and the wrestler's hands, and have them push. The paper doesn't move, so its velocity is zero and its acceleration is zero. Newton's second law of motion, the law of acceleration, tells you the net external force must be zero because the acceleration is zero. Gemma is pushing on the paper one way and the wrestler is pushing on the paper in the opposite direction, and the forces must cancel, so the forces are equal in magnitude and in the opposite directions.

Bumping with the contact force

The *contact force* is the force between two objects when they're in contact. Contact forces are everywhere and play an important role in biophysics. Look around you and everywhere you see two objects touching there is a contact force. For example, you hold your biophysics book with your hands, joggers push off the ground, people hold hands, a virus attaches to a cell, or a molecule attaches to an enzyme. Even your own body has many contact forces. The contact forces arise in the ligaments connecting one bone to another, tendons connecting muscles to bone, and the cartilage connecting the bones in the joints. In biophysics, the fundamental force that produces the contact force is the electromagnetic force between the protons and electrons within the two objects.

You may be wondering about gravity. You're in contact with the earth right now (unless you like to read your biophysics book while skydiving), but gravity isn't a contact force because it pulls on you independent of whether you're touching the ground or skydiving. On the other hand, the previous example where Gemma is trying to push the sumo wrestler over is a contact force. The force between Gemma's hands and the sumo wrestler's hands stops as soon as Gemma breaks contact.

Pointing in a specific direction — the normal force and tangential force

A contact force that is present in many biomechanical situations is the contact between an object and a surface. In many applications of biomechanics, splitting the contact force with a surface into its components is convenient. The two components are

- ✔ **A normal force perpendicular to the surface:** The *normal force* is the part of this contact force that points perpendicular to the surface. It prevents gravity and other forces from sucking the object into the (surface) ground. For example, if Harry jumps off the top of his kitchen table, gravity accelerates him downwards until he hits the ground, and then the normal force produced by contact with the ground stops his downward motion. *Remember:* Normal means perpendicular.

- ✔ **A tangential force parallel to the surface:** The *tangential force* plays an important role in biomechanics as well, from everyday activities to sports. For example, runners, sprinters, joggers, and power walkers push the ground backwards, and according to Newton's third law of motion, the law of action and reaction, the ground is pushing them forward. Yes, everyone walks by using the reaction force from the ground. The tangential force in the case of these athletes is *static friction* between their feet and the ground, which allows them to move forward.

Meeting Conservative Forces — No Tea Party Folks Here

A *conservative force* is path independent, which means that if the force does work on an object moving it from some point to another, then you must do the same amount of work against that force to move it back to the original point. For example, if Isha's biophysics book slips out of her hands, gravity will pull the book downwards and it will fall onto the table. Isha has to reach down and lift the book up against gravity to bring it back to its starting position.

The force does work, and this work can be associated with a change in energy. *Potential energy* is the potential of a conservative force to do work on an object. Every conservative force has a corresponding potential energy. To continue Isha's example, the book has a large amount of *gravitational potential* when she is holding it, but the book loses potential energy as gravity does work pulling the book down toward the table. The book regains its gravitational potential energy as Isha lifts it back up.

These sections discuss two conservative forces: the force of a spring (Hooke's law) and the force of gravity (Newton's universal law of gravity) along with their corresponding potential energies.

Hooking into Hooke's Law

Hooke's law describes the force of a spring. If you were to take a spring and compress it, the spring would push back against you. Also, if you pull the spring and stretch it, it would fight back and pull in the opposite direction. Springs like to be a specific length and don't want to be compressed or stretched.

The force of a spring equals the displacement times the spring constant. The force is in the direction opposite to the displacement from equilibrium. The mathematical expression for the force in Hooke's law is

$$\vec{F}_H = -k_H \left(\vec{s} - \vec{s}_{eq} \right)$$

$\vec{F}$ is the force applied to an object by the spring, k_H is the spring constant, $\vec{s}$ is the position of the object in contact with the spring, and $\vec{s}_{eq}$ is the equilibrium position, which is where the object needs to be for the spring to produce no force on the object. Hooke's spring constant is a measure of the spring's stiffness. In other words, the larger the Hooke's spring constant becomes, the more force is required to move it from its equilibrium position.

Meanwhile, the potential energy equals the magnitude of the displacement squared times half the spring constant. The mathematical expression for the potential energy in Hooke's law is

$$E_{P,H} = \frac{k_H}{2} \left| \vec{s} - \vec{s}_{eq} \right|^2$$

In this formula, $E_{P,H}$ is the potential energy of the spring, k_H is the spring constant, $\vec{s}$ is the position of the object in contact with the spring, and $\vec{s}_{eq}$ is the equilibrium position, which is where the object needs to be for the spring to produce no force on the object.

Most scales that measure weight work on a spring. When you step on the scale, your weight compresses the scale and the spring inside the scale supplies a countering force (this is Newton's third law of motion, the law of action and reaction). This countering force is what you measure on the scale.

For example, suppose James weighs 150 pounds (667 newtons) and the bathroom scale compresses 0.25 inches (0.64 centimeters) when he stands on it. What is the value of the spring constant? What is the potential energy stored in the spring while he is standing on it?

You can find the spring constant from Hooke's law. You know the force of the spring equals the force of James's weight. (The acceleration is zero, so the net force is zero from Newton's second law, the law of acceleration.) You also know the displacement, so Hooke's spring constant is

$$k_H = \frac{F_H}{\Delta y} = \frac{150 \text{ lb}}{0.25 \text{ in}} \left(\frac{12 \text{ in}}{1 \text{ ft}} \right) = 7200 \frac{\text{lb}}{\text{ft}} \left(1.05 \times 10^5 \frac{N}{m} \right)$$

Now that you have Hooke's spring constant, you can find the potential energy:

$$E_{P,H} = \frac{k_H}{2} |\Delta y|^2 = \frac{1}{2} \left(7200 \frac{\text{lb}}{\text{ft}} \right) \left(0.25 \text{ in} \frac{1 \text{ ft}}{12 \text{ in}} \right)^2 = 1.56 \text{ foot pounds } (2.12 \text{ joules})$$

Getting heavy with the effect of gravity

Weight is the force of gravity between an object and the earth, and *gravity* is the force that causes you to hit the floor when you fall out of bed. Gravity keeps you stuck to the earth so you don't fly off into space. The force of gravity keeps the earth orbiting the sun. The force of gravity is an important force in some biophysical situations, such as cliff diving and skydiving. Also, weightlessness in space is very hard on the human body; your body is designed to live in a world with gravity.

The force of gravity is an attractive force between two objects with mass. The force is equal to Newton's gravitational constant times the mass of the first object times the mass of the second object divided by the distance between the objects squared. Newton's gravitational constant is $G = 3.44 \times 10^{-8}$ pound feet squared/slug squared (6.67×10^{-11} newton meters squared/kilogram squared). The mathematical expression for the force acting on the second mass (M_2) is

$$\vec{F}_{g,2} = -\vec{F}_{g,1} = G M_1 M_2 \frac{\left(\vec{s}_1 - \vec{s}_2 \right)}{\left| \vec{s}_1 - \vec{s}_2 \right|^3}$$

In this formula, $\vec{s}_1$ is the position of the first mass (M_1) and $\vec{s}_2$ is the position of the second mass (M_2). I include the force acting on the first mass because Newton's third law of motion, the law of action and reaction, states that the force on mass 1 is equal and opposite.

The force of gravity is a conservative force, so the work done by gravity can be associated with a change in the potential energy. The potential energy of gravity equals Newton's gravitational constant times the mass of the first object times the mass of the second object divided by the distance between the objects. Newton's gravitational constant is $G = 3.44 \times 10^{-8}$ foot pounds squared/slug squared (6.67×10^{-11} newton meters squared/kilogram squared). The mathematical expression for the potential energy is

$$E_{P,g} = \frac{G M_1 M_2}{\left| \vec{s}_1 - \vec{s}_2 \right|}$$

For example, suppose Merlin with a mass of 5.5 slugs (80.2 kilograms) is dancing with Kaitlin whose mass is 4.5 slugs (65.7 kilograms), and they're separated by a distance of 1 foot (0.305 meters). The force of gravitational attraction between them is pulling them towards each other, so what is the magnitude of this force?

You can find the answer by substituting these numbers into the equation for the gravitational force:

$$\left| \vec{F}_g \right| = \frac{G M_1 M_2}{\left| \vec{s}_1 - \vec{s}_2 \right|^2} = \left(3.44 \times 10^{-8} \frac{\text{lb ft}^2}{\text{slug}^2} \right) \frac{(5.5 \text{ slugs})(4.5 \text{ slugs})}{(1 \text{ ft})^2} =$$

$$8.51 \times 10^{-7} \text{ pounds} \left(3.79 \times 10^{-6} \text{ newtons} \right)$$

This is an extremely small force, so in biophysics, you can ignore the gravitational force between the objects except with the earth, which has a very large mass. The gravitational force between an object and the earth is called the *weight* of the object. At the earth's surface, the weight of an object is

$$\vec{F}_g = \vec{g} \, m$$

In this formula, g is the magnitude of acceleration due to gravity and is equal to G times the mass of the earth divided by the radius of the earth squared. The direction of the force is straight down toward the center of the earth. The *average value of g* is 32.2 feet per second squared (9.81 meters per second squared).

The gravitational potential energy close to the surface of the earth is

$$E_{P,g} = -m \, \vec{g} \bullet \vec{s} = m \, g \, y$$

The magnitude of the weight of Merlin is 177 pounds (788 newtons) and of Kaitlin is 145 pounds (645 newtons), by using mg, which means the force of attraction between them (8.51×10^{-7} pounds $= 3.79 \times 10^{-6}$ newtons) is about a billion times smaller than their weights.

Knowing the difference between weight and mass

Mass is a physical property of an object; it depends only on the object. *Weight* is the gravitational force of attraction between the object and the earth, which depends on the location on the Earth and the altitude.

Many people talk about weight and mass in terms of pounds or kilograms. You very rarely hear the terms *slugs* or *newtons* when talking about the mass or weight of objects. The reason is the slug is very insensitive to change with a mass of 1 slug having a weight of 32 pounds. The newton is the opposite; it's too sensitive of a scale for the weight of everyday objects.

For instance, an object with a weight of 1 newton has a mass of 0.102 kilograms. If you drink a cup of water, then your weight has suddenly increased by 2.45 newtons. The human body can fluctuate a few pounds during a 24-hour period. An athlete can lose several pounds during a competition. Saying a person's weight is 788 newtons doesn't make much sense when his or her weight could be 775 newtons in the morning and 800 newtons just 12 hours later. Instead of using kilograms for weight (which isn't wise), I like to think of it in terms of dekanewtons (daN); so an object with a mass of 1 kilogram has a weight of 0.981 daN or 2.20 pounds.

The weight of an object depends on where it's located on earth because the mass inside the earth isn't distributed uniformly, and the earth looks like a ball that someone has sat on. The earth's diameter from the North Pole to the South Pole is 7,901 miles (12,714 kilometers), and the acceleration due to gravity at the poles is 32.26 feet per second squared (9.832 meters per second squared). The earth's average diameter at the equator is 7,928 miles (12,756 kilometers), and the acceleration due to gravity at the equator is 32.20 feet per second squared (9.814 meters per second squared). If you went to the equator and weighed yourself, you would find the *effective acceleration due to gravity* of 32.09 feet per second squared (9.781 meters per second squared). If you're interested in effective weight, check out Chapter 6.

The easiest weight-loss program is to move closer to the earth's equator or move to higher altitudes. If you have the money to visit the international space station, then you would lose all your effective weight (and some of your weight). Of course, doing so would have no effect on your mass. Weight-loss programs should actually be called mass-loss programs.

Recognizing the Nonconservative Forces: No Bleeding Hearts Here

A *nonconservative force* is a force that is path dependent, which means that the amount of work done by the force on an object while moving the object from some point to another depends on how the object moves between the two points. *Dissipative* forces (friction) and applied forces are nonconservative forces.

You can't describe nonconservative forces with a potential energy. Only conservative forces have a potential energy. These sections discuss a few nonconservative forces that arise in biomechanics.

Walking in the park — static friction

Static friction is a contact force between two objects that aren't moving relative to each other and prevents the two surfaces from sliding apart. The magnitude of the static friction is between zero and some maximum value depending on the applied forces.

If an applied force exceeds the maximum magnitude of the static friction, then the two surfaces will break apart and slide. The maximum static friction is in the direction opposite to the direction the object is trying to move in, and the magnitude of the maximum static friction is equal to the coefficient of static friction times the magnitude of the normal force. The mathematical description for the magnitude of the maximum static friction is

$$\left|\overrightarrow{F}_{sf}^{max}\right| = \mu_s \left|\overrightarrow{F}_N\right|$$

Allow me to analyze the activities of walking, jogging, and running as an example. If I'm standing still on flat ground, there is no static friction; otherwise, the friction force would make me accelerate across the ground. If I'm standing on a hill, then static friction is holding me up; otherwise, I would start to slip down the hill. If the hill has wet grass or mud, then I could still slide downwards because of the insufficient amount of friction between my shoes and the ground. This example tells me that the coefficient of static friction depends on the type of shoes I'm wearing and the type of ground I'm standing on. The easiest method for me to find a numerical value for the coefficient of static friction between my shoes and the ground is to perform an experiment.

I'm walking on flat ground, so I lift a foot and bring it forward while pushing backwards on the ground with the other foot. Newton's third law of motion, the law of action and reaction, says the reaction force from the earth pushes on my foot. The contact force between my foot and the ground has two parts:

- ✔ **The normal force pushing the ground downward:** The normal force is pushing the foot (and my body) upward.

- ✔ **The tangential (static friction) force is pushing the earth backward:** The static friction is pushing my foot and body forward.

Notice the static friction is the reaction force that is pushing my body forward. When you're walking, you're trying to make your foot move backward, and static friction prevents your foot from sliding backward. Static friction also prevents your front foot from sliding forward when you step down. If you make too big of a step in shoes with a slippery heel, you'll end up doing the splits.

The acceleration of a body is limited by the maximum value of the static friction, which is why it's so hard to run on slippery surfaces like ice. Most animals have a maximum horizontal acceleration that is less than the acceleration of gravity, $g = 32.2$ feet per second squared (9.81 meters per second squared), even on rough surfaces.

Hurting in the joints when moving — kinetic friction

Kinetic friction is a contact force between two objects that are sliding against each other. The kinetic friction is in the direction opposite to the direction the object is moving in, and the magnitude is equal to the coefficient of kinetic friction times the magnitude of the normal force. The mathematical description for the magnitude of the kinetic friction is

$$\left|\vec{F}_{kf}\right| = \mu_k \left|\vec{F}_N\right|$$

The electromagnetic force is trying to bind the two objects together and the roughness of the surfaces (ridges and valleys on both surfaces) rub together and act like brakes stopping the sliding motion. The kinetic friction is converting the object's mechanical energy into heat energy.

The coefficient of kinetic friction is usually less than the coefficient of static friction, which is why it's harder to get an object to start moving than to keep it moving.

In many situations within biophysics the kinetic friction force is an unwanted force that is working against the system, but nature is smart in reducing this friction. For example, in synovial joints, such as the knee, shoulder, and hip, the two bones don't rub against each other; they're separated by cartilage and synovial fluid. In normal joints, the coefficient of static friction is about 0.01, whereas the coefficient of kinetic friction can be as low as 0.003. A normal joint has very little friction, which means the joints don't heat that much during activities. In an arthritic joint, the coefficient of kinetic friction can be 100 times larger, which can generate a lot more heat and pain.

Identifying other nonconservative forces

Other nonconservative forces play a role in biomechanics, especially in sports activities. I briefly mention three in these sections.

Rolling with rolling resistance

If you push a wheel and let it roll along a flat surface, it will eventually stop because of the *rolling resistance*. The mathematical description for this force is

$$\left|\vec{F}_{rf}\right| = \mu_r \left|\vec{F}_N\right|$$

The mathematical form is similar to the frictional forces that I discuss in this chapter. The coefficient of rolling resistance depends on many factors and properties of the road and the rolling object. Reducing this type of force is important for the fuel economy of your vehicle and in many sports such as bike races. You aren't going to win the Tour de France if your bike has a large rolling resistance.

Resisting is futile — fluid resistance

Fluid resistance on an object moving through a liquid and air resistance on small particles floating in the air, such as dust, is linearly dependent upon the speed of the object. Fluid resistance is also known as the *drag force* and the motion of objects through fluids with high viscosity (see Chapter 10). The mathematical description for fluid resistance is

$$\left|\vec{F}_{ff}\right| = \mu_f \left|\vec{v}\right|$$

The fluid resistance is an important force in swimming competitions. Many international competitions are won by a few hundredths of a second, and fluid drag force can make all the difference in a race. One way swimmers reduce drag is by covering their head or shaving it bald. Males and females also remove the rest of their body hair to reduce drag. Technological swimwear also has reduced fluid drag.

Skydiving — air resistance is a good thing

Air and other gases aren't as dense as fluids, so their resistance isn't dependent upon the speed but is dependent on the speed squared. The mathematical description of air resistance is

$$\left|\vec{F}_{af}\right| = \tfrac{1}{2}D \rho A v^2$$

In this formula, D is the drag coefficient, ρ is the air density, A is the frontal cross-sectional area, and v is the speed. The *drag coefficient* is a measure of how aerodynamic the object is. For example, a person skydiving will wear regular clothes that will flap in the air and create a large amount of drag, whereas a cliff diver will wear a swimming suit that will create very little drag.

The *frontal cross-sectional* is the area of the object that is pushing against the air. The cliff diver does a beautiful swan dive, and only the head and shoulders push against the air; the cliff diver has a very small cross-sectional area. On the other hand, the skydiver does a belly flop, and the air pushes against his face, chest, midsection, and legs. The frontal cross-section in this case is the entire body, which is very large. The weight density of air is 0.0749 pounds per cubic foot at a temperature of 70 degrees Fahrenheit and 14.70 pounds per square inch atmospheric pressure. The mass density of air is 1.2041 kilograms per cubic meter at a temperature of 20 degrees Celsius and 1 atmosphere (101325 Pascals). (You can get these values in the CRC Handbook of Chemistry and Physics.)

In biomechanics, air resistance can be a good thing or a hindrance. Skydivers and base jumpers appreciate air resistance. Many other sports, such as auto racing and downhill skiing, focus a lot of training on reducing air resistance. On the other hand, birds and aircraft need the air for lift, but the air drag wastes a lot of energy.

Thinking Green — Conservation Is Good; So Is Energy, Work, and Power

Work is related to force, and the change in the energy is related to the work, and power is related to work as well. Studying work, energy, and power tells you a lot about a system's biomechanics. It's equivalent to using forces and Newton's laws of motion. In addition, conservation laws tell you what does not change in time. The concepts of energy, work, power, and conservation laws are very important in biophysics. Some situations in biophysics can be solved easier using energy and work as opposed to Newton's laws and forces. Also, knowing what doesn't change in a biological system is very important to understanding the system. The following sections introduce these concepts and discuss some of the applications to biomechanics.

Conserving momentum

Including all interacting objects within your system makes the total momentum of the system a constant over time. According to Newton's second law of motion, the law of acceleration, the total *momentum* of the system changes over time when the system interacts with external sources, which means that the total momentum of the universe is conserved and doesn't change for the entire life span of the universe (the universe is an isolated system), unless the universe interacts with other universes.

Strictly speaking, everything within the universe is interacting with your biological system, but most of those interactions can be ignored. For example, if you're studying a person doing the high jump, you don't need to consider the gravitational attraction of the high jumper with the sports announcer sitting in the top balcony of the stadium.

The next two sections look at conservation of linear momentum and conservation of angular momentum.

Conservation of linear momentum

The total *linear momentum* of the biological system is conserved for all time if the net external force acting on the system is zero. In plain English, it means the momentum is a constant as long as there is no net force acting on the object according to Newton's second law of motion, the law of acceleration. The mathematical condition from Newton's second law of motion, the law of acceleration, is

$$\vec{F}_{NET} = \hat{x}\, F_{x,NET} + \hat{y}\, F_{y,NET} + \hat{z}\, F_{z,NET} = \vec{0}$$

$$F_{x,NET} = \sum F_x = 0, \ F_{y,NET} = \sum F_y = 0, \ F_{z,NET} = \sum F_z = 0$$

The force in each direction must be zero. As a consequence, the total momentum in each direction is a constant, and the mathematical formula for total linear momentum is

$$\vec{p}_{TOTAL} = \hat{x}\, p_{x,TOTAL} + \hat{y}\, p_{y,TOTAL} + \hat{z}\, p_{z,TOTAL}$$

$$p_{x,TOTAL} = \sum p_x, \ p_{y,TOTAL} = \sum p_y, \ p_{z,TOTAL} = \sum p_z$$

Conservation of angular momentum

The total *angular momentum* of the biological system is conserved for all time if the net external torque acting on the system is zero. The mathematical condition from Newton's second law of motion, the law of acceleration, is

$$\vec{N}_{NET} = \hat{x}\, N_{x,NET} + \hat{y}\, N_{y,NET} + \hat{z}\, N_{z,NET} = \vec{0}$$

$$N_{x,NET} = \sum N_x = 0, \ N_{y,NET} = \sum N_y = 0, \ N_{z,NET} = \sum N_z = 0$$

The angular momentum ($\vec{L}$) is equal to the moment of inertia (I) times the angular velocity ($\vec{\omega}$), and the total angular momentum is

$$\vec{L}_{TOTAL} = \hat{x}\, L_{x,TOTAL} + \hat{y}\, L_{y,TOTAL} + \hat{z}\, L_{z,TOTAL}$$

$$L_{x,TOTAL} = \sum L_x, \; L_{y,TOTAL} = \sum L_y, \; L_{z,TOTAL} = \sum L_z$$

For example, consider Omar diving off the 10-meter platform into the swimming pool below. As Omar jumps off the edge of the platform in an upright position, he starts spinning his body, so his body has angular momentum. (His body will spin while he drops toward the water.)

In order to spin faster, Omar has to pull his body into a tuck position because of conservation of angular momentum. When Omar goes into the tuck position, he has decreased the moment of inertia of his body, which causes his angular velocity to increase so the angular momentum doesn't change.

The reason angular momentum is conserved is because the net torque acting on Omar is zero. Air resistance can be ignored, and the only force acting on him is gravity, which is pulling his body downward. The force of gravity doesn't produce any torque on Omar, so the net torque is zero and his angular momentum is conserved.

Calculate Omar's initial angular velocity and estimate his minimum speed off the platform in order to do 2.5 rotations before hitting the water. Before Omar climbs the ladder to the diving board, you make some measurements. He weighs 175 pounds (778 newtons) or his mass is 5.43 slugs (79.3 kilograms), and he is 6 feet (1.83 meters) tall while in the upright position. When he is diving in a tuck position, he has a radius of 1.5 feet (0.457 meter). During his dive, he is timed to take 1.50 seconds to hit the water after leaving the platform, and he's in the tuck position for 1.30 seconds.

Follow these steps to solve this problem:

1. Determine the physics of the problem and the formulas needed.

Remember, average angular velocity equals angular displacement divided by elapsed time. You know how long it takes (elapsed time) Omar to do his rotations while in the tuck position (angular displacement).

Omar has conservation of angular momentum ($L_{initial} = L_{final}$) while falling. The angular momentum is $L = I\,\omega$, where L is the angular momentum, I is the moment of inertia, and ω is the angular velocity, so you can find $\omega_{initial}$ from conservation of angular momentum once you know the other three. I ignore the vector arrow on the angular variables because they're one-dimensional.

Omar's speed is equal to his initial angular velocity times half the length of his body. The problem asks you to estimate Omar's speed off the platform. At the moment Omar jumps off the platform, his feet are momentarily stationary with respect to the platform and his *center of mass* is moving with his body speed. (The center of mass of the human body is slightly above the hips.) According to the center of mass, his feet and the platform are moving backwards with the speed of his body. Chapter 6 discusses center of mass more.

2. **Find the numbers necessary to calculate Omar's final angular velocity.**

   You know you need the time and angular displacement:

   The elapsed time to do the rotations is Δt = 1.30 seconds.

   The angular displacement is $\Delta\theta$ = (2.5 revolutions)(2π radians per 1 revolution) = 5π radians.

3. **Calculate Omar's final angular velocity.**

   The magnitude of the final angular velocity is

   $$\omega_{final} = \frac{\Delta\theta}{\Delta t} = \frac{5\pi \text{ rads}}{1.30 \text{ s}} = 12.1 \frac{\text{radians}}{\text{second}}$$

4. **Find the numbers so you can calculate Omar's initial angular velocity.**

   You need to use conservation of angular momentum to calculate the initial angular velocity, so you need to find numbers for the initial moment of inertia and final moment of inertia and the final angular velocity:

   Omar is an athletic diver, so he looks more like a 6-foot pole instead of a sphere or disk rotating around his center as he jumps off the platform. His initial moment of inertia is approximately that of a rod, so look up the formula for the moment of inertia of a rod: $I_{initial} = mL^2/12$ = (5.43 slugs)(6 feet)^2/12 = 16.3 slug feet squared (22.1 kilogram meters squared).

   When Omar is in a tucked position, he looks like a rotating disk. His final moment of inertia is approximately that of a disk, so look up the formula for the moment of inertia of a disk: $I_{initial} = mR^2/2$ = (5.43 slugs)(1.50 feet)^2/2 = 6.11 slug feet squared (8.29 kilogram meters squared).

   The final angular velocity from Step No. 3 is ω_{final} = 12.1 radians per second.

5. **Apply the conservation of angular momentum formula and solve for the initial angular velocity.**

   $$\omega_{initial} = \frac{L_{initial}}{I_{initial}} = \frac{L_{final}}{I_{initial}} = \frac{I_{final}\omega_{final}}{I_{initial}} = \frac{\left(6.11 \text{ slug ft}^2\right)\left(12.1 \text{ rads/s}\right)}{\left(16.3 \text{ slug ft}^2\right)} = 4.54\frac{\text{radians}}{\text{second}}$$

6. **Find the numbers, so you can calculate Omar's initial speed.**

   You know Omar's initial angular velocity is $\omega_{initial}$ = 4.54 radians per second.

   You know half his height is R = 3.00 feet (0.914 meters).

7. **Calculate the initial speed.**

   $v_{initial} = R\,\omega_{initial}$ = (3 ft)(4.54 rads/s) = 13.6 feet per second (4.14 meters per second).

Moving energy and work

This section illustrates the concepts of energy, work, and their relationship to each other. They're very powerful concepts in biophysics and physics. Understanding these concepts is critical to understanding biomechanics and bioenergetics. Here I show how the two are connected, their importance in conservation laws, and how they're used in biomechanics. Chapter 3 includes a more detailed definition of the two (energy and work).

Examining the work-energy theorem

The *work-energy theorem* states the change in the total kinetic energy of a biological system is equal to the net work done on that system by external forces. In other words, a force acting on a moving object does work on the object. When you add up all the work done on the object, it equals the net work done on the object. The change in the object's kinetic energy is equal to this net work done on the object. The mathematical expression for this theorem is

$$\Delta E_K = E_{K,final} - E_{K,initial} = W_{NET}$$

For example, when you strike a baseball with a bat, the bat first does work, stopping the ball going over the plate, and then it does more work, giving the ball kinetic energy so it flies into the field.

Conservation of mechanical energy

Another way of interpreting the work-energy theorem is the change in the total mechanical energy of a biological system is equal to the work done by nonconservative forces such as applied forces and friction. *Conservation of total mechanical energy* means the total kinetic energy plus the total potential energy doesn't change in time. This means that the total work done by nonconservative forces is zero.

The mathematical expression for the work-energy theorem written in terms of the kinetic energy, potential energy, and the work done by dissipative forces is

$$\Delta E_M = \Delta E_K + \Delta E_P = W_{NCF}$$

In this expression, ΔE_M is the change in the total mechanical energy of the biological system, ΔE_K is the change in the total kinetic energy of the biological system, ΔE_P is the change in the total potential energy of the biological system, and W_{NCF} is the work done against nonconservative forces.

For example, Paige is riding a bike along a flat, horizontal road at constant velocity. The velocity is constant, so the kinetic energy isn't changing ($\Delta E_K = 0$), and she is riding along a horizontal road, so the gravitational potential energy isn't changing ($\Delta E_P = 0$), so the total mechanical energy of the system is conserved ($\Delta E_M = 0$). However, Paige is doing work against the dissipative forces, so the total work done by nonconservative forces is zero ($W_{NCF} = 0$). Yes, Paige's muscles are a nonconservative force according to the bike. The dissipative forces she is doing work against are friction in the bike's bearings, rolling resistance, and air resistance.

When the net work done by nonconservative forces is zero and the potential energy is proportional to the mass, such as the gravitational potential energy, then the equation becomes independent of the mass. The mathematical formula for conservation of mechanical energy and the gravitational potential energy is

$$\frac{v_{final}^2}{2} + g y_{final} = \frac{v_{initial}^2}{2} + g y_{initial}$$

The conservation of mechanical energy combined with the gravitational potential energy formula is very useful in biomechanics. It gives the minimum speed (v) necessary to achieve a certain vertical height (h) against gravity, which is important in several sports such as basketball. The formula for this specific situation is

$$v = \sqrt{2 g h}$$

A 3-foot (0.914 meter) vertical jump requires an initial vertical velocity of 9.48 miles per hour (15.2 kilometers per hour) upward.

Here is an example.

BRAINTEASER

> The outdoor world record for the pole vault is 20.1 feet (6.14 meters). What was the minimum horizontal speed the athlete required to make this jump?
>
> In this event, the athlete runs horizontally and then sticks the pole in the ground. The athlete changes his kinetic energy into work on the pole, which is stored as potential energy within the pole as it bends. This potential energy is converted back into work on the athlete, which changes the athlete's kinetic energy, but now in the vertical direction. This kinetic energy does work against gravity, and the energy is converted into gravitational potential energy within the athlete as the athlete moves vertically upwards.

To solve this problem, follow these steps:

1. **Determine the physics of the problem and the formulas needed.**

   The formula needed is the change in the mechanical energy equals the work done by nonconservative forces.

   The minimum amount of kinetic energy required corresponds to the situation when there are no nonconservative forces present, which is equivalent to saying the pole vaulter has conservation of mechanical energy.

2. **Find the numbers needed to solve the problem.**

   All the mechanical energy is kinetic energy just prior to the pole hitting the ground, so $y_{initial}$ = 0 feet (0 meters).

   At the highest height all the mechanical energy is gravitational potential energy, so v_{final} = 0 feet per second (0 meters per second), and y_{final} = 20.1 feet (6.14 meters).

3. **Solve the problem using conservation of mechanical energy.**

   Rearrange the conservation of mechanical energy equation and solve for $v_{initial}$:

   $$v_{initial} = \sqrt{v_{final}^2 + 2gy_{final} - 2gy_{initial}} = \sqrt{0 + 2\left(32.2\frac{\text{ft}}{\text{s}^2}\right)(20.1\text{ ft}) - 0} =$$

   $$36.0\frac{\text{feet}}{\text{second}}\left(11.0\frac{\text{meters}}{\text{second}}\right)$$

Working with energy and power

Energy, work, and power are very powerful tools in biomechanics. You may encounter five types of energy (six if you include dark energy), which are kinetic energy, potential energy, heat energy, radiation energy, and mass energy. These sections show you how to use concept of conservation combined with energy, work, and power to study many different situations in biomechanics.

Conservation of total energy

Conservation of total energy means you can't create or destroy energy; you can only change it from one form to another. In other words, the sum of all the kinetic energies, potential energies, heat energy, radiation energy, and mass energies within an isolated biological system is a constant. If the biological system isn't isolated, the change in this total energy of the system is equal to the amount of heat energy entering the system plus the work done on the system by external sources. (The heat energy is negative if it's flowing out of the system, and work done is negative if the biological system is doing work on its surroundings.)

You can only convert energy from one form to another, which is true on all levels, including the cellular level with the transport of energy, the interaction of the sun with life on earth through photosynthesis, and even athletes competing.

Conservation of power and work

Conservation of total energy is always true, but sometimes in biomechanics it's better to think of the problem in terms of the work (power) and energy put into the system and the work (power) and energy output by the system. They have to be the same unless the biological system is storing the energy or using internal energy that was already stored in the system.

For example, suppose Quintin is riding his bike. The power from his pedaling is transferred into the pedals, through the chain, and then to the back wheel, which pushes the ground backward with the same amount of power that Quintin puts into pedaling. Newton's third law states that this force is equal to the force the earth pushes back on the bike with and propels the bike forward. Three things can happen:

✔ The pedaling does work (power in equals work in divided by time) which is greater than the work done against dissipative forces (power out equals work out divided by time), and the excess work in does work on the ground and accelerates the bike forward.

✔ The pedaling does work (power in equals work in divided by time), which is the same as the work done against dissipative forces (power out equals work out divided by time), and the bike moves at a constant speed.

✔ Quintin's pedaling does work (power in equals work in divided by time), which is less than the work done against dissipative forces (power out equals work out divided by time), and the bike slows down. The bike is using kinetic energy to do work against the dissipative forces. The kinetic energy comes from the linear forward motion of the bike plus the rotational kinetic energy of the wheels.

In all three cases, the power in equals the power out and nothing is lost. You may have a couple of questions pop into your mind, such as:

✔ **What are the dissipative forces?**

The wheels, the pedals, the sprockets, and the derailleur (gear changer) have frictional force within their bearings.

Rolling resistance depends on the tires and the type of surface Quintin is riding on.

Air resistance can play an important role (try riding a bicycle through a hurricane), depending on how fast he is riding, his sitting position, and the type of bike he is riding.

Gravity if Quintin is going uphill. If he is going downhill, then it can help him. (This is not a real dissipative force.)

✔ **Why is it easier to pedal in certain gears?**

To answer this question, follow these steps to see how work is transferred through the bike:

1. **Find the work done on the pedal by Quintin.**

   Quinton's feet do work on the pedals, which in mathematical notation is $W_{foot} = N_{foot} \, \Delta\theta_{front} = F_{foot} \, r_{pedal} \, \Delta\theta_{front}$.

   N_{foot} is the torque his foot produces on the pedal. F_{foot} is the force he applies to the pedal. r_{pedal} is the distance from the center of the front sprocket to his foot. $\Delta\theta_{front}$ is the angle he moved the pedal and the front sprocket.

2. **Use the conservation of mechanical energy between the pedal and the chain.**

   The work Quintin does on the pedal is equal to the work the front sprocket does on the chain. In terms of mathematical notation, it's $W_{front} = W_{foot} \, (F_{front} \, r_{front} \, \Delta\theta_{front} = F_{foot} \, r_{pedal} \, \Delta\theta_{front})$.

F_{front} is the force applied to the chain by the front sprocket. r_{front} is the distance from the center of the front sprocket to the chain. $\Delta\theta_{front}$ is the angle he moved the pedal and the front sprocket.

3. **Find the motion of the chain across both sprockets.**

   The chain moves both sprockets the same amount. The chain is a solid loop, and the amount the chain moves at the back must equal the amount it moves at the front. Mathematically the condition is $r_{front}\,\Delta\theta_{front} = r_{back}\,\Delta\theta_{back}$.

4. **Use the conservation of mechanical energy between the back sprocket and the chain.**

   The work the chain does on the back sprocket is the same amount of work the front sprocket does on the chain. ($W_{back} = W_{front}$) In terms of mathematical notation: $F_{back}\,r_{back}\,\Delta\theta_{back} = F_{front}\,r_{front}\,\Delta\theta_{front} = F_{foot}\,r_{pedal}\,\Delta\theta_{front}$.

   F_{back} is the force applied to the back sprocket by the chain. r_{back} is the distance from the center of the back sprocket to the chain. $\Delta\theta_{back}$ is the angle the back sprocket and back tire move.

5. **Find the conservation of mechanical energy between the back sprocket and the tire on the road.**

   The work done on the back sprocket is the same amount of work done on the road by the tire ($W_{tire} = W_{back}$). In terms of mathematical notation: $F_{road}\,r_{tire}\,\Delta\theta_{back} = F_{back}\,r_{back}\,\Delta\theta_{back} = F_{front}\,r_{front}\,\Delta\theta_{front} = F_{foot}\,r_{pedal}\,\Delta\theta_{front}$.

   F_{road} is the static friction applied to the road by the back tire. r_{tire} is the distance from the center of the back sprocket to the road. $\Delta\theta_{back}$ is the angle the back sprocket and back tire move.

6. **Realize that the work done by Quintin is the same as the work done by the tire on the road (conservation of mechanical energy).**

   All the works calculated are the same, so match the first and last expression for the work ($F_{road}\,r_{tire}\,\Delta\theta_{back} = F_{foot}\,r_{pedal}\,\Delta\theta_{front}$), and use the expression for the chain movement ($r_{front}\,\Delta\theta_{front} = r_{back}\,\Delta\theta_{back}$) to obtain:

   - The chain moves the back sprocket the same amount as the front sprocket: $\Delta\theta_{front} = \Delta\theta_{back}\,(r_{back}/\,r_{front})$

   - The work done by Quintin on the pedal equals the work done by the back tire on the road: $F_{foot} = F_{road}\,(r_{tire}/\,r_{pedal})(r_{front}/\,r_{back})$

   These two equations are based on conservation of energy and work and tell you a couple of interesting things:

   The ratio (r_{front}/r_{back}) is the gear ratio. Each link in the chain is the same length, so the distance between the teeth on the sprockets must be the same. Therefore, the number of teeth on a sprocket tells you the size (and radius) of the sprocket.

If the front sprocket is large and the back sprocket is small, then to maintain a certain amount of force on the road requires more force from Quintin's feet, but the pedals move less; whereas if the front sprocket is small and the back sprocket is large, then the force required is smaller but the pedals need to move a lot. The first situation is good for flat roads at high speeds, and the second situation is good for hills or rough terrain.

The bike is one of the most energy-efficient forms of transportation with almost all of your energy going into your pedals, which reaches the road propelling you forward. You can travel very large distances without much expenditure of energy.

Colliding objects

The conservation of momentum is the key to understanding collisions. Objects are always colliding in biomechanics, especially within the body and in sports.

In the "Stopping me requires force — Newton's second law of motion" section earlier in this chapter, I discuss Newton's second law, the law of acceleration. Even though the internal forces and torques during a collision can be quite large, if the external forces or external torques are zero (or small along with the elapsed time being small, then they can be ignored), and you can use conservation of momentum with the velocities of the objects just prior to the collision and just after the collision. In the case when the external forces or external torques are small (and the elapsed time is small), using conservation of momentum is a reasonable approximation during the collision.

The total energy is always conserved, but keeping track of all the different forms of energy during a collision can be difficult. Instead, the collisions are split into two different kinds depending on the kinetic energy.

Inelastic collisions

If kinetic energy is lost during the collision, it's said to be *inelastic*. The greatest loss of kinetic energy occurs when the two objects stick together, which is called a *completely inelastic collision*. Most collisions in biomechanics fall in this category. If the motion is linear, then you need to use conservation of linear momentum. The mathematical formula for two objects colliding is

$$\vec{p}_{1,initial} + \vec{p}_{2,initial} = \vec{p}_{1,final} + \vec{p}_{2,final}$$

If the motion is circular, then you need to use conservation of angular momentum. The mathematical formula for two objects colliding is

$$\vec{L}_{1,initial} + \vec{L}_{2,initial} = \vec{L}_{1,final} + \vec{L}_{2,final}$$

Elastic collisions

If kinetic energy is conserved during the collision, then the collision is said to be *elastic*. A few collisions in biomechanics can be approximated as an elastic collision. The classic example is the collision between pucks on an air hockey table.

In the case of linear motion, you have conservation of linear momentum and conservation of kinetic energy. The two conservation laws for two particles colliding are mathematically, represents by:

$$\vec{p}_{1,initial} + \vec{p}_{2,initial} = \vec{p}_{1,final} + \vec{p}_{2,final}$$

$$E_{K,1,initial} + E_{K,2,initial} = E_{K,1,final} + E_{K,2,final}$$

In the case of circular motion, you have conservation of angular momentum and conservation of kinetic energy. The mathematical representation of these two laws for two particles colliding is

$$\vec{L}_{1,initial} + \vec{L}_{2,initial} = \vec{L}_{1,final} + \vec{L}_{2,final}$$

$$E_{K,1,initial} + E_{K,2,initial} = E_{K,1,final} + E_{K,2,final}$$

If a problem doesn't state what type of collision you have to solve for, then assume it's inelastic, which is the most common type of collision in biophysics. The problem should state whether the collision is completely inelastic or whether the collision is elastic.

The mathematical expressions for the momenta and energies are

$$\vec{p} = m\vec{v};\ \vec{L} = I\vec{\omega};\ E_{K,linear} = \frac{mv^2}{2};\ E_{K,rotational} = \frac{I\omega^2}{2}$$

Raj passes a soccer ball to Sarah, who then kicks it back towards Raj without stopping the ball first. Assume the ball is initially moving in the negative *x*-axis direction before the collision, the collision is elastic, and all the motion is one-dimensional. Find the motion of Sarah's foot and the ball's motion after the kick. Ignore the spin (rotational kinetic energy) of the soccer ball.

The mathematical representation of conservation of momentum and conservation of kinetic energy is

conservation of momentum: $m_{foot}v_{foot,initial} - m_{ball}v_{ball,initial} =$

$m_{foot}v_{foot,final} + m_{ball}v_{ball,final}$

conservation of kinetic energy: $\dfrac{m_{foot}v_{foot,initial}^2}{2} + \dfrac{m_{ball}v_{ball,initial}^2}{2} =$

$\dfrac{m_{foot}v_{foot,final}^2}{2} + \dfrac{m_{ball}v_{ball,final}^2}{2}$

The first line in this formula is the conservation of momentum in the direction of the ball's motion. The second line is the conservation of kinetic energy. Sarah's initial momentum is positive representing the motion to the right, whereas the ball's initial momentum is negative representing its motion to the left. The final momenta will either be negative (left traveling) or positive (right traveling), depending on the solution to these equations and the numerical values.

Sarah's final speed isn't important here, so solve the momentum equation for her final speed, substitute it into the kinetic energy equation, and solve for the ball's final speed. There are two different solutions; the first solution will correspond to Sarah's missing the ball and both Sarah and the ball keep moving at their initial velocities. (Sarah is a better soccer player than that, so the second solution is the desired solution.) The second solution is where she makes contact with the ball. My solution to the equations is

$$v_{foot,final} = v_{foot,initial} - \frac{m_{ball}}{m_{foot}}\left(v_{ball,initial} + v_{ball,final}\right)$$

$$v_{ball,final} = \frac{\left(m_{foot} - m_{ball}\right)}{\left(m_{foot} + m_{ball}\right)}v_{ball,initial} + \frac{2m_{foot}}{m_{foot} + m_{ball}}v_{foot,initial}$$

Even though these solutions are for an elastic collision, they are still a good approximation for kicking a soccer ball (football). One thing to note from these solutions is the ball will have a much larger final velocity and travel farther if it is moving towards Sarah rather then sitting on the pitch stationary when she kicks it. As a comment, m_{foot} isn't just the mass of the foot. Good soccer players are able to transfer a lot more momentum to the ball than the mass of the foot times the foot's speed.

I have left the formulas in terms of symbols instead of using numbers because they're important. The first formula by itself is the solution for any one-dimensional inelastic collision with conservation of linear momentum. The two formulas combined are the solution for any one-dimensional elastic collision with conservation of linear momentum and conservation of kinetic energy. Remember the directions I had selected for the initial velocities. In two dimensions, the formulas change, but you now know how to solve these problems.

Chapter 5

Sitting with Couch Potatoes — Static Equilibrium

· ·

In This Chapter

▶ Talking translational equilibrium

▶ Ruminating on rotational equilibrium

▶ Forming static equilibrium of rigid bodies

· ·

A *static* object isn't moving — it has no linear velocity and no angular velocity. An object in *static equilibrium* has no linear acceleration, no angular acceleration, and no motion. For example, while you're reading this biophysics book, your head isn't moving and it's in static equilibrium. Even though your head isn't moving, several interesting biophysical things are going on, such as a multitude of electrical signals is going into the brain from your five senses, thoughts are racing through your brain as you think, a multitude of signals is leaving your brain to control your body, blood is flowing through your brain supplying energy, and forces are pulling your head in different directions so you can balance it on your neck and keep it in static equilibrium.

In this chapter, I discuss the static equilibrium of biological systems composed of *rigid bodies,* meaning the objects in the biological system won't break or deform under the forces acting on them. For example, my kitchen table is a rigid body. If I place food, plates, and utensils on the table, these applied forces don't change the table's shape. However, even rigid bodies have limits, and if I place an elephant on my table, the table will change its shape. (Luckily, I don't have an elephant hiding in my fridge, and I consider only forces that don't change the object's shape.) I show you how to solve problems of rigid biological systems in static equilibrium. The *rigid biological systems* usually consist of a combination of animals and inorganic objects, such as duct-taping your friend to the ceiling (c'mon, you've all been there).

Static equilibrium can be split into static translational equilibrium and static rotational equilibrium. The fact that stationary biological systems are in static translational equilibrium and static rotational equilibrium isn't boring, contrary to what you may think. These balances provide a lot of information about the biological system in addition to the fact that the velocity, angular velocity, acceleration, and angular acceleration are all zero. You may be surprised how important static equilibrium is. By the end of this chapter, you'll be a pro at static equilibrium!

Understanding Static Translational Equilibrium

A biological system in static equilibrium is in *static translational equilibrium,* which means that the linear acceleration and the linear velocity of the biological system are both zero. According to Newton's second law of motion, *the law of acceleration,* the net external force acting on the system must be zero if the acceleration is zero. For an example of zero net force, think of your car at a traffic light. Your car isn't in static translational equilibrium when the traffic light turns from red to green and you step on the gas pedal. At that precise moment, the car isn't moving (so it's static), but it has an acceleration forward. But if you step on the brake and the gas at the same time and do a brake-stand, then the car's velocity and acceleration are zero, and it's in static translational equilibrium. The engine is trying to make the car move forward and the brakes are preventing the car from moving, so forces are acting on the car, but the net force is zero.

Meanwhile, *net force* means you have to add all the forces together. Forces are vectors with both direction and magnitude, so you need to use a compass (coordinate system) and add the forces together in each of the three directions (width, depth, and height) separately.

In the next sections, I discuss how to solve a static translational equilibrium problem, explain the free-body diagram, and examine the applications that illustrate how to put all the pieces together and apply them to biological systems.

Solving static translational equilibrium problems

Solving translational equilibrium problems involves satisfying these three equations. It sounds simple enough. The mathematical formulas for translational equilibrium are

$$\sum F_x = 0, \quad \sum F_y = 0, \quad \sum F_z = 0$$

Before you solve a static translational equilibrium problem, you need to know two things:

- The direction of the force is just as important as its magnitude. You can't add forces as if they're scalars. (A *scalar* has only magnitude and is just a number with possibly physical dimensions (units) too.) Chapter 3 discusses scalars.

- You don't need to consider the internal forces of the biological system when you're calculating the net force.

    Newton's third law of motion says that the action-reaction pair of forces cancels out for a pair of internal forces. For example, if you pull a trailer with your car, the car pulls the trailer with a force and the trailer pulls back on the car with the same force. When you treat the car and trailer as a single object, these forces cancel out and your trailer doesn't fly off from the car in some random direction (good thing too!).

By following these steps, you can solve any static translational equilibrium problem:

1. **Draw a free-body diagram of the biological system.**

    I explain how to do these diagrams in the next section. This is the most important step but the one people often ignore.

2. **Calculate each of the external forces, using your free-body diagram as a guide.**

3. **Use this formula, which adds all the scalar components of the forces together in each direction and sets them equal to zero.**

    $$\sum F_x = 0, \quad \sum F_y = 0, \quad \sum F_z = 0$$

These three equations remind you that you must split up the forces into scalar components parallel to the x-axis (East–West direction), the y-axis (North–South direction), and the z-axis (Up–Down direction), and then add each set of components together as scalars.

4. **Solve for the unknown forces acting on your biological system.**

Drawing free-body diagrams

A *free-body diagram* is a picture of an object with all the surrounding objects removed and all the external forces acting on the biological system added. For example, suppose you're sitting in your favorite chair, reading your favorite biophysics book, and holding the book in the air with both hands. The free-body diagram for the book shows only the book with three forces acting on it — the force of gravity, the contact force of your left hand, and the contact force of your right hand.

Drawing a free-body diagram is the most important part of understanding and solving problems with static translational equilibrium for a couple of reasons:

   ✔ It contains all the information about the system.

   ✔ It lets you visualize what's going on. A picture is worth a thousand words after all!

Despite these advantages, a free-body drawing is the part everybody wants to skip over. I make it easy for you by splitting the free-body diagrams into easy-to-follow steps:

1. **Draw a diagram of the biological system as if you have taken a picture of it.**

   Figure 5-1a shows a diagram of a person lying on a table on a slope.

Figure 5-1: A person lying on an inclined table over a swimming pool (a) and a free-body diagram of that person (b).

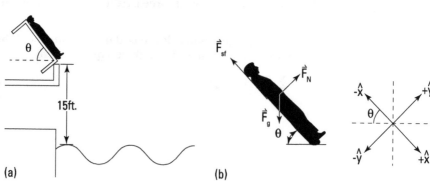

(a)   (b)

2. Draw a free-body diagram that includes all the external forces acting on the system without the other objects in the picture.

Take your time with the free-body diagram to get it right. Be sure to include all the forces and their directions. If the free body diagram has errors, then there's a high probability you'll incorrectly solve the problem. Continuously ask yourself if the diagram matches the biological system and the information given.

Pick a point of view that shows all the forces most clearly. If you're in doubt about the correct perspective, draw three pictures: one from above, one from the side, and one from the front.

Figure 5-1b is the free-body diagram of the sample sloping person from Step 1. I use a side view because that's the best view of the external forces acting on the person:

- Gravity (obviously)

- The normal force

- The static friction between the person and the table

You usually need to split the contact force between the biological system and an object (such as the table) into a normal force and a tangential force. (*Normal force* is the force required to prevent the object from falling through the surface [in this case, the table]; the *tangential force* is trying to accelerate or decelerate the object along the surface.) For example, suppose you're at a water park on the waterslide. The force between you and the slide is a contact force, which you can split into the normal force (the slide holding you up) and the *kinetic friction* (tangential force) trying to pull your swimming suit over your head.

The direction an external force is acting on a rigid biological system is very important for static translational equilibrium. In order to calculate the forces, you need to know the direction it's applied relative to some reference.

3. Draw a compass on your free-body diagram, which shows your directions for the axes, as in Figure 5-1b.

The compass is very important because the forces split into components parallel to each axis given by the compass. You then add force components together that are parallel to each other.

Usually, you want to choose one of the axes in the direction of an unknown force to simplify the problem.

This choice of axes defines your coordinate system for the problem. I show the *x*-axis and *y*-axis in Figure 5-1b. The *z*-axis isn't shown and sticks straight out of the figure toward you. You can think of the positive *x*-axis as being north, the negative *x*-axis as being south, the positive *y*-axis as being up, and the negative *y*-axis as being down. This choice then makes the positive *z*-axis east and the negative *z*-axis west.

When you finish your free-body diagram, you're ready to solve your translational equilibrium problem for the unknown forces.

Finding forces with static translational equilibrium

This section provides information on how you apply static translational equilibrium to biophysical problems. If you want to use static translational equilibrium to study forces in a static biological system, read on.

In typical biophysical problems dealing with static translation equilibrium, usually the only thing that changes from one problem to the next is the wording and the numbers. Textbooks usually give the solution as a formula, and students substitute the numbers into the formula without understanding where the formula came from because they don't understand the problem. I'm here to see that this doesn't happen to you.

You need to understand when formulas are valid and how they're derived. To help you do so, I present two similar problems and their solutions. The setup is the same for each problem:

You're hosting your weekly Saturday biophysics party and a friend invites a non-biophysics friend to the party (Larry the lawyer). He assumed the party involved alcohol, came very intoxicated (c'mon, you've all been there), and shortly after arriving passes out on a table. You and your physics friends immediately think of a biophysics experiment: You want to calculate the coefficient of static friction between the table and Larry's polyester suit. (I discuss static friction in Chapter 4.) You measure Larry's weight as 145 pounds (65.8 kilogram mass) on the bathroom scale. You notice that the top of your balcony is 15 feet (4.57 meters) above a swimming pool, so you and your friends move the table to the balcony and start to tip it over. At an angle of 35 degrees above the horizontal, Larry slides off the table and down into the swimming pool (don't worry, he wakes up then and knows how to swim).

At this point you're probably saying, "This problem is hard! How do I solve it without the Internet or formulas?" A problem like this usually contains important facts that aren't explicitly stated, which is what makes them interesting. In this problem, the four facts not stated explicitly are as follows:

✔ You need to figure out what physics rules are needed to solve the problem.

Note that if the angle is less than or equal to 35 degrees, then Larry continues to lie on the table and doesn't move, which means he is in static equilibrium, and you can use static translational equilibrium. (At any angle greater than 35 degrees, he'll slide off the table, which is what you want him to do.)

✔ At 35 degrees, the magnitude of the static friction is at its *maximum value* (the coefficient of static friction times the magnitude of the normal force).

✔ You should choose the direction negative *x*-axis parallel to the static friction in the free-body diagram.

You want to solve for the coefficient of static friction, and this choice makes it easier to isolate the friction force and solve.

✔ The height of the balcony is irrelevant for this problem.

You can solve the problem without having Larry slide off the table. Of course, you'll want to make him slide off the table as part of the experiment.

You're ready to solve the problem:

1. **Draw your picture.**

   Check out Figure 5-1a for my picture.

2. **Draw your free-body diagram.**

   This is the important part. Figure 5-1b is my free-body diagram. How does my diagram compare to yours?

   The angle θ in Figure 5-1b is 35 degrees. The force $\vec{F}_g$ is Larry's weight (mg), $\vec{F}_{sf}$ is the static friction, and $\vec{F}_N$ is the normal force.

3. **Draw a compass on your free-body diagram, which shows your directions for the axes, as in Figure 5-1b.**

4. **Calculate each of the forces, using the free-body diagram.**

A *force* is a vector with both magnitude and direction. It's represented by a symbol with an arrow on top. The magnitude of a force doesn't have an arrow. The weight always points toward the center of the earth, and the normal force points away from the surface and is perpendicular to it.

The free-body diagram (refer to Figure 5-1b) helps you to rewrite the forces parallel to the x-axis and the y-axis:

$$\vec{F}_g = \left(+\hat{x}\right)F_g \sin(\theta) + \left(-\hat{y}\right)F_g \cos(\theta)$$

$$\vec{F}_N = \left(+\hat{y}\right)F_N$$

$$\vec{F}_{sf} = \left(-\hat{x}\right)\mu_s F_N$$

5. **Solve for static translational equilibrium with the formula:**

$$\sum F_x = 0, \quad \sum F_y = 0, \quad \sum F_z = 0$$

You substitute the three forces into these equations. I obtained the following equations:

x-axis: $F_g \sin(\theta) + 0 - \mu_s \ F_N = 0$

y-axis: $-F_g \cos(\theta) + F_N + 0 = 0$

1. Solve for the normal force in the y-axis equation.

Why start with the y-axis? You want to solve for the coefficient of static friction and it appears only in the x-axis equation, but you don't know the normal force yet, so you use the y-axis equation to find the normal force.

$$F_N = F_g \ \cos(\theta) = 145 \text{ lb } \cos\left(35°\right) = 119 \text{ pounds } \left(529 \text{ newtons}\right)$$

A common mistake students typically make is using this formula to calculate the normal force no matter what the problem is. This equation is valid only if no other forces are in the direction perpendicular to the surface. Always start with the free-body diagram.

2. Find the coefficient of static friction, using the x-axis equation:

$$\mu_s = \frac{F_g \sin(\theta)}{F_N} = \frac{145 \text{ lb } \sin\left(35°\right)}{119 \text{ lb}} = 0.700$$

The coefficient of static friction between Larry's suit and the table is 0.700.

To help you understand why you should start with the diagrams and not the formulas, I change the problem to make the two previous formulas invalid:

You and your biophysics friends notice that the top of your balcony is 15 feet (4.57 meters) above a swimming pool, so you and your friends move the table to the balcony and tip the table to an angle of 10 degrees above the horizontal. You tie a rope to Larry's body and bring the free end of the rope to a friend's balcony directly across the swimming pool so that they can pull the rope horizontally across the pool. Figure 5-2 shows this scenario. Larry's weight is still 145 pounds (65.8 kilogram mass). You measure the tension (force) in the rope to be 67.5 pounds (300 newtons) just prior to Larry's sliding off the table and into the pool. Your mission is to find the coefficient of static friction between the table and Larry's polyester suit.

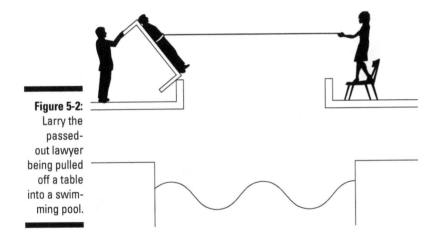

Figure 5-2:
Larry the passed-out lawyer being pulled off a table into a swimming pool.

The problem sounds similar to the one in which you slide Larry off the table, but it also sounds a lot harder because the problem has an extra force (the rope). You can solve it though, if you follow the steps for solving for static translational equilibrium — you just don't use the formulas. You're ready to solve the problem, so just follow these steps:

1. **Draw your picture.**

   Figure 5-2 shows a picture of my interpretation of the problem.

2. **Draw your free-body diagram.**

   From the picture, you can draw a free-body diagram; mine is shown in Figure 5-3. How does your figure compare?

   The angle θ is 10 degrees. The force $\vec{F}_g$ is Larry's weight (mg), $\vec{F}_{sf}$ is the static friction, $\vec{F}_N$ is the normal force, and $\vec{F}_r$ is the tension in the rope.

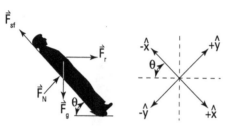

Figure 5-3: Free-body diagram of Larry being pulled off a table.

3. **Draw a compass on your free-body diagram, which shows your directions for the axes.**

   My choice is shown in Figure 5-3. I chose the direction parallel to the table surface as my x-axis and the perpendicular direction as my y-axis. This is because I want to isolate for the static friction.

4. **Calculate each of the forces using the free-body diagram in Figure 5-3.**

   $$\vec{F}_g = \left(+\hat{x}\right)F_g \sin(\theta) + \left(-\hat{y}\right)F_g \cos(\theta)$$
   $$\vec{F}_N = \left(+\hat{y}\right)F_N$$
   $$\vec{F}_{sf} = \left(-\hat{x}\right)\mu_s F_N$$
   $$\vec{F}_r = \left(+\hat{x}\right)F_r \cos(\theta) + \left(+\hat{y}\right)F_r \sin(\theta)$$

   You may have noticed that the forces look the same except for an extra line for the rope's tension (force).

5. Solve for static translational equilibrium with the formula:

$$\sum F_x = 0, \quad \sum F_y = 0, \quad \sum F_z = 0$$

1. You substitute the four forces into these equations:

   x-axis: $F_g \sin(\theta) + 0 - \mu_s \, F_N + F_r \cos(\theta) = 0$

   y-axis: $-F_g \cos(\theta) + F_N + 0 + F_r \sin(\theta) = 0$

   You can see why the formulas in the previous problem don't work. The problems look very similar and the steps used to solve the problems are the same, but the formulas are different.

2. Solve for the normal force in the y-axis equation.

   $$F_N = F_g \cos(\theta) - F_r \sin(\theta) = 145 \text{ lb } \cos(10°) - 67.5 \text{ lb } \sin(10°)$$

   $$= 131 \text{ pounds } (583 \text{ newtons})$$

   Note that the rope reduces the magnitude of the normal force. The table doesn't need to apply as much force to hold Larry up because the rope is partially holding up Larry. This is the same as people who hang from a bar over their head and slowly lower their body onto the bathroom scale and stop dropping when they get to the weight they want to be.

3. Find the coefficient of static friction, using the x-axis equation:

   $$\mu_s = \frac{F_g \sin(\theta) + F_r \cos(\theta)}{F_N} = \frac{145 \text{ lb } \sin(10°) + 67.5 \text{ lb } \cos(10°)}{131 \text{ lb}} = 0.699$$

The coefficient of static friction between Larry's suit and the table is still about 0.7 (0.699).

If your friends produce a tension greater than 67.5 pounds (300 newtons) in the rope, then Larry will slide off the table and fall into the swimming pool below. A 10-pound fishing line wouldn't work, so hopefully you have a bigger rope in your home.

Static systems are in static translational equilibrium. No matter how different a problem may look from the solutions in this section, if a system is in static translational equilibrium, then the biophysics is the same and the steps involved in solving it are the same.

Turning to Static Rotational Equilibrium

A biological system that's in static equilibrium, in addition to not moving and being in static translational equilibrium (which I discuss in the previous section), is also in static rotational equilibrium. *Static rotational equilibrium* means that the angular acceleration of the biological system is zero, and according to Newton's second law, the net torque acting on the system must be zero.

Think of changing a flat tire: You place the tire wrench on the lug nut and pull, producing a torque on the lug nut. (*Torque* tries to make an object spin faster or slower — see Chapter 4 for a complete discussion.) Unfortunately, the lug nut has rusted to the hub, and the friction force produces a countering torque in the opposite direction. The two torques add up to zero, and the lug nut doesn't move.

An example of a stationary object not in rotational equilibrium is you on a swing moving back and forth. When the swing reaches its highest position, you stop momentarily. You and the swing are stationary, but gravity is producing a torque, which causes you to start swinging in the opposite direction.

Solving rotational equilibriums

Solving static rotational equilibrium problems also involves satisfying these three equations. The mathematical formula for static rotational equilibrium is

$$\sum N_x = 0, \quad \sum N_y = 0, \quad \sum N_z = 0$$

Some facts about static rotational equilibrium to keep in mind are

- ✔ Torques are vectors with both direction and magnitude, so you need to add the components of the torques together in each of the three directions (width, depth, and height).

- ✔ Here I discuss biological systems in static rotational equilibrium that are rigid. *Rigid* means the system doesn't deform from the torques applied to it. If the system weren't rigid, it would bend, twist, or fold from the applied torques.

- ✔ All static equilibrium systems are in both static rotational equilibrium and static translational equilibrium, and this is always true.

  This isn't the case for moving objects in equilibrium, such as the hands on a grandfather or wall clock, which are in dynamical rotational equilibrium (angular velocity is not zero) but not in dynamical translational equilibrium.

The main steps involved in solving problems with static rotational equilibrium are as follows:

1. **Draw a picture of the system.**

   A picture is worth a thousand words and helps with the second step.

2. **Draw a free-body diagram of the biological system showing where all the external forces are acting on it.**

   This step is the most important step.

3. **Draw an axis of rotation and a compass on your free-body diagram, which shows your directions for the axes.**

4. **Calculate each of the external forces and their corresponding torques using the free-body diagram.**

   The direction of the torque is given by the right-hand rule. I discuss the direction of the torque in Chapter 4, which involves three simple steps:

   1. Place your right-hand wrist at the axis of rotation.

   2. Point your right hand toward the location where the force is being applied (without moving your wrist).

   3. Curl your fingers in the direction the force is pointing. The right-hand thumb is pointing in the same direction as the torque.

   The magnitude of the torque is calculated using the formula:

   $$N = r\,F\,\sin(\theta) = r\,F_{\perp} = r_{\perp}F$$

   N is the magnitude of the torque, r is the distance from the axis of rotation to where the force F is applied, and θ is the angle between them. $F_{\perp}$ is the component of $\vec{F}$ perpendicular to $\vec{r}$, and $r_{\perp}$ is the component of $\vec{r}$ perpendicular to $\vec{F}$.

   The right-hand rule and the formula for calculating the magnitude of the torque can be combined by the cross-product formula:

   $$\vec{N} = \vec{r} \times \vec{F}$$

5. **Add all the components of the torques together and set the sums equal to zero, using the formulas:**

   $$\sum N_x = 0, \quad \sum N_y = 0, \quad \sum N_z = 0$$

6. **Solve for the unknown torque acting on your system.**

Doing static rotational free-body diagrams

The most important part of understanding and solving problems with static rotational equilibrium is the free-body diagram because it helps you calculate the torque produced by a force. Just follow these three steps:

1. **Draw a picture of the biological system.**

2. **Draw a free-body diagram of the object only, showing where all the external forces are acting on it.**

   The location where an external force acts on a rigid biological system is very important as well as the direction of the force. In order to calculate the torque, you need to know where the force is relative to the axis of rotation.

3. **Select a location for the axis of rotation.**

   The *axis of rotation* is the axis an object rotates around. An axis of rotation can be real, such as the axis of a bike wheel, or it can be virtual, such as a ballerina spinning in a circle whose axis of rotation is the line of molecules in the body that don't move while the rest of her body moves in a circle around these molecules. The earth's axis of rotation is a line that goes from the North Pole to the South Pole.

   The object isn't moving so it doesn't matter where you place the axis of rotation, so put it where there are many unknown forces.

   Suppose your problem has several unknown forces. The magnitude of the torque is proportional to the force times the distance from the axis of rotation to where the force is applied. If you place the axis of rotation where the unknown forces are applied, then their torques will be zero, which makes the problem simple to solve.

   Many people have a hard time choosing an axis of rotation, especially if the biological system has a natural axis. As an example, consider a person standing still with her feet in contact with the ground and an arm stretched out horizontally holding onto a pole — it could be a ballerina that was spinning on her toes and used the pole to stop spinning, or it could be a child who used his feet to stop spinning around the pole. As long as the person is stationary, you don't know which situation is correct and what is the true axis of rotation. It doesn't matter as long as it is static.

Bending to the will of static rotational equilibrium

In this section, I go through the steps of applying static rotational equilibrium to biophysical problems — specifically to a human's spinal muscles. Try to solve this problem:

You and your biophysics friends are shopping for supplies for your weekly Saturday biophysics party when a colleague, Allie, sees you approach. She has just purchased new shoes and shorts and wants to show them off. She bends over with her legs vertical and her back horizontal and attempts to lift a box that weighs 48.2 pounds (21.9 kilogram mass) as shown in Figure 5-4. You and your friends see Allie and get really excited, drop everything you're carrying, pull out your notebooks, and do some biophysics calculations. You want to calculate the increase in the tension of the spinal muscles.

Figure 5-4:
Allie lifting a
box.

First, you and your friends must estimate some of her physical properties while she's bent over. (I will help with this part.) The four forces acting on a person's back in this posture are:

✔ The weight of the upper body is $\overline{F}_{g,U}$. This is her total weight minus the weight of her legs and feet (about 30 percent of her total weight). You assume the magnitude to be 140 − 42 = 98 pounds. (A mass of 44.5 kilograms.)

✔ The weight of the box being lifted is $\overline{F}_{g,L}$. In this case, its magnitude is 48.2 pounds. (A mass of 21.9 kilograms.)

✔ The tension in the spinal muscles is $\vec{F}_T$.

✔ The strain in the sacroiliac joints is $\vec{F}_S$.

You and your friends assume her upper body's center of mass is located approximately x_{SU} = 10 inches (25 centimeters) horizontally from the sacrum. The box is located a horizontal distance of x_{SL} = 30 inches (76 centimeters) from the sacrum. The spinal muscles are holding the body up and are attached to the back approximately x_{ST} = 15 inches (37.5 centimeters) from the sacrum. In this position, the spinal muscles are pulling toward the sacrum at an angle of approximately θ = 12 degrees above the horizontal. (In a normal problem, all this information is usually given to you.)

The sacrum is the triangular bone at the bottom of the spine. It connects to the hip bones (iliac bones) at the sacroiliac joint. The tip of the triangle is the tail bone (coccyx).

Second, you solve the problem by following the steps list at the beginning of the section, which is applicable to any object in static rotational equilibrium:

1. **Draw a picture of the biological system.**

   My picture is shown in Figure 5-4.

2. **Draw the free-body diagram of the back and show all the forces acting on the back.**

   The spinal muscles are an internal force within the body, so to use static rotational equilibrium, you have to remove the legs (only virtually) as shown in Figure 5-5. Doing so creates two new external forces in the free-body diagram: the contact force between the hips and upper body at the sacrum and the tension in the spinal muscles.

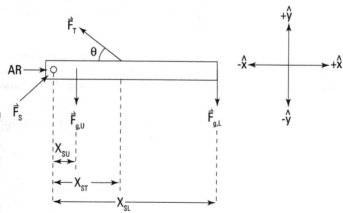

Figure 5-5: Free-body diagram of Allie lifting a box.

3. Select a location for the axis of rotation.

Your best choice for the axis of rotation is the sacrum. Choosing a different axis of rotation means you need to consider static translational equilibrium and the strain in the sacrum as well. Plus, you need to choose a coordinate system (compass) for the force calculations. This makes the problem a lot harder than it needs to be. My advice: Stick with the sacrum.

My free-body diagram is shown in Figure 5-5. AR marks the point (circle) where my axis of rotation cuts through the sacrum.

4. Draw a compass on your free-body diagram, which shows your directions for the axes.

5. Calculate each of the external forces and their corresponding torques using the free-body diagram.

In this problem, the torques are one-dimensional vectors perpendicular to the plane of Figure 5-5, the z-axis. (Don't forget your right-hand rule.)

TIP

Forces that try to produce counterclockwise rotations around the axis of rotation are positive torques, and forces that try to produce clockwise rotations are negative torques.

$$\vec{N}_S = \hat{z} \; x_{SS}F_s \sin(\phi) = \hat{z}(0 \text{ ft})F_s \sin(\phi) = \hat{z} \;(0 \text{ lb ft})$$

$$\vec{N}_T = \hat{z} \; x_{ST}F_T \sin(\theta) = \hat{z} \;(1.25 \text{ ft}) \; F_T \sin(12°)$$

$$\vec{N}_U = -\hat{z} \; x_{SU}F_{g,U} = -\hat{z}(0.833 \text{ ft})(98 \text{ lb}) = -\hat{z}(81.7 \text{ lb ft})$$

$$\vec{N}_L = -\hat{z} \; x_{SL}F_{g,L} = -\hat{z}(2.50 \text{ ft})(48.2 \text{ lb}) = -\hat{z}(120 \text{ lb ft})$$

6. Add all the components of the torques together and set equal to zero using the formulas.

$$\sum N_x = 0, \quad \sum N_y = 0, \quad \sum N_z = 0$$

Substitute the torques into these equations. All the torques are parallel to the z-axis, so the first two equations are 0=0. We are left with only one equation:

$$z\text{-axis} : 0 \text{ lb ft} + (1.25 \text{ ft}) \; F_T \sin(12°) - 81.7 \text{ lb ft} - 120 \text{ lb ft} = 0 \text{ lb ft}$$

7. Solve for the unknown torque that is acting on your system.

The tension in the spinal muscles is

F_T = 314 pounds (1,398 newtons) if there is no box

or

F_T = 776 pounds (3,452 newtons) if you include the box.

The tension in the spinal muscles has to increase by 462 pounds (2,055 newtons) or 147 percent in order to lift a 48.2 pound (21.9 kilogram mass) box.

Breaking Rigid Bodies with Static Equilibrium

A biological system in static equilibrium is in both static translational equilibrium and static rotational equilibrium. In the previous two sections I focused on how to solve a problem with just static translational equilibrium or static rotational equilibrium. In this section, I put it all together and outline the steps involved in solving problems involving static equilibrium.

1. **Determine what forces or distances you want to calculate.**

   This is the "setting it up" and "understanding what's going on" part of the problem when you determine what you know and what information you need.

2. **Draw your picture.**

   Most people find visualizing the problem very helpful.

3. **Draw your free-body diagrams.**

   If you have to calculate internal forces, distances, or torques, then you need to split up your biological system. *Note:* You may still have to break up the biological system in some problems where all the forces you want to calculate are external. Don't be afraid to split up the system, especially if you know the relative positions of the forces.

4. **Solve with static rotational equilibrium.**

   This requires knowing the relative positions of the forces so you can calculate the torques.

5. **Solve any remaining forces with static translational equilibrium.**

In the following sections, I apply these steps to solve two problems.

Applying static translational equilibrium multiple times — break a leg

Larry the lawyer who ended up in the swimming pool in earlier sections awakens upon impact with the water and swims to the side. He wants revenge on the biophysicists, so in anger, he runs along the wet deck, slips, and breaks a leg. You rush him to the hospital where they put the leg in a cast and traction.

The leg has a weight of 27.5 pounds (12.5 kilogram mass). The doctor elevates the leg to an angle of 20 degrees above the horizontal. The leg is pulled lengthwise toward the foot with a force of 6 pounds (26.7 newtons) by a pulley system to keep the parts of the bone straight and aligned while the bone mends. Figure 5-6 shows the leg and pulley system.

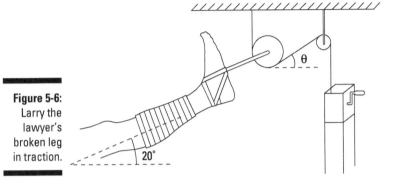

Figure 5-6:
Larry the
lawyer's
broken leg
in traction.

You want to figure out what tension and direction the rod connecting the foot to the pulley system must have. What tension is needed in the cable and what angle θ is required to hold the leg in place? Assume the cable and pulleys are massless and frictionless.

Follow these steps to solve:

1. **Determine what forces or distances you want to calculate.**

   Your first step is to figure out what's going on so that you can set up the problem. Facts to consider include

   - The leg isn't moving, so the hip is pulling back with a force of $F_H = 6.00$ pounds (26.7 newtons) along the length of the leg.

   - The leg has a weight of $F_g = 27.5$ pounds (12.5 kilogram mass) straight downwards.

   - The leg makes an angle of 20 degrees relative to the horizontal.

   - You need to find the tension in the cable, F_T, and the angle, θ. The magnitude of the tension in the cable is the same on both the right side and left side of the pulley because the pulley is massless and frictionless.

   - You have to split the system apart because the contact force, $\vec{F}_c$, is an internal force and we need it to be an external force.

2. **Draw your picture.**

   Figure 5-6 shows my picture.

3. **Draw your free-body diagrams.**

   Figures 5-7 and 5-8 show mine. I made one for the leg and one for
   the pulley. I don't need to make one for the pulley and leg combined
   because the two diagrams allow me to solve for all the unknown quantities. You can also use the leg plus pulley in Figure 5-6 as a free-body diagram. The only difference between Figure 5-6 and Figure 5-8 is that the
   hip force and weight of the leg replaces the contact force.

   In Figure 5-7: $\vec{F}_H$ is the force applied to the leg by the hip, $\vec{F}_g$ is the
   weight of the leg, and $\vec{F}_c$ is the contact force between the leg and the
   pulley.

   In Figure 5-8: $\vec{F}_{TL}$ is the force of the left-side cable pulling the pulley
   upwards, $\vec{F}_{TR}$ is the force of the right-side cable pulling the pulley
   upwards and to the right, and $-\vec{F}_c$ is the contact force between the leg
   and the pulley.

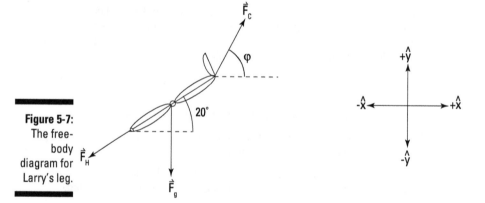

Figure 5-7:
The free-
body
diagram for
Larry's leg.

4. **Try using static rotational equilibrium,**

   $$\sum N_x = 0, \quad \sum N_y = 0, \quad \sum N_z = 0$$

   Static rotational equilibrium requires calculating the torques, but you
   don't know the location of the forces relative to the axis of rotation, so
   you need to use static translational equilibrium instead.

5. **Solve any remaining forces with static translational equilibrium:**

   $$\sum F_x = 0, \quad \sum F_y = 0, \quad \sum F_z = 0$$

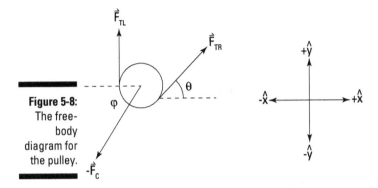

Figure 5-8:
The free-body diagram for the pulley.

1. Calculate the forces. I have drawn a compass in Figure 5-7:

$$\vec{F}_C = \left(+\hat{x}\right)F_c\cos(\phi)+\left(+\hat{y}\right)F_c\sin(\phi)$$

$$\vec{F}_H = \left(-\hat{x}\right)F_H\cos(20°)+\left(-\hat{y}\right)F_H\sin(20°)$$

$$\vec{F}_g = \left(-\hat{y}\right)F_g$$

2. Substitute into the static translational equilibrium equations:

x-axis: $F_C\cos(\phi)-F_H\cos(20°)+0=0$

y-axis: $F_C\sin(\phi)-F_H\sin(20°)-F_g=0$

3. Solve for ɸ.

$$\tan(\phi)=\frac{F_C\sin(\phi)}{F_C\cos(\phi)}=\frac{F_H\sin(20.0°)+F_g}{F_H\cos(20.0°)}=\frac{6\text{ lb }\sin(20.0°)+27.5\text{ lb}}{6\text{ lb }\cos(20.0°)}$$

The contact force makes an angle of 79.2 degrees relative to the horizontal.

4. Substitute the angle of 79.2 degrees into the x-axis equation and solve for F_c = 30.1 pounds (134 newtons).

5. You need to repeat the calculations for the tension in the cable. I have drawn a compass in Figure 5-8. You should have one in yours as well. The forces from Figure 5-8 using my compass are

$$-\vec{F}_C = \left(-\hat{x}\right)F_c\cos(\phi)+\left(-\hat{y}\right)F_c\sin(\phi)$$

$$\vec{F}_{TL} = \left(+\hat{y}\right)F_T$$

$$\vec{F}_{TR} = \left(+\hat{x}\right)F_T\cos(\theta)+\left(+\hat{y}\right)F_T\sin(\theta)$$

6. Substitute the forces into the static translational equilibrium equations,

x-axis: $F_T \cos(\theta) + 0 - F_c \cos(\phi) = 0$

y-axis: $F_T \sin(\theta) + F_T - F_c \sin(\phi) = 0$

7. Combine these equations and solve for F_T:

$$F_T^2 = \left[F_T \sin(\theta) \right]^2 + \left[F_T \cos(\theta) \right]^2 = \left[-F_T + F_c \sin(\phi) \right]^2 + \left[F_c \cos(\phi) \right]^2$$
$$= F_T^2 - 2F_T F_c \sin(\phi) + F_c^2$$

I use the identity $\cos^2 + \sin^2 = 1$, and I can solve for F_T.

The tension in the cable is

$$F_T = \frac{F_c}{2 \sin(\phi)} = \frac{30.1 \text{ lb}}{2 \sin(79.2°)} = 15.3 \text{ lb}$$

Substitute the tension into the x-axis equation and solve for the angle:

$$\cos(\theta) = \frac{F_c \cos(\phi)}{F_T} = \frac{30.1 \text{ lb } \cos(79.2°)}{15.3 \text{ lb}} = 0.369$$

The angle is $\theta = 68.4$ degrees.

The rod has to hold up the leg plus provide 6.00 pounds (26.7 newtons) of force along the length of the leg, which is why the rope has a tension of 30.1 pounds (134 newtons) and is pulling upwards at an angle of 79.2 degrees above the horizontal. The cable is pulling the pulley and rod. The cable is pulling on both sides of the pulley and shares the force between the cable on the left and right sides of the pulley. This makes the tension in the cable much smaller at only 15.3 pounds (68.1 newtons). The right cable is set an angle of 68.4 degrees above the horizontal. This will allow Larry's leg to heal properly so he can chase biophysicists in the future.

For biophysical problems like this you need to recognize that you can't use static rotational equilibrium unless you're given the locations of the forces or you need to find the location where the force is being applied. When dealing with internal forces, you need to split up the system to figure out what forces are involved.

Applying static rotational and static translational equilibrium — the iron cross

The problem in this section is a combination of all aspects of static equilibrium biomechanics and illustrates how to implement the steps given at the beginning of the "Breaking Rigid Bodies with Static Equilibrium" section.

Because static equilibrium objects are in both static translational and static rotational equilibrium, you need to solve for both of them as well as you usually have to split the system apart (like Larry, his leg, and the pulley).

The iron cross is a gymnastic exercise on the rings that requires a lot of strength. The gymnast hangs on a set of rings with her arms held out horizontally from her body. In this case, here I switch things around and have the person upside down.

Figure 5-9 shows Helga of the women's Olympic Team who weighs $F_{g,B} = 220$ pounds (100 kilogram mass) performing an upside-down iron cross. The deltoid muscles (located at the top of the shoulder) hold her entire body in the air. The problem is to calculate the tension in her deltoid muscle and the strain in her shoulder joint.

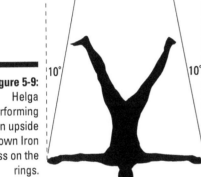

Figure 5-9:
Helga
performing
an upside
down Iron
Cross on the
rings.

For biophysical problems like this, the trick is to know how to break the system apart and figure out what forces are involved. In biological systems, the main forces in these types of problems are the muscles, the contact forces in the joints, and gravity.

1. **Determine what forces or distances you need for the calculations and what you want to calculate.**

   To calculate the tension in the deltoid muscle, you need to know what external forces are acting on Helga's arm:

   The strain in the shoulder joint is $\vec{F}_{S,S}$. You don't know the magnitude or the direction of this force, but it's located at the shoulder joint.

   The tension in the deltoid muscle is $\vec{F}_{T,D}$. The deltoid muscle is attached to the humerus bone $x_{SD} = 6$ inches (0.152 meters) horizontally from the shoulder joint. The force is directed toward the body at an angle of 20 degrees below the horizontal.

The weight of the upper arm is $\vec{F}_{g,U}$. It has a magnitude of 6 pounds (mass of 2.72 kilograms) and is located at x_{SU} = 5.25 inches (0.133 meters) horizontally from the shoulder joint. The force is directed downwards toward the center of the earth.

The weight of the lower arm and hand is $\vec{F}_{g,L}$. It has a magnitude of 4.75 pounds (mass of 2.16 kilograms) and is located at x_{SL} = 16.5 inches (0.419 meters) horizontally from the shoulder joint. The force is directed downwards toward the center of the earth.

The tensions in the ropes are $\vec{F}_{T,RL}$ and $\vec{F}_{T,RR}$. The ring in the hand is located at x_{SR} = 26.0 inches (0.660 meters) horizontally from the shoulder joint. The force is directed upwards and 10 degrees toward the center of the body.

The external forces acting on Helga's body are

> The weight of the body is $\vec{F}_{g,B}$.

> The tension in the right rope is $\vec{F}_{T,RR}$.

> The tension in the left rope is $\vec{F}_{T,RL}$.

2. **Draw your free-body diagrams.**

Figure 5-10 shows my free-body diagrams for Helga's body and arm. Figure 5-10a is the free-body diagram for Helga's body. Figure 5-10b is the free-body diagram for Helga's arm.

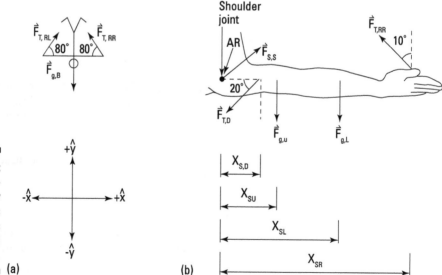

Figure 5-10: The free-body diagrams of Helga's body (a) and her arm (b).

(a)

(b)

3. **Solve the tension in the deltoid with static rotational equilibrium on the arm.**

$$\sum N_x = 0, \quad \sum N_y = 0, \quad \sum N_z = 0$$

Figure 5-10b shows three unknown forces acting on Helga's body (deltoid muscle, strain in the shoulder joint, and the tension in the rope). You have to find one of these three forces some other way before you can solve this problem. Luckily, you can find the tension in the ropes by considering the whole system (Helga's body). You don't know the relative locations of the ropes and her center of mass, so you need to use static translational equilibrium instead of rotational equilibrium.

4. **Solving for the tension in the rope using static translational equilibrium on the whole body.**

$$\sum F_x = 0, \quad \sum F_y = 0, \quad \sum F_z = 0$$

1. Calculate the forces acting on the body.

    In Figure 5-10a, I show the three forces and my compass (+y-axis is up and +x-axis is north):

$$\vec{F}_{T,RR} = \left(-\hat{x}\right) F_{T,R} \sin\left(10°\right) + \left(+\hat{y}\right) F_{T,R} \cos\left(10°\right)$$

$$\vec{F}_{T,RL} = \left(+\hat{x}\right) F_{T,L} \sin\left(10°\right) + \left(+\hat{y}\right) F_{T,L} \cos\left(10°\right)$$

$$\vec{F}_{g,B} = \left(-\hat{y}\right) F_{g,B}$$

2. Substitute into the static translational equilibrium equations

    x-axis: $- F_{T,R} \sin\left(10°\right) + F_{T,L} \sin\left(10°\right) + 0 = 0$

    y-axis: $F_{T,R} \cos\left(10°\right) + F_{T,L} \cos\left(10°\right) - F_{g,B} = 0$

These equations are solved by the tension in the ropes. The x-axis equation shows that the magnitude of the two forces must be the same. The tension in each of the ropes from the y-axis equation is

$$F_{T,R} = F_{T,L} = \frac{F_{g,B}}{2\cos\left(10°\right)} = \frac{220 \text{ lb}}{2\cos\left(10°\right)} = 112 \text{ pounds } \left(498 \text{ newtons}\right)$$

You can now find the tension in Helga's deltoid and the strain in her shoulder joint by solving the static equilibrium problem for the arm. You start with static rotational equilibrium:

$$\sum N_x = 0, \quad \sum N_y = 0, \quad \sum N_z = 0$$

1. **Choose an axis of rotation for the torque calculations.**

   My free-body diagram is shown in Figure 5-10b, where AR marks the point where the axis of rotation cuts through the figure. I picked my axis of rotation to pass through the shoulder joint.

   Your best choice for the axis of rotation is the shoulder joint because the torque produced by the strain in the shoulder joint is zero. Choosing a different axis of rotation means you need to consider static translational equilibrium and the strain in the shoulder joint simultaneously with the tension in the deltoid. This makes the problem a lot harder than it needs to be. My advice: Stick with the AR at the shoulder joint.

2. **Calculate the torques.**

   (Don't forget your right-hand rule.) You can use the Figure 5-10a compass for Figure 5-10b too.

   $$\vec{N}_{S,S} = \hat{z} \, x_{SS} F_{S,S} \sin(\phi_{S,S}) = \hat{z}(0 \text{ ft}) F_{S,S} \sin(\phi_{S,S}) = \hat{z} \, 0 \text{ lb ft}$$

   $$\vec{N}_{T,D} = -\hat{z} \, x_{SD} \, F_{T,D} \sin(20°)$$

   $$\vec{N}_{g,U} = -\hat{z} \, x_{SU} \, F_{g,U}$$

   $$\vec{N}_{g,L} = -\hat{z} \, x_{SL} \, F_{g,L}$$

   $$\vec{N}_{T,RR} = \hat{z} \, x_{SR} \, F_{T,R} \sin(80°)$$

3. **Add all the torques together and set equal to zero.**

   $$\sum N_x = 0, \quad \sum N_y = 0, \quad \sum N_z = 0$$

4. **Solve for the unknown torque that is acting on your system.**

   Substitute the torques into these equations (there is only one non-zero equation):

   $$z\text{-axis: } 0 + x_{SD} F_{T,D} \sin(20°) - x_{SU} F_{g,U} - x_{SL} F_{g,L} + x_{SR} F_{T,R} \sin(80°) = 0$$

   Solve for the tension in the deltoid muscle:

   $$F_{T,D} = \frac{-x_{SU} F_{g,U} - x_{SL} F_{g,L} + x_{SR} F_{T,R} \sin(80°)}{x_{SD} \sin(20°)}$$

   $$= \frac{-(5.25 \text{ in})(6 \text{ lb}) - (16.5 \text{ in})(4.75 \text{ lb}) + (26 \text{ in})(112 \text{ lb})\sin(80°)}{(6 \text{ in})\sin(20°)}$$

   $$= 1344 \text{ pounds } (5980 \text{ newtons})$$

5. **Solve any remaining force with static translational equilibrium.**

   $$\sum F_x = 0, \quad \sum F_y = 0, \quad \sum F_z = 0$$

The only force left to solve for is the strain in the shoulder joint.

1. Calculate the forces in Figure 5-10b. My compass shows the x-axis (north–south) and the y-axis (up–down).

$$\vec{F}_{S,S} = \left(+\hat{x}\right)F_{SS,x} + \left(+\hat{y}\right)F_{SS,y}$$

$$\vec{F}_{T,D} = \left(-\hat{x}\right)F_{T,D}\cos\left(20°\right) + \left(-\hat{y}\right)F_{T,D}\sin\left(20°\right)$$

$$\vec{F}_{g,U} = \left(-\hat{y}\right)F_{g,U}$$

$$\vec{F}_{g,L} = \left(-\hat{y}\right)F_{g,L}$$

$$\vec{F}_{T,RR} = \left(-\hat{x}\right)F_{T,R}\sin\left(10°\right) + \left(+\hat{y}\right)F_{T,R}\cos\left(10°\right)$$

2. Substitute into the static translational equilibrium equations.

x-axis: $F_{SS,x} - F_{T,D}\cos\left(20°\right) + 0 + 0 - F_{T,R}\sin\left(10°\right) = 0$

y-axis: $F_{SS,y} - F_{T,D}\sin\left(20°\right) - F_{g,U} - F_{g,L} + F_{T,R}\cos\left(10°\right) = 0$

3. Solve for the strain in the shoulder joint.

$$F_{SS,x} = F_{T,D}\cos\left(20°\right) + F_{T,R}\sin\left(10°\right) = 1344 \text{ lb}\cos\left(20°\right) + 112 \text{ lb}\sin\left(10°\right)$$

$$= 1282 \text{ pounds}\left(5700 \text{ newtons}\right)$$

and

$$F_{SS,y} = F_{T,D}\sin\left(20°\right) + F_{g,U} + F_{g,L} - F_{T,R}\cos\left(10°\right)$$

$$= 1344 \text{ lb}\sin\left(20°\right) + 6 \text{ lb} + 4.75 \text{ lb} - 112 \text{ lb}\cos\left(10°\right)$$

$$= 360 \text{ pounds } \left(1600 \text{ newtons}\right)$$

In biophysics, you usually need to calculate the magnitude of the force and the direction (angle). The magnitude of the strain and the angle are calculated from these forces using Pythagoras' theorem and trigonometric properties.

$$F_{SS} = \sqrt{F_{SS,x}^2 + F_{SS,y}^2} = \sqrt{\left(1282 \text{ lb}\right)^2 + \left(360 \text{ lb}\right)^2} = 1{,}330 \text{ pounds } \left(5{,}920 \text{ newtons}\right)$$

and

$$\phi = \tan^{-1}\left[\frac{F_{SS,y}}{F_{SS,x}}\right] = \tan^{-1}\left[\frac{360 \text{ lb}}{1282 \text{ lb}}\right] = 15.7°$$

You see that Helga is very strong. Each of her deltoid muscles is producing a force of 1,340 pounds (5,980 newtons) at an angle of 20 degrees below the horizontal. The strain in her shoulder joint has a magnitude of 1,330 pounds (5,920 newtons) at an angle of 15.7 degrees above the horizontal.

Chapter 6

Building the Mechanics of the Human Body and Animals

*T*here are many situations in biophysics when studying the setup is just as interesting and important as studying the motion of the objects. This chapter focuses on understanding some of the biomechanics without worrying about if the system is moving or not. An example is the stability of an athlete on the balance beam. It doesn't matter if she's moving along the beam or stationary; the principle of not falling off the balance beam is the same.

You discover more about statics in this chapter, specifically the center of mass and stability and how well the body holds up to accelerations and forces. You also find out about the six basic machines and the mechanical advantage and applications within your body. In addition, objects aren't perfectly rigid but can bend and twist, so I introduce the concepts of stress and strain. The last section of this chapter is about scaling and dimensional analysis, and I discuss their importance in biophysics.

Getting Down with Gravity

Gravity is a force biological organisms can't escape. Most people don't think about it, but it plays an important role with all life on the planet. People who have it foremost on their mind are probably skydivers when they jump out of a plane and mountaineers when climbing a cliff. Gravity is important in sports and animal mechanics, so that is the focus of this section.

I focus on the concept of center of mass, which is a special point in objects where gravity thinks all the mass is located. I also discuss stability, which is a pretty important concept in life too. I show when you should expect someone to be stable or whether the person will fall on his or her face.

Shifting to the center of mass

Gravity thinks all the mass in your body is located at a single point called the *center of mass*. The location of your center of mass plays a critical role in bio-mechanics, such as in many sports from surfing to skateboarding to wrestling to biking and so on. You can add almost any sport you can think of to this list. (Center of mass isn't too critical in checkers though.)

To understand the center of mass, take a uniformly proportioned stick and balance it on one finger at each end. If you remove the left hand finger, the stick will rotate around the right hand finger and fall down. Now, place your right hand finger in the center of the stick, and you should be able to hold up the stick with a single finger. You can think of there being the same amount of mass on both sides of the finger, or you can think of all the mass within the stick as being located at a single point, which is above your finger. The external forces acting on the stick are gravity and the normal force from your finger. According to Newton's second law, the law of acceleration, the two forces must add to zero so there's no acceleration. The two forces must be located at the same place as well, so the net torque is also zero, and there's no angular acceleration.

You can calculate and use the idea of center of mass with single objects such as the stick or with multiple objects. An example where you may be interested in the location of the center of mass of multiple objects is the tightrope walker. The system consists of multiple objects: a long pole to help with stability, the mass of each lower leg, the mass of each upper leg, the mass of the torso, the mass of each forearm, the mass of each upper arm, and the mass of the head. In this example, you want to know where the center of mass of the entire system is as the person swings a leg in a semi-circle to bring the leg from behind the other leg to in front of it.

The formula for calculating *center of mass* is

$$m_{total} \, x_{CM} = m_1 \, x_1 + m_2 \, x_2 + \ldots$$
$$m_{total} \, y_{CM} = m_1 \, y_1 + m_2 \, y_2 + \ldots$$
$$m_{total} \, z_{CM} = m_1 \, z_1 + m_2 \, z_2 + \ldots$$

If you multiply this equation by the acceleration due to gravity, *g*, which equals 32.2 feet per second squared (9.81 meters per second squared), this changes the masses to the magnitude of the weights and the equations become the formula for calculating the *center of gravity*.

 Think of the formula for center of gravity as the sum of the torques produced by the individual objects' weight equal to the torque produced by the total weight located at the center of gravity.

 The previous equation is for a collection of discrete objects. When the object is continuous, the formula uses calculus:

$$\vec{s}_{CM} = \frac{\int \vec{s} \, dm}{\int dm}$$

Here $\vec{s}$ is the position vector and *dm* is an element of mass located at this position, which equals the mass density times a volume element.

 It's Saturday night and you're having a biophysics party. You want to calculate the location of your body's center of mass. I don't know your dimensions, so I use Sam's numbers instead. When you're looking over my calculations, you should substitute your own body's numbers into the calculations. It's more fun knowing your own center of mass.

Follow along in these steps:

1. **Set up the experiment and measure the center of mass/gravity of the table.**

   Ask some of your closest biophysics friends to bring their bathroom scales to the party. You also need to use a couple of broom handles and your kitchen table. Then, follow these steps:

   1. Place the scales on the kitchen floor.

   2. Place the broom handles on the scales.

   3. Flip your kitchen table upside down and place it on the broom handles.

4. Measure the readings on the bathroom scales and the distance from the end of the table to each of the broom handles.

I have made measurements on my kitchen table:

$\vec{F}_{N,1} = +\hat{y}\, 27$ pounds (120 newtons), $\vec{F}_{N,2} = +\hat{y}\, 23$ pounds (102 newtons), $x_1 = 1$ foot (0.305 meters) and $x_2 = 8.5$ feet (2.59 meters).

Figure 6-1 illustrates the kitchen setup. Does your setup look similar? The figure also shows what I mean by x_1 and x_2.

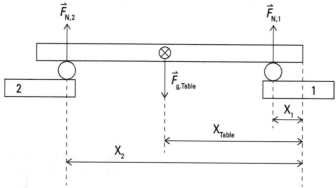

Figure 6-1:
A diagram for the kitchen table and scales showing the forces acting on the table.

Three forces act on my table: the gravitational force (weight) and the normal force at each broom handle. The table isn't moving so the acceleration of the table is zero and according to Newton's second law, the law of acceleration, the net force acting on the table must be zero. This means the weight of the table cancels the normal forces pushing the table up as such:

$$\vec{F}_{g,Table} = -\vec{F}_{N,1} - \vec{F}_{N,2} = -\hat{y}\, 50 \text{ pounds } (222 \text{ newtons})$$

2. **Calculate the center of mass/gravity of the table.**

You can find the center of gravity (mass) by using either the weights or the normal forces because the normal forces are countering the weight and the forces and torques of the system must balance:

$$x_{Table} = \frac{|\vec{F}_{N,1}|x_1 + |\vec{F}_{N,2}|x_2}{|\vec{F}_{g,Table}|} = \frac{(27 \text{ lb})(1 \text{ ft}) + (23 \text{ lb})(8.5 \text{ ft})}{50 \text{ lb}} = 4.45 \text{ feet}(1.36 \text{ meters})$$

3. Measure the center of mass/gravity of the table with you on it.

You now know where the center of gravity of the table is and can calculate Sam's center of mass (gravity). You can calculate your center of mass (gravity) using your numbers. Lie down on the table and have a biophysics friend measure the location of your feet and the force on the bathroom scales. Sam's values are

$$\vec{F}_{N,1} = +\hat{y}\,120 \text{ pounds } (534 \text{ newtons}), \vec{F}_{N,2} = +\hat{y}\,100 \text{ pounds } (445 \text{ newtons}) \text{ and}$$

$$x_{feet} = 1.5 \text{ feet } (0.457 \text{ meters}).$$

Figure 6-2 illustrates the setup in my kitchen of the table with Sam lying on it. Does your setup look similar?

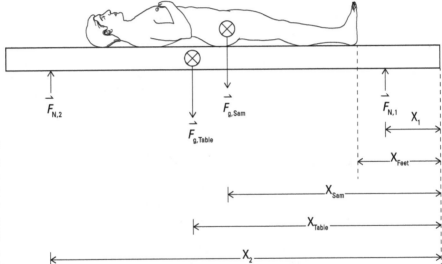

Figure 6-2: The free-body diagram for the kitchen table and Sam.

REMEMBER

The reason you want to measure the location of your feet is very simple. Suppose you're walking down the street and a stranger asks you, "Where's the location of your center of mass?" You can immediately respond that your center of mass is so many inches above the ground instead of "I don't know."

Four forces are acting on the system: the gravitational force acting on the table, the gravitational force acting on Sam, and the normal force at each broom handle. The table and Sam aren't moving so the acceleration of the table is zero, and according to Newton's second law of motion, the law of acceleration, the net force acting on them must be zero. This means Sam's weight plus the table's weight must cancel the normal forces pushing the table up. You now know Sam's weight:

$$\overline{F}_{g,Sam} = -\overline{F}_{g,Table} - \overline{F}_{N,1} - \overline{F}_{N,2} = -\hat{y}\,170 \text{ pounds } (757 \text{ newtons}).$$

4. **Calculate your center of mass/gravity.**

You can find your center of mass (gravity) by using this formula (I calculate Sam's center of mass):

$$x_{Sam} = \frac{\left|\overline{F}_{N,1}\right| x_1 + \left|\overline{F}_{N,2}\right| x_2 - \left|\overline{F}_{g,Table}\right| x_{Table}}{\left|\overline{F}_{g,Sam}\right|} =$$

$$\frac{(120 \text{ lb})(1 \text{ ft}) + (100 \text{ lb})(8.5 \text{ ft}) - (50 \text{ lb})(4.45 \text{ ft})}{170 \text{ lb}}$$

Sam's center of mass (gravity) is located at 4.40 − 1.5 = 2.90 feet (0.883 meters) above his feet.

Women typically find their center of mass at hip level whereas men find their center of mass to be slightly above the hips. The lower center of mass in the female body makes it more stable than the male body, but I discuss stability in the next section.

Staying stable and balanced

An animal or object is stable when its center of mass is over its base. The base is the area on the ground enclosed by the locations of the normal forces. If the center of mass of an object isn't over the base, then the object will fall over. For example, a three-legged table can have normal forces at the three legs and the base is a triangle enclosed by the three-legs. For a human standing on his feet, the base is from the toes to the heels and from the outer edge of the right foot to the outer edge of the left foot. As long as the person's center of mass (gravity) is over that base, he won't fall over.

As an example, Sam's center of mass is located 2 feet 11 inches (88 meters) above his feet, half way between the left foot and right foot and approximately half way between the heel and ball of the foot. Now suppose Sam puts on an extra 100 pounds (445 newtons) of weight (45.3 kilogram mass). For many humans, excess fat builds up on the torso and mostly on the front. How far has Sam's center of mass shifted forward if the center of mass of the fat is 8 inches (20 centimeters) in front of the original center of mass?

Use this formula to solve the problem:

$$x_{CM,new} = \frac{\left|\vec{F}_{g,Sam,old}\right| x_{Sam,old} + \left|\vec{F}_{g,fat}\right| x_{fat}}{\left|\vec{F}_{g,CM,new}\right|} = \frac{(170 \text{ lb})(0 \text{ in}) + (100 \text{ lb})(8 \text{ in})}{270 \text{ lb}}$$

Sam's new center of mass is located 3 inches (7.62 centimeters) in front of the original center of mass. This means Sam has to arch his back backward so his new center of mass is shifted backward 3 inches (7.62 centimeters) and remains over his feet (base). This arching isn't natural and can lead to back problems and discomfort for people that are obese. This can also occur for pregnant women.

Feeling the Effects of Acceleration

The human body is designed for a gravitational force (mass times acceleration) due to gravity (g = 32.2 feet per square second = 9.81 meters per square second) and small accelerations. (*Acceleration* is a measure of how fast the velocity is changing. Hitting a brick wall at 100 miles per hour [160 kilometers per hour] definitely isn't a small acceleration.) The development of technology has made it easier for the human body to undergo extreme accelerations. These sections look at some of the effects of these accelerations on the human body.

Noticing the physiological effects of acceleration

Newton's second law of motion, the law of acceleration, states that the acceleration equals the net external force divided by the mass of the object. Therefore, the larger the acceleration, the larger the net force is for a given mass. A larger force means more potential to damage the biological system. Four primary factors are key to determining the danger of the net force acting on a biological system:

✔ **The type of biological system:** It's suggested that bacteria can survive huge forces and extreme conditions to the point that they can survive interstellar travel between planets. Experimental measurements have shown that fleas, ticks, and other bugs of this type can subject their own bodies to accelerations that could kill a human.

✔ **The magnitude of force being applied to the biological system:** Dropping your biophysics book on your foot isn't too bad, but having a bus drive over your foot really hurts.

✔ **The duration of force being applied to the biological system:** During a car accident, a person hitting the windshield stops in a couple of milliseconds, and depending on the car's initial speed may not survive. Another person in the car wearing a seatbelt takes about 100 times longer to stop and has a much smaller average acceleration. The seat belted person has a much greater chance of survival.

An airbag in a car subjects the occupant to an acceleration of up to about 60g but only for about 1/30 of a second; whereas a trained jet pilot can withstand an acceleration of 5g for several seconds before passing out. Serious health risks exist if the duration and/or force were increased.

✔ **The *pressure* or more precisely the area over which the force is being applied:** In the example of the car accident and the person without a seatbelt, a very large force is needed to stop the person in a couple of milliseconds. In addition, the force is applied to just the small area of the forehead, which creates a pressure (force divided by area) large enough to fracture the bone (skull).

Gaining a hold of effective weight — blackouts and redouts

Blackouts occur when you are accelerating upwards. An upward acceleration means the effective weight is larger and the heart has to work harder. At an acceleration of about *3g* ($F_{g,eff} = 4mg$), the body starts to experience blackout effects. The retina in the eye starts to become oxygen deprived and the vision blackouts. Jet pilots are trained to withstand this blackout, and even at *5g* ($F_{g,eff} = 6mg$), they can hold it off for several seconds with proper equipment. At an acceleration of *6g* ($F_{g,eff} = 7mg$), the pilot will completely black out (become unconscious) from the reduced supply of blood to the brain. ***Note:*** An upward acceleration of *5g* is equivalent to 161 feet per second squared (49.1 meters per second squared).

Redouts are the opposite of blackouts, which means that the heart can't effectively pump the blood out of the brain and replace it with oxygen rich blood. In this case, the acceleration causes too much blood to flow into the head. This forces the blood vessels to enlarge and the body then reduces the activity of the heart, which in turn reduces the oxygen supply to the retina and brain. The buildup of blood in the retina causes the vision to turn red, and you'll shortly pass out afterwards from a lack of oxygen. You start to experience a redout with a downward acceleration of −*3g* ($F_{g,eff} = 2mg$ upside down).

A person's *effective weight* is expressed as

$$\vec{F}_{g,eff} = \vec{F}_g - \vec{F}_{NET}$$

This formula states that your weight changes if you're accelerating. To verify this, take your bathroom scale to a high speed elevator, stand on the scale in the elevator, and press a few buttons. If you're stationary, the scale reads your weight. If you're accelerating upwards, it feels like your stomach is in your shoes and the scale reads a force greater than your weight. The reason is the scale is providing the normal force, which is countering your weight and providing the force necessary to accelerate you upwards. The bathroom scale is measuring the normal force.

The common experiences of effective weight that people notice occur in elevators, cars, and amusement parks when it feels like your stomach has moved into your toes or throat. However, everybody is experiencing an effective weight because of the rotation of earth, but no one notices it because it's a very small effect compared to gravity.

For example, suppose Tabitha has a mass of m = 3.75 slugs (54.7 kilograms). The average acceleration of gravity on earth is g = 32.2 feet per second squared (9.81 meters per second squared). The magnitude of Tabitha's nominal weight is $m\,g$ = 121 pounds (537 newtons). Suppose Tabitha moves to the equator. The average acceleration due to gravity at the equator is 32.20 feet per second squared (9.814 meters per second squared), which gives Tabitha a weight that's the same as the nominal weight.

Mass has units of slugs (kilograms) where 1 slug = 1 pound square second per foot. 1 slug = 14.59 kilograms. The magnitude of the weight has units of pounds (newtons) where 1 newton = 1 kilogram meter per square second. 1 pound = 4.448 newtons. Personally, I think we should all use dekanewtons (daN) for weight, and then Tabitha would have a weight of 53.7 daNs.

But at the equator, Tabitha is traveling in a big circle every 23 hours and 56 minutes and according to Newton's first law of motion, the law of inertia, objects travel in straight lines unless subjected to a net external force. If Tabitha has a net force acting on her, there must be an acceleration according to Newton's second law of motion, the law of acceleration. When standing, smelling the roses, you don't think of your body as having acceleration, but it does.

The *radial (centripetal) acceleration* is the acceleration needed to travel in a circle. The radial acceleration at the equator is

$$a_r = r\,\omega^2 = 4.19 \times 10^7 \, \text{ft} \left(\frac{2\pi \ \text{radians}}{8.616 \times 10^4 \, \text{s}} \right)^2$$

The acceleration, a_r, is the radial (centripetal) acceleration, and it's directed toward the center of the circle. The subscript on the a is to remind you that this is the radial acceleration, which has the special form shown in the equation. The r is the circle's radius, which is the radius of earth at the equator in this case. The earth's average radius at the equator is 7,928 miles (12,756 kilometers) or 4.19×10^7 feet (1.28×10^7 meters). ω is the angular velocity.

In this example, the angular velocity is approximately a constant, and you can use the average angular velocity, which is the angular displacement divided by the elapsed time. The person travels in a full circle (2π radians) in 23 hours 56 minutes, which is 8.616×10^4 seconds. (Earth is orbiting the sun, so it takes the earth an extra four minutes of spinning to bring the sun to the same point in the sky, which is why the day is 24 hours.)

I can now calculate Tabitha's effective weight by the following equation:

$$\vec{F}_{g,\text{eff}} = m\left[\vec{g} - \vec{a}_r\right] = (3.75 \text{ slugs})\left[-\hat{y}32.20 \text{ ft/s}^2 - \left(-\hat{y}0.223 \text{ ft/s}^2\right)\right] =$$
$$(3.75 \text{ slugs})\left[-\hat{y}32.0 \text{ ft/s}^2\right]$$

The effective acceleration due to gravity is $g_{\text{eff.}} = 32.0$ feet per second squared (9.75 meters per second squared) downward, and the magnitude of Tabitha's effective weight is 120 pounds (533 newtons). If Tabitha originally lived at the South Pole and then moved to the equator, she would've lost 1.1 pounds (4.7 newtons) of weight just by moving — the quickest weight-loss program around. Of course, there was zero mass lost in the process.

Perceiving angular momentum and balance

The ear does more than let you hear. The inner ear plays an important role in balance. The inner ear is a bony labyrinth of hollow cavities within the temporal bone of the skull. The bony labyrinth can be split into three parts:

- **Cochlea:** The part of the ear dedicated to hearing (covered in more detail in Chapter 15)
- **Semicircular canals:** Parts of the ear dedicated to the detection of motion and balance

    The human inner ear has three semicircular canals:

    - **Lateral semicircular canal (or the horizontal semicircular canal):** This semicircular canal is for the detection of *yaw* rotation. This corresponds to sitting on a chair that rotates and you start spinning around in circles.

- **Anterior semicircular canal (or the superior semicircular canal):** This semicircular canal is for the detection of *pitch* rotation. This corresponds to bending over to tie your shoes.

- **Posterior semicircular canal:** This semicircular canal is for the detection of *roll* rotation. This corresponds to sitting in a canoe and tipping over.

✔ **Vestibule:** The central part of the inner ear close to the cochlea and where the semicircular canals connect; works with the semicircular canals to detect motion and help with balance; also works with eyes to help you focus on objects while in motion.

Look at your coffee cup and start spinning the cup (not the coffee). What happens if you don't spill the coffee? At first there's an angular acceleration and then constant angular velocity. During the initial angular acceleration, the coffee can't keep up with the cup, but as you continue to spin the coffee cup at a constant angular velocity, the coffee eventually catches up and spins at the same angular velocity.

The semicircular canals are filled with a fluid called *endolymph*. When you suddenly move your head, it's the same as when you start spinning the coffee cup. The endolymph doesn't move in the canal and applies pressure to the cupula, within the ampulla, in the opposite direction. (The *ampulla* is an enlarged part of the semicircular canal near the vestibule. The nerve cells end in hair cells that are connected to the *cupula* [a gelatinous material] within the ampulla. A signal is sent to the brain when the fluid changes the pressure applied to the cupula.)

After the head stops accelerating, it takes the endolymph about a second to catch up with the semicircular canal's motion. It then takes the cupula about a third of a minute to relax after that. After the cupula relaxes, the inner ear can't detect the motion.

The signal from the cupula to the brain not only tells you that your head is accelerating, but also it triggers the *nystagmus reflex* in the eye. This is a flicking of the eyes from one object to the next. The eyes focus on an object for a moment and then fix on another object, focus, and then move to the next. This reflex stops as soon as the cupula stops sending signals to the brain. A lot people then see the objects streaking past and lose their balance.

Test the nystagmus reflex by placing a chair in the middle of the room that can spin freely. Before you begin, make sure there's nothing close to the chair in case someone flies off the chair. The first victim sits on the chair, and you spin it for a minute. A constant angular velocity of 2π radians per second should be fine. You want to avoid any angular acceleration so the cupula can relax. After one minute, you bring it to an abrupt stop and have the person immediately walk across the room to the other side and pick up a glass of water. Now, you and the rest of your friends can try it.

Your observations include the following:

✔ Was she losing her balance and turning green while in motion? People can train their eyes not to see the streaking motion and maintain balance.

✔ Was she staggering when she left the chair and walking toward the glass of water? During the stopping motion, the nystagmus reflex is making the eye focus on the moving objects and treating the stationary objects as if they were moving. If the stopping and walking are done quickly, the brain has trouble making the adjustment and hence the unbalance and staggering.

Floating in space and the effects of weightlessness

Weightlessness occurs when the effective weight is zero. There is a force of gravity everywhere, but you can have a zero effective weight. For example, even on the International Space Station, the acceleration due to gravity is 28.7 feet per square second (8.73 meters per square second), so a person who weighs 150 pounds (667 newtons) on earth will weigh 134 pounds (594 newtons) on the International Space Station.

The effective weight is zero when the net force acting on the object is equal to the weight. This is indeed the case for the people on the International Space Station. They're falling toward the center of the earth due to a net force equal to their weight. The only reason they don't crash into the earth is because the tangential velocity is too large, and they keep missing the ground. Falling and missing the ground is usually referred to as *being in orbit*.

The human body is designed for an effective weight equal to the weight of the body at the surface of earth and not approximately zero, so one out of every two astronauts will get space sickness, which includes nausea, vomiting, vertigo, and headaches. *Space sickness* is a short-term illness, which lasts up to three days. *Long-term effects* can occur, as well, and can be split into three categories: severe, moderate, and light. Some of the worst problems that arise from long-term weightlessness include the following:

✔ Severe

• **Muscle atrophy:** The muscles are always at work, holding the body in place against the forces acting on it. In weightlessness, muscles aren't needed for standing, running, or lifting heavy objects and other similar activities. The muscles start to shrink, and the body's muscles mass drops.

- **Osteopenia:** The skeleton no longer has to work against gravity to maintain the body's shape, and bone mass is lost.

✔ Moderate

- Balancing problems

- A weaker immune system

- A reduced red blood cell count

- A slowing of the cardiovascular system

Rising of the Machines — The Bio-Terminator

There are six kinds of basic machines, which are important in understanding the biomechanics of animal motion. In this section, I define the mechanical advantage, which measures the amount of applied force required against a load force. You also get the six different machines and discover the human body in terms of these machines.

Marching to the mechanical advantage

With machines, you apply a force to the machine, and the machine applies a force to some load. The *mechanical advantage* of a machine is defined as the magnitude of the output (load) force divided by the magnitude of the input (applied) force. Mathematically, mechanical advantage's formula is

$$MA = \frac{\left|\vec{F}_{out}\right|}{\left|\vec{F}_{in}\right|}$$

The mechanical advantage is used with conservation of energy to determine forces, displacements, or speeds. Of course, in real machines, including animal bodies, there's energy lost through dissipative forces, such as friction. The *efficiency* (η) of the machine is defined as the power out (P_{out}) divided by the power in (P_{in}):

$$\eta = \frac{P_{out}}{P_{in}}$$

Machines do work, but they need energy to do the work. The formula tells you how good the machine is at using the energy to do work on its surroundings. The efficiency of a realistic machine is between 0 and 1. If the efficiency is one, the machine is perfect, and it obeys conservation of energy, which means work in is equal to work out, or power in is equal to power out. You can view the human body as a machine where you use energy to do work on your surroundings. For most activities, the human body has an efficiency less than 10 percent ($\eta = 0.10$). The body is good at a few things. For example, with walking the human body has an efficiency around 0.2 (20 percent). The *total efficiency* of a system with multiple machines connected in a sequence is the product of the efficiencies:

$$\eta_{total} = \eta_1 \eta_2 \eta_3 \ldots$$

Combining mechanical advantage with conservation of energy and the definition of work gives you the following useful relationships for the ideal system:

$$MA = \frac{F_{out}}{F_{in}} = \frac{\Delta x_{\|,in}}{\Delta x_{\|,out}} \text{ for linear work}$$

$$MA = \frac{F_{out}}{F_{in}} = \frac{r_{\perp,in}\Delta\theta_{in}}{r_{\perp,out}\Delta\theta_{out}} \text{ for angular work}$$

Δx is the displacement parallel to the force, the $\Delta\theta$ is the angular displacement vector parallel to the torque, the $r_{\perp}$ is the component of the displacement from the axis of rotation (think of the line from the earth's North Pole to the South Pole) to the location where the force is applied, which is perpendicular to the force and the direction of the axis, and $r_{\perp}$ times F is the torque (N).

Suppose you're doing arm curls with 50 pound (222 newton) weights in each hand. With your biceps muscle in the vertical direction and your forearm horizontal, what's the force supplied by your biceps muscles, and what's the mechanical advantage?

The distance from the elbow joint to where your biceps muscle is attached to the forearm is 2.0 inches (5.1 centimeters), and the distance from the elbow joint to the weight is 13.0 inches (33.0 centimeters). $\Delta\theta$ is the same for both the biceps muscle and the weight. Substitution into the equation for the rotational mechanical advantage above gives the following:

$$MA = \frac{r_{\perp,in}\Delta\theta_{in}}{r_{\perp,out}\Delta\theta_{out}} = \frac{(2 \text{ in})\Delta\theta}{(13 \text{ in})\Delta\theta} = 0.154; \text{ and}$$

$$F_{in} = \frac{F_{out}}{MA} = \frac{50 \text{ lb}}{0.154} = 325 \text{ pounds}(1{,}450 \text{ newtons})$$

The force in each biceps muscle is 325 pounds (1,450 newtons). The mechanical advantage of the arm is 0.154. A mechanical advantage less than 1 means more force is needed for a given output load, but the load will move farther. If the mechanical advantage is greater than 1, then the input force is less than the output force.

Pulleys and levers are the most common applications of mechanical advantage.

Perusing the machines within biomechanics

Welcome to the machines. A *machine* is a device that changes the direction and/or magnitude of a force. It has a mechanical advantage. Machines, including your entire body, can be decomposed into six basic types of machines, which these sections discuss.

Slipping on slopes — the inclined plane

The inclined plane is a flat surface that allows you to change the rate at which vertical displacement changes relative to the horizontal displacement. A couple of examples include the following:

- **The stairs in an apartment building:** It's a lot easier to climb the stairs to the third flood than to scale the outside of the building to the third floor balcony.

- **A wheelchair ramp:** For the disabled, a ramp reduces the incline so the chair can make it up and down the change in the vertical distance.

Dividing things — the wedge

A *wedge* is a triangular shaped tool used to separate two objects, to split a single object into two pieces, or to hold two objects together. For example, a wood axe is used to split firewood. Also, metal or wooden wedges can be driven into the axe handle to make the handle wider and hold the axe head in place.

Rolling — the wheel and axle

The wheel and axle allows a force applied to one to be transferred to the other so it can do work. An example would be the engine and transmission in a car or the chain on a bike turning the axle, which in turn moves the wheel. Alternatively, a force can rotate the wheel such as in a windmill, watermill, wind turbine, and water turbine. The wheel then turns the axis, which then does work.

Screwing — the screw

A screw converts torque into force and rotational motion into linear motion. Some screws have threads such as wood screws. The threads drive the wood screw linearly into the object. A tire jack uses a screw mechanism with threads as well. As you turn the screw, the threads pull the ends of the jack together, and it lifts the car. Other screws use a helical motion, such as a corkscrew, instead of threads.

Pulling — the pulley

The pulley is used to transfer power, change the direction of a force, and reduce the force needed to counter a load. The pulley consists of an axle with wheels attached to the axle that are free to rotate. A cable is then placed around the wheel. In Figure 6-3, you see some pulleys with mechanical advantages of 1, 2, 3, 4, and 5 lifting a box labeled W. If you notice, there's a pattern for the pulleys and the mechanical advantage. The mechanical advantage equals the number of times the rope passes between the pulleys. The pulleys for mechanical advantages of 6 and higher follow this same pattern.

TIP

If you want to determine the mechanical advantage of a pulley for yourself, follow these steps:

1. **Draw a free-body diagram of the pulleys with the load when it is stationary.**

   For example, consider the pulley system in Figure 6-3 that has a mechanical advantage of 3. Try drawing its free-body diagram. After you're finished, you can compare it to my free-body diagram. My free-body diagram in Figure 6-4 shows the bottom group of pulleys (1 pulley in this problem). Notice my free-body diagram has a compass included to help with directions.

2. **Calculate the forces in the free-body diagram with the help of compass.**

   The four forces in the example are

   $$\vec{F}_{rL} = +\hat{y}\,F_T;\ \vec{F}_{rC} = +\hat{y}\,F_T;\ \vec{F}_{rR} = +\hat{y}\,F_T;\ \text{and}\ \vec{F}_W = -\hat{y}\,W$$

   F_T is the tension in the rope, and W is the load on the pulley (the magnitude of the load force).

3. **Apply Newton's second law of motion, the law of acceleration.**

   Assume the acceleration of the system is zero, so the net force is zero, which gives a relationship between the input force and the output force and hence the mechanical advantage.

In the example shown in Figure 6-4, this gives $3F_T - W = 0$, and the mechanical advantage is W divided by F_T, which is 3.

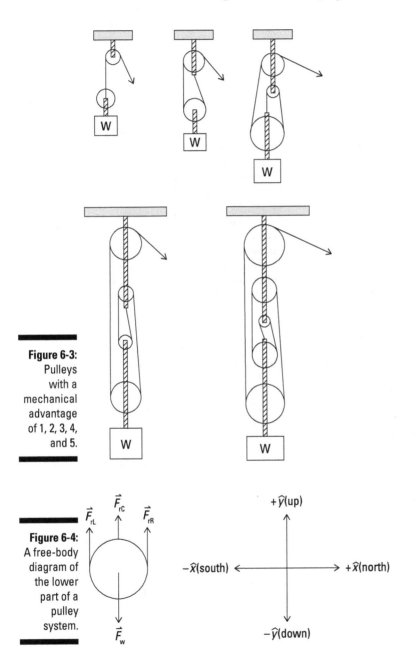

Figure 6-3:
Pulleys
with a
mechanical
advantage
of 1, 2, 3, 4,
and 5.

Figure 6-4:
A free-body
diagram of
the lower
part of a
pulley
system.

As an example, suppose you want to lift your 1,000-pound (453-kilogram mass) piano from the ground up to your third floor apartment balcony because it doesn't fit through the front doors. You're supervising so your friends have to lift the piano 27 feet (8.23 meters). The advantages of using pulleys are

- ✔ You can transfer the power from your friends' muscles to power lifting the piano against gravity.

- ✔ The pulleys change the direction of the force so your friends can stand on the ground beside the piano as they lift it up instead of standing on the roof of the apartment building; and you can set up the pulleys so your friends don't feel the full weight of the piano.

If you use a pulley system with a mechanical advantage of 5, how much force must your friends apply and how far do they need to pull the rope?

In the section, "Marching to the mechanical advantage," I mention the relationship between mechanical advantage, the ratio of the forces, and the ratio of the displacements. If you use the pulley system in Figure 6-3 with a mechanical advantage of 5, your friends have to lift a weight of only $F_{in} = F_{out}/MA = 1,000$ lbs/5 = 200 lbs (890 newtons). On the down side, your friends have to pull $x_{in} = MA\, x_{out} = 5$ (27 ft) = 135 feet (41.1 meters) of rope to lift the piano 27 feet (8.23 meters).

Lifting — the lever (Class 1, 11, and 111)

The last machine is the lever, which is very important in biomechanics. Levers consist of a *fulcrum*, an *applied (input) force*, and a *load (output force)*. The fulcrum is the point around which the lever rotates. It can be a ball in a socket like the elbow joint or the *axis of rotation,* such as the axis of a wheelbarrow or a teeter-totter. The orientation of the fulcrum and the two forces places the lever into one of three types:

- ✔ **Class I lever:** The fulcrum is in the middle with the two forces on either side.

- ✔ **Class II lever:** The output force is in the middle with the input force and fulcrum on the outsides.

- ✔ **Class III lever:** The input force is in the middle with the output force and fulcrum on the outsides.

The three levers are shown in Figure 6-5.

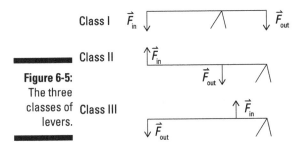

Figure 6-5:
The three classes of levers.

Working with your body

The body is loaded with machines such as the following:

- ✔ **The wheel and axle:** The spinal column

- ✔ **Screw:** The biceps (biceps brachii) muscle when it rotates the forearm and hand

- ✔ **Pulley system:** The trochlea and the superior oblique muscle

    The *trochlea* is a cartilaginous pulley, and the superior oblique muscle starts behind the eye, goes through the trochlea, and turns sharply to attach to the eye. The muscle pulls the eye downwards and toward middle.

- ✔ **Class I lever:** Your head while reading your biophysics book with the weight pulling downwards and the muscles holding your head up

- ✔ **Class II lever:** The foot when you stand up on your toes

    The weight of the body is pushing downward on the anklebone and the Achilles' tendon is pulling the foot upward.

- ✔ **Class III lever:** The lower jaw when chewing food, the spine when you're bent over, and the forearm when you're performing biceps muscle curls.

Responding to Biological System's Forces

Apply sufficient force to anything and the object will bend, twist, stretch, compress, or break. There's always some limit to what can be done to a biological system before it changes its shape. The limit varies depending on what the biological system is and how the force is being applied. For example, some objects can withstand a large linear force but can't withstand twisting forces such as the twist-off caps on some types of bottles.

In this section, you look at the concepts of elasticity, stress, and strain and how biological systems behave to linear forces. I also cover the response of biological systems to torques.

Grasping elasticity, stress, and strain

If you pull the skin on your cheek, it will stretch, and when you let go, it will bounce back in place. You can do the same thing with a rubber band, except the rubber band will probably stretch farther than your cheek. Everything is elastic to some extent and each object will behave differently to stresses. *Stresses* are forces (interactions) that cause an object to change its shape. The amount of change is the *strain*. This section introduces and defines the concepts of elasticity, stress, and strain.

Here are some clear definitions in plain English that can help you understand the following sections:

Elasticity is the ability of solid objects to change shape without becoming permanently changed or broken. The force must be sufficiently small and depends on the material. For example, if I hit my front door with a pillow, I won't notice the deformation to the door, but if I use a hammer on the door, there will be a permanent deformation to the door.

The *elastic limit* or *limit point* is the maximum amount of force that can be applied to the object and the object will return to its original form after the force is removed. The range of forces from zero to this maximum value is known as the *elastic range*, whereas forces beyond this maximum value is known as the *plastic range*. If the force is in the plastic range, the material undergoes *plastic deformation*.

Fatigue is when the forces binding the material together weaken. It occurs when the forces that cause plastic deformation go through many cycles. As an example, take one of your spoons and bend it; now bend it back; bend it again; bend it back; and keep repeating. Eventually your spoon will break.

Stressing out over force

Stress is defined as the total force divided by the area of the object over which the force is applied as shown in Figure 6-6. In Figure 6-6a and 6-6b, the area is the side of the box where the force is applied, whereas in Figure 6-6c, it's the top of the box. Mathematically, the stress is written as

$$\sigma = \frac{|\vec{F}|}{A}$$

σ is the symbol for stress, *F* is the magnitude of the force, and *A* is the area of the surface over which the force is applied.

To understand the formula and what it means, consider the following example. Suppose you're shooting peas across a cafeteria with an elastic band. If you pull on an elastic band with a certain amount of force, it will stretch a given amount. If you now place a second identical elastic band on top of the first elastic band and pull both of them at the same time, the stress σ hasn't changed because this is a property of the material. But, the area *A* has doubled. Therefore, you'll need twice as much force to stretch the elastic bands the same distance as before. The area *A* and the magnitude of the force *F* both doubled but the stress remains unchanged.

Three kinds of stress include the following:

- ✔ **Tensile (tension) stress, σ_T:** This is the stress in the object when the force is trying to pull the object apart as shown in Figure 6-6a. *L* is the width and the area is the depth times the height of the box in Figure 6-6a. The force has caused the object to stretch by a distance Δ*L*.

- ✔ **Compressive (compression) stress, σ_C:** This is the stress in the object when the force is trying to crush the object as shown in Figure 6-6b. *L* is the width and the area is the depth times the height of the box in Figure 6-6b. The force has caused the object to be compressed by a distance Δ*L*.

- ✔ **Shear stress, σ_S:** This is the stress in the object when the force acts like scissors on the object — for example, when someone comes and rips my elastic band in two as shown in Figure 6-6c. *L* is the height and the area is the depth times the width of the box in Figure 6-6c.

Materials have an elastic range and for larger stresses they have a plastic range. The size of these ranges depends on the properties of the material. You can classify materials depending on these ranges. Three classifications for materials are

- ✔ *Brittle* materials have a very short range of force beyond the elastic limit before the object breaks.

- ✔ *Ductile* materials deform in the plastic range for a large range of tensile stress.

- ✔ *Malleable* materials deform in the plastic range for a large range of compression stress.

Straining and deforming

Strain is a measure of the amount of deformation that occurs within the material, and it's defined as the change in the length divided by the total length of the object. Mathematically, it's written as

$$\varepsilon = \frac{\Delta L}{L}$$

ε is the strain in the material caused by a stress, ΔL is the amount the material moves because of the stress, and L is the original length of the material.

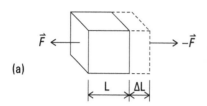

(a)

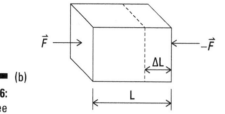

(b)

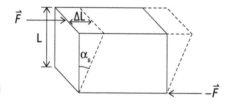

(c)

Figure 6-6: The three kinds of stresses and strains: tensile (a), compressive (b), and shear (c).

The three kinds of strain are

- **Tensile (tension) strain, ε_T:** This is the strain in the object when the length of the object is increased by ΔL as shown in Figure 6-6a.

- **Compressive (compression) strain, ε_C:** This is the strain in the object when the length of the object is decreased by ΔL, as shown in Figure 6-6b.

- **Shear strain, ε_S:** This is the strain in the object when the length of the object is shifted as a function of height as shown in Figure 6-6c.

Seeing the link between stresses and strains

Stresses cause strains and the relationship between the two is very complicated. Some general properties do exist for different types of material. I want to discuss some of the general features of the stresses and strains for ductile and brittle materials.

Figure 6-7a shows the tensile stress versus strain curve for a balloon, which is an example of ductile material. In Figure 6-7a, from the origin to No. 1, the stress is directly proportional to the strain. This corresponds to the linear limit of the elastic range. No. 2 in Figure 6-7a is the elastic limit. If you blow only a little bit of air into a balloon and then let it out, the balloon will contract back to its original shape. If you blow too much air into the balloon and let it out, the balloon won't deflate to its original shape, which means you have passed the elastic limit and entered the plastic range. As you continue to blow up the balloon, you notice that it becomes easier to blow it up. You have passed the maximum or ultimate tensile strength (the No. 3 in Figure 6-7a). Now blowing up the balloon is easier; it keeps getting larger and larger until it bursts in your face (which is the No. 4 in Figure 6-7a).

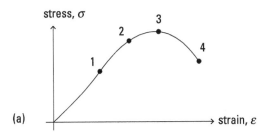

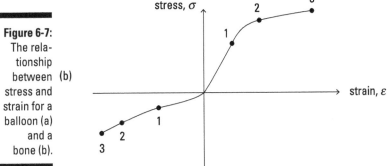

Figure 6-7: The relationship between stress and strain for a balloon (a) and a bone (b).

Figure 6-7b shows the tensile (right of the vertical axis) and compressive (left of the vertical axis) stress versus strain curve for a bone, which is an example of a brittle material. The No. 1 corresponds to the linear limit of the elastic range. From the origin to these points the stress is directly proportional to the strain. No. 2 is the elastic limit where the curve passes from the elastic range to the plastic range. No. 3 is the point at which the bone breaks.

Bending, buckling, and compressing

If you're standing on the end of the spring diving board, you'll notice that it has bent under your weight. If you rotate the board on its side and walk on it, it won't bend but buckle to the side. If you replace the board with a wooden beam, it will still compress a little as you walk on it. Nothing in biophysics is rigid, so this has applications in all areas of biophysics and is the focus of this section. To start this section though, I need to introduce several definitions so you understand them more clearly.

Young's modulus is the ratio of the stress divided by the strain in the linear region of the elastic range. It measures the amount of force needed to cause a change in the length of the object. The mathematical expression is

$$E = \frac{\sigma}{\varepsilon}$$

E is Young's modulus, σ is the stress, and ε is the strain. The slope of the stress versus strain curve at any arbitrary point is called the *tangent modulus* and is equal to Young's modulus in the linear region. In both Figure 6-7a and 6-7b, the Young's modulus is the slope of the curve from the origin to the No. 1. Young's modulus is one of the *moduli of elasticity* or *elastic moduli*. Two other common elastic moduli are the *shear modulus* and the *bulk modulus*. The bulk modulus is similar to Young's modulus, except the force is applied uniformly from all directions instead of along a single direction.

In the elastic range, the material will return to its original shape after the stress is removed, which is similar to a spring, which means *Hooke's law* is obeyed in the linear region of the elastic range. Hooke's law states that the force applied to a mass by a spring is proportional to the displacement from the equilibrium position. The mathematical expression for the ratio of the applied force divided by the deformation becomes

$$\frac{\left|\vec{F}\right|}{\Delta L} = \frac{\sigma A}{\varepsilon L} = \frac{EA}{L}$$

This formula starts on the left with the applied force divided by the deformation. By definition, the applied force is equal to the stress times the area and the deformation is equal to the strain times the length L, but the definition of stress divided by the strain is Young's modulus, which is the far right term. Hooke's law states that the force divided by the deformation is equal to Hooke's spring constant. Based on the expression in the formula, this is

Hooke's law with Hooke's spring constant $k_H = EA/L$. (For more information on Hooke's law, see Chapter 5.)

You can use these concepts for this example problem.

Calculate the change in length of the femur subjected to a 10-pound (44.5 newton) tensile force and then the compressive force. The minimum cross-sectional area of a femur is 0.9 square inches (6×10^{-4} square meters) and its length is 1.5 feet (0.46 meters). Along the axis of a bone, the tensile Young's modulus is $E_T = 2.32 \times 10^6$ pounds per square inch (1.60×10^{10} newtons per square meter) while the compressive Young's modulus is $E_C = 1.3 \times 10^6$ pounds per square inch (0.90×10^{10} newtons per square meter).

This problem has multiple steps:

1. **Solve for the stress.**

   Substitute the numbers into the equation for stress.

   $$\sigma = \frac{|\vec{F}|}{A} = \frac{10 \text{ lb}}{0.9 \text{ in}^2} = 11.1 \frac{\text{pounds}}{\text{inch}^2} \left(1,600 \frac{\text{pounds}}{\text{foot}^2} = 7.66 \times 10^4 \text{pascals} \right)$$

2. **Calculate the corresponding strains using Young's modulus and the stress.**

   $$\varepsilon_T = \frac{\sigma_T}{E_T} = \frac{11.1 \text{ psi}}{2.32 \times 10^6 \text{psi}} = 4.78 \times 10^{-6}; \text{ and } \varepsilon_C = \frac{\sigma_C}{E_C} = \frac{11.1 \text{ psi}}{1.30 \times 10^6 \text{psi}} = 8.54 \times 10^{-6}$$

3. **You now know the length and the strain, so calculate the change in the length of the femur.**

   $$\Delta L_T = \varepsilon_T L = \left(4.78 \times 10^{-6} \right) 1.5 \text{ ft} = 7.17 \times 10^{-6} \text{feet} \left(2.19 \times 10^{-6} \text{meters} \right)$$
   $$\Delta L_C = \varepsilon_C L = \left(8.54 \times 10^{-6} \right) 1.5 \text{ ft} = 1.28 \times 10^{-5} \text{feet} \left(3.90 \times 10^{-6} \text{meters} \right)$$

The calculations are in the linear region of the elastic range. The change in the bone's length is very small. If the force is too large, the bones will break because they have a maximum (ultimate) strength before they snap. Along the axis of the bone, the maximum (ultimate) tensile strength is $\sigma_T = 1.7 \times 10^4$ pounds per square inch (2.4×10^6 pounds per square foot = 1.2×10^8 newtons per square meter) and the maximum (ultimate) compressive strength is $\sigma_C = 2.5 \times 10^4$ pounds per square inch (3.6×10^6 pounds per square foot = 1.7×10^8 newtons per square meter).

Based on the maximum strengths of bones along the axis, how much force is required to break the femur? The problem told me the minimum cross-sectional area of the femur is 0.9 square inches (6.3×10^{-3} square feet $= 5.8 \times 10^{-4}$ square meters). Now use the formula for stress to calculate the force required to break the femur:

$$\left| \vec{F} \right|_T = \sigma_T A = \left(1.7 \times 10^4 \, \text{psi} \right) 0.9 \, \text{in}^2$$

The tensile (tension) force required to break the femur is 1.5×10^4 pounds (6.7×10^4 newtons), and the compressive force required to break the femur is 2.2×10^4 pounds (9.8×10^4 newtons). This is a huge force!

Your bones are very strong along the axis and well designed to hold up your body, but what about force perpendicular to the length? *Bending* occurs when the object is subjected to a directional force, which is zero at one or more points and is countered by the internal forces within the object throughout the rest of the object. An example is your balcony sticking out of the apartment wall with gravity trying to pull it downward. The balcony has a very slight bend to it, which you may not be able to notice. In one place I lived, the balcony had a very noticeable bend to it, especially when I stood on the end.

Bending strength is defined as the torque from the internal forces to counter an external force trying to bend the object. Mathematically, the net torque is

$$\left| \vec{N} \right| = \frac{E \, I_A}{R}$$

E is *Young's modulus*, R is the *radius of curvature*, and I_A is the *area moment of inertia*.

The area moment of inertia is a measure of the object's ability to resist a force, which is trying to cause the object to bend. If you draw the bent object in a full circle, the radius of that circle is called the *radius of curvature*. When an object bends, part of it is being compressed and the other part is under tension. The boundary between the two regions is undergoing no change in length and is called the *neutral surface*. Figure 6-8 shows the radius of curvature and the neutral surface.

Three area moments of inertia that are common in biophysics include the following:

$$I_A = \frac{\pi r_{outer}^4}{4} \text{ for a solid cylinder}$$

$$I_A = \frac{\pi \left(r_{outer}^4 - r_{inner}^4 \right)}{4} \text{ for a hollow cylinder}$$

$$I_A = \frac{\pi r_{\parallel}^3 r_{\perp}}{12} \text{ for a rectangle}$$

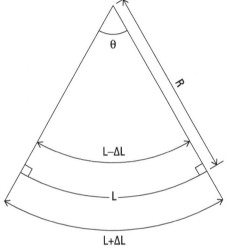

Figure 6-8:
The neutral surface and radius of curvature for a bent object.

The two r's for the rectangular object are the two thicknesses of the bar; the thickness ($r_{\parallel}$) parallel to the radius of curvature (R) shown in Figure 6-8, and the thickness perpendicular ($r_{\perp}$), which is sticking out of the page in Figure 6-8.

Strictly speaking, calculus is needed to couple the torque to the area moment of inertia, even if E and R are constants. Let x be the position relative to the neutral surface and parallel to R shown in Figure 6-8. The torque is as follows:

$$\left| \overline{N} \right| = \iint \frac{E\, x^2}{R}\, dA$$

N is the magnitude of the torque, E is Young's modulus, R is the radius of curvature, dA means an integral over the cross-sectional area of the beam (or bone), and x is the distance from the neutral surface parallel to R.

Bending usually occurs when the stress is perpendicular to the long axis of the material. In the case of forces parallel to the long axis there is another possibility. *Buckling* occurs when the object is subjected to a compression (compressive) force, and it can't be compressed anymore. The buckling occurs when there's *structural failure* in the object.

Buckling can also occur from a bending force if the object is too thin in the perpendicular direction. A competition exists between the bending and the buckling, which arises in biophysics, physics, and engineering. To see this, suppose you purchased a 12-foot long 1-inch by 10-inch wooden board so you can practice your balance beam routine. You place the board on two supports, one support at each end, and the rest of the board is in the air with the 1 inch thickness in the vertical direction and the 10 inch width being the width of the board. As soon as you walk on the board, you notice the board has a very large bend in it. To prevent the board from bending, you flip the board on its side so the 10-inch width is the vertical height and the 1-inch thickness is the width. As soon as you step on it you notice the board hardly bends at all, but it has a tendency to twist to the side (buckle).

Hollow objects are better at withstanding perpendicular forces compared to a solid object if they have the same length and weight. But this weakens the object's ability to withstand buckling. In animals, bones are approximately hollow because they're stiffer than solid bones for the same amount of weight. The human femur, for example has an outer radius that is twice the inner radius. Bird bones are very thinned-walled to reduce the bird's weight. The inner radius of the bone in a bird can be as high as 90 percent the outer radius. The bones are still very strong against tensile and compressive types of break. You have to fall from a very large height to cause a break this way. The majority of the breaks in bones are caused by torques and twists.

Shearing and twisting

A force parallel to the surface produces *shear stress* on an object. The force can produce a torque and twisting motion of the object, such as turning a round-handle doorknob. Figure 6-6c shows the distortion of the object caused by the tangential force. The shear stress is equal to the tangential force divided by the area. The deformation caused by the tangential force is the shear strain, which is related to the *shear angle* (twisting angle):

$$\varepsilon_S = \frac{\Delta L}{L} = \tan(\alpha_S)$$

ε_S is the shear strain, $\Delta L/L$ is the deformation per unit length, and α_S is the shear angle. The easiest way to see this relationship is to look at Figure 6-6c, where you can see the tangential force, the deformation, and the shear angle α_S. The shear modulus is

$$G = \frac{\sigma_S}{\varepsilon_S}$$

G is the shear modulus, σ_S is the shear stress, and ε_S is the shear strain. In the linear elastic range, the amount of strain is linearly proportional to the shear stress, and as soon as the stress is removed, the material goes back to its original shape. Hooke's law describes a linear relationship between force and linear displacement; therefore, Hooke's law for a shear force is

$$\left|\vec{F}\right| = \sigma_S A_S = G\varepsilon_S A_S = \frac{GA_S}{L}\Delta L = GA_S \tan(\alpha_S)$$

F is the shear force, σ_S is the shear stress, A_S is the area over which the force is applied, G is the shear modulus, ε_S is the shear strain, ΔL is the amount of deformation per unit length, and α_S is the shear angle. Comparison with Hooke's law shows that the Hooke's constant is $k_H = G\,A_S/L$. Shear forces occur in many areas of biophysics. Examples within the body are the forces acting on joints and the vertebrae of the spine.

Another example of shear stresses in biophysics is most broken legs in skiing sports. These types of breaks occur when different parts of the legs have forces going in different directions. A *couple* is a pair of forces with equal magnitude but that act on different locations of an object in opposite directions. Examples of couples include turning the steering wheel with both hands, turning a round-door handle, moving the driveshaft in a car, or the forces acting on your limbs in many sports. Couples produce no net force; they're shear forces like in Figure 6-6c. If the forces are localized to one part of the object, they produce a net *twisting torque* on the object. The twisting torque is

$$\left|\vec{N}\right| = G\,I_p\frac{\alpha_S}{L}$$

I_p is the polar moment of inertia and is a measure of the object's ability to resist a twisting torque. L is the distance along the object from where it's not twisted to where is has been twisted by an angle of α_S.

For example, suppose Vicky is downhill skiing and falls causing her leg to twist, and she breaks her tibia. What twisting angle was produced on her leg just as it broke?

Some information about Vicky's tibia you need includes the following:

✔ The torque needed to break the tibia is 73.8 pound foot (100 newton meters).

✔ The length of the tibia is 1 foot (0.3048 meters).

✔ The shear modulus of bone is $G = 1.5 \times 10^6$ pounds per square inch (2.15×10^8 pounds per square foot $= 1.0 \times 10^{10}$ newtons per square meter).

✔ The polar moment of inertia is $I_p = 0.11$ inches to the fourth power (4.4×10^{-8} meters to the fourth power).

Vicky's tibia broke when it was twisted by an angle of

$$\alpha_S = \frac{L|\vec{N}|}{GI_p} = \frac{(12\ \text{in})(73.8\ \text{lb ft})(12\ \text{in}/1\text{ft})}{(1.5\times10^6\ \text{lb}/\text{in}^2)(0.11\ \text{in}^4)} = 0.0644\ \text{radians}$$

The angle is 0.064 radians = 3.7 degrees. It doesn't take much to break a leg by twisting it.

Defining Scaling: No Scales Required

Scaling is a technique where dimensional analysis is used to study characteristics that change with size. It assumes the characteristics change in a simple way where the size of the biological system can be labeled by a simple parameter. Scaling was started when Galileo noticed that bones didn't scale at the same rate as the size of the animal. Scaling has been used in studying many aspects of animal physiology. In this chapter, I introduce the scaling length and explain why trees and animals can get only so big and apply the concepts of scaling to properties and situations within the human body.

Growing cows and trees

A *scaling length* is a parameter that tells you how a quality will differ when you vary the spatial dimension. Trees can grow only so large before their weight is too large to support. The scaling comes from the following: mass (density times volume) of the tree must be proportional to the cross-sectional area of the tree trunk (support against buckling). The volume (L^3) scales as the area (r^2) or $L \sim r^{2/3}$. This indicates that the tree trunk must grow faster in width than in height. The maximum height a tree can grow is proportional to this scaling:

$$H_{max} = c\,r^{2/3}$$

c is a constant and r is the radius of the tree.

A few interesting things about c in the equation include

✔ If the column is assumed to be uniform (cylinder) and it's bending slightly under its own weight, then

$$c = \left(\frac{2E}{g\rho}\right)^{1/3}$$

E is Young's modulus for the material, g is the acceleration due to gravity, and ρ is the mass density of the material. If the cylinder exceeds H_{max}, then it will buckle under its own weight.

✔ $c = 34.9$ is the best-fit experimental value for trees.

✔ None of the experimental heights of trees reaches the maximum height if c is calculated for a tapered column. The ideal situation can't be attained in real life.

This relationship can be used for the scaling of animal sizes as well; except now, r is the radius of the leg, and H becomes L, a scaling length for the size of the animal. The relationship is

$$r \sim L^{3/2}$$

This scaling relationship shows that the size of an animal's legs must grow faster than the characteristic length of its body, which means there's a limit to how large an animal can be. You may be thinking, "Why are humans not as tall as trees then?" (I was thinking it.) A few reasons include

✔ Humans walk, jog, and run with usually one leg in the air. Combined with the impact with the ground and motion, the normal force on a single leg is usually much greater than the weight of the body.

✔ As the animal gets bigger, the mass of the animal's bones scales faster than the animal's total mass. If the animal becomes large enough, it becomes all bone with no organs or skin, which puts a limit on how large an animal can become. Galileo first noticed it in the bones of large animals compared to small animals and lead to him developing the scaling method.

✔ The power output required to walk and do activities doesn't scale linearly. The work done to walk will eventually become greater than the amount of power produced by the muscles.

Scaling in the body

An introduction to scaling in biophysics usually assumes that each dimension scales as the characteristic length to the first power. This isn't always the case. There are many examples in biophysics and nonlinear physics where the exponent isn't 1. In fact, in the preceding section, I use an example where the scaling exponent was ⅔ because of the buckling strength required in the animal's leg or in a tree's trunk. If I assume that animals are a collection of cylinders (legs, torso, arms) and the cylinders are restricted by the buckling strength, then this will introduce a different scaling instead of the exponent 1.

I present the scaling of different quantities for the body with both assumptions: 1-to-1 (the first scale) and the buckling strength scale of $2/3$ (in square brackets).

A few scaling rules that apply to the physical properties of the body are

- **Volume:** $V \sim R^2 L \sim L^3 \, [L^4]$

  For specific shapes the volumes are L^3 for a cube, $\pi \, L \, R^2 \sim L^3$ for a cylinder, and $1/3 \, \pi \, R^3 \sim L^3$ for a sphere. All of them scale as the length cubed. [The volume of a cylinder is $\pi \, R^2 \, L \sim (L^{3/2})^2 \, L = L^4$]

- **Body surface area:** $A_{sur} \sim RL \sim L^2 \, [L^{5/2}]$

- **Cross-sectional area:** $A_{csa} \sim R^2 \sim L^2 \, [L^3]$

- **Density:** $\rho \sim$ constant [constant]

  People are mostly water, carbon, and air, and the density is approximately a constant at about 80 percent of the density of water.

- **Mass:** $M = \rho \, V \sim L^3 \, [L^4]$

  A common scaling parameter is the mass of the animal, so the previous quantities become

    Length: $L \sim M^{1/3} \, [M^{1/4}]$

    Width: $R \sim M^{1/3} [\sim L^{3/2} \sim M^{3/8}]$

    Volume: $V \sim L^3 \, [L^4] \sim M \, [M]$

    Body surface area: $A_{sur} \sim L^2 \sim M^{2/3} \, [\sim RL \sim L^{5/2} \sim M^{5/8}]$

    Cross-sectional area: $A_{csa} \sim R^2 \sim L^2 \sim M^{2/3} \, [\sim R^2 \sim L^3 \sim M^{3/4}]$

  A log-log plot of the body surface area of mammals as a function of their mass is approximately a straight line with a slope of 0.63. Note that the body surface area above gives $2/3 = 0.667$ and $5/8 = 0.625$, which are both close to the experimental value for animals.

- **Muscle stress:** $\sigma \sim$ constant [constant]

  The force per unit area is the same in all muscles in all mammals. The maximum stress is approximately 60 pounds per square inch (40 newtons per square centimeter).

- **Muscle force:** $F_M = \sigma \, A_{csa} \sim R^2 \sim L^2 \sim M^{2/3} \, [\sim L^3 \sim M^{3/4}]$

  If a weightlifter has muscles with twice the radius of your muscles, you would expect that the weightlifter can lift $(2)^2 = 4$ times as much weight as you can.

- **Muscle power:** $P_M = F_M v = \sigma \, A_{csa} \, v \sim R^2 \sim L^2 \sim M^{2/3} \, [\sim L^3 \sim M^{3/4}]$

  The velocity, v, of voluntary muscle fiber contraction is the same for all muscles in all mammals.

✔ **Bones:** $\sim A_{csa} \sim R^2 \sim L^2 \sim M^{2/3}$ $[\sim L^3 \sim M^{3/4}]$

The weight a bone can support is proportional to the cross-sectional area of the bone.

✔ **Limbs:** $\sim L$.

Limbs act like levers, which scale as their length.

Here are four physiological applications of scaling that build on the previous list:

✔ **Drug dosage ~ mass of the body:** The drug is absorbed by the body and spreads throughout the volume. The effectiveness of medication is proportional to its concentration, so the dosage of the drug scales as the volume, which scales as the mass of the person.

You want the medicine to be effective, but you don't want to take too much, which is harmful to the body. This is important for children whose weight changes a lot as they grow up. If you look at the directions for children's medicines, some of them state to take a specified amount per unit of weight. Adult prescriptions drugs are sometimes prescribed per unit of weight as well.

✔ **Jumping: Energy output per muscle mass ~ constant:** If you jump up to dunk a basketball, then your muscles have to do work: $W = F\,\Delta x = \sigma A_{csa}\,\Delta x \sim R^2 L \sim L^3 \,[L^4] \sim M$. Therefore, the energy per unit mass is a constant, which is about the same for all mammals. If this work goes mostly into kinetic energy and very little is lost to dissipative forces and gravitational potential energy ($M\,g\,\Delta x$), the kinetic energy will equal the gravitational potential energy (mgh) at the top of the jump. This means the work done by your muscles (W) scales as the mass (M) equals M g h; therefore, the height you jumped is approximately independent of M and L, which means all mammals of the same type will jump approximately to the same height.

✔ **Walking: Energy output per muscle mass ~ constant:** If you're walking, your muscles have to do work: $W_M = F\,\Delta x = \sigma A_{csa}\,\Delta x \sim R^2 L \sim L^3 \,[L^4] \sim M$. To move, you have to do work that's approximately equal to your weight times the distance. For one step, $W_{step} = F L \sim M L \sim M^{4/3}\, [M^{5/4}]$.

In order for you or any animal to move, its muscles must be capable of moving its body. The work done for one step divided by work done by the muscles is: $W_{step}/W_M \sim M^{1/3}\, [M^{1/4}]$. Therefore, cows can get only so big before they can't move and become trees because the work done to take a step grows faster than the work your muscles can do as the mass increases.

✔ **Metabolic properties of the body:** The muscle power output is $P_M \sim M^{2/3}\, [M^{3/4}]$. The heat production of the body is proportional to the power output and the rate of metabolic heat generation $\sim M^{2/3}\, [M^{3/4}]$. Experimentally, the heat production of mammals scales as $M^{0.75}$.

The heart muscle power output is $P_M \sim M^{2/3}$ $[M^{3/4}]$ and does work $W_{heart} = P_M \Delta t \sim M^{2/3}$ $[M^{3/4}]$. The blood flowing through the lungs is absorbing oxygen, so the surface area in the lungs should scale as $\sim M^{2/3}$ $[M^{3/4}]$. Experimentally, the surface area of the lungs in mammals scales as $M^{0.75}$.

The volume of blood pumped through the heart times the pulse rate is proportional to the power output of the heart. Therefore, the pulse rate scales as $\sim M^{2/3}/V$ $[M^{3/4}/V] \sim M^{-1/3}$ $[M^{-1/4}]$. This means that larger animals have a slow pulse rate.

Chapter 7

Making The World Go Round with Physics — Dynamics

. . . .

In This Chapter

▶ Understanding motion in a straight line

▶ Grasping circular motion

. . . .

*D*ynamics means to study the causes of motion and changes in the motion. The focus is to study how forces combine to create a net external force and/or a net torque acting on a biological system. In dynamics, you're also asking why did the things move or why did their motion change? For example, a figure skater standing in one spot and spinning can change the speed at which she is spinning. This chapter tries to figure out how that happens. Or you may be interested in sports, so how does a baseball curve to the right or left? Dynamics deals with the why.

This chapter starts with one-dimensional linear motion; you may be interested in the motion of birds, fish, or mammals. If you're interested in linear motion in two or three-dimensions, then don't worry. This chapter also focuses on motion in a circle. This motion is two dimensional, but you can treat circular motion as one-dimensional motion if you look at the motion in terms of the angle (angular position). (The radius of the circle is a constant.) The centrifuge involves circular motion as well as the figure skater spinning in a circle.

Reducing Motion to a Straight Line

In many situations within biomechanics, you can ignore the activity in two directions, focusing your analysis of the motion in a straight line. Consider a sprinter. All the motion is in a straight line from the starting blocks to the finish line. In some situations, it's necessary to consider the forces perpendicular to the motion, such as the sprinter's weight and the normal force acting on the sprinter's feet, but the motion is still in a straight line. A *straight line* means the direction of the acceleration and the direction of the velocity are constants that don't change.

In dynamics, answering the why is important, so understanding the motion of biological systems and how the motion changes are the focus of this section. The understanding involves several ordered steps, which I lay out for you. If you follow these steps for each dynamical problem given to you, you should have no problems:

1. **Determine all forces acting on the biological system and all sources of energy within the system.**

2. **Draw a free-body diagram.**

   Visualizing what's going on is important to solving many problems. The free-body diagram includes the biological system and all external forces acting on it. I like to include a compass in my free-body diagram so I know my directions. (This is a common source of error for most people.) I also like to show the direction of the acceleration. When I do a vector sum of all the forces, it has to produce a net force that points in the same direction as the acceleration; if it doesn't, then there's a mistake somewhere.

   It's necessary to split the biological system into multiple parts within the free-body diagram when you need to consider the internal forces.

3. **Calculate the forces by using the free-body diagram as a guide.**

4. **Apply the laws of physics.**

   Chapter 5 discusses the laws to consider. The hard part is to figure out, which law is the best one to use. For example, you need to ask yourself, should I use Newton's laws of motion, the work-energy theorem, or conservation of energy?

Reading steps is easy. The big question is how do you apply them to real problems? This section illustrates how to apply these steps.

Riding my bike

Imagine you want to study the biomechanics of riding your bike down the road. (If you don't ride a bike, then you can think of me on my bike or someone in a bike race like the Tour de France. All you have to change is the numbers to the appropriate values. Be creative and have fun — I promise biophysics can be fun.)

The road has an incline of 4.50 degrees (5 percent grade), and you want to calculate your maximum acceleration up the hill. In addition, you can calculate the maximum speed you can ride your bike up the hill. The maximum speed exists because you're including air resistance, which is proportional to the speed squared. If you don't include a dissipative force that is speed dependent (air resistance), there's no maximum speed. The maximum speed is called *terminal velocity*.

The following example shows the steps involved in solving these types of problems, as well as the power of these concepts and how to use them in any biomechanical problem.

Step No. 1: Determine the forces and sources of energy

You can feel your muscles work as you ride your bike up the hill. Your muscles are one of the sources of energy. Some sources are going to do work on the system and supply it with mechanical energy (muscles); other sources are going to work against the system and change the mechanical energy into heat energy (friction — dissipative forces).

To help you with this step, keep the sources of the forces acting on the bike organized. I find the easiest way of doing that is by making a list. The following is my list:

- When riding your bike, your muscles work on the system to make the bike go forward.

- You have to do work against gravity because you're going up a hill at an angle of 4.50 degrees. Include your weight and the bike's weight. If you don't know the bike's weight, it's usually much smaller than your body weight, so you can ignore it. The total weight of my bike and me is 198 pounds (881 newtons) downwards, and the total mass is 6.15 slugs (90.0 kilograms). (Notice I add the direction downward to the weight because it's a force that has both direction and magnitude.)

- The normal force is holding you and the bike up and is pointing perpendicular to the road. The amount of normal force on the back wheel and front wheel is important and depends on your riding position. People usually sit with most of the weight over the back wheel. You can do a measurement or calculation to determine your weight distribution or you can guess. In my case, two-thirds of my weight (including the bike's weight) is over the back wheel and one-third is over the front wheel. These are related to the two normal forces acting on the bike wheels.

- The rolling resistance force is a dissipative force, which is equal to the coefficient of rolling resistance times the total normal force. You can estimate your rolling resistance by making a few measurements of how fast you slow down. My coefficient of rolling resistance is 0.01.

- The back wheel pushes the earth backwards with the static friction force and the reaction force pushes the bike forward. The coefficient of static friction between my back wheel and the road is 0.80. The maximum force you can apply with your muscles between the tire and the road is the coefficient of static friction times the magnitude of the normal force on the back wheel. If you pedal harder so you apply a force to the ground greater than the static friction, then the back wheel will spin on the ground.

✔ Air resistance can play an important role, especially if you like to ride a bike through a hurricane. This force is also dissipative, which is sucking energy out of the system. You will have to make some measurements and estimate your drag coefficient and frontal cross-sectional area (the area of your body pushing against the wind.) My drag coefficient is 0.8 and I am sitting upright, so the wind is pushing against my face, upper torso and the front of my legs, so my frontal cross-sectional area is 6.46 square feet (0.600 square meters). The frontal cross-sectional area of most bikes is very small and can be ignored. The weight density of air is 0.0749 pounds per cubic foot (11.8 newtons per cubic meter) and the mass density of air is 0.00233 slugs per cubic foot (1.2041 kilograms per cubic meter). The mathematical formula for air resistance is covered in Chapter 5, but here is the formula for you:

$$\left| \vec{F}_{af} \right| = \frac{1}{2} D \rho A v^2$$

F_{af} is the magnitude of the air resistance force, D is the drag coefficient, ρ is the density of air, and v is the speed of you and the bike.

Step No. 2: Draw a free-body diagram

In this step, you draw a free-body diagram that shows the forces acting on you and the bike. The diagram should also include a compass that shows directions.

Figure 7-1 shows a free-body diagram of myself. In the figure, I also show the external forces, my compass, and the direction of the maximum acceleration. I ignore the internal forces, such as me pushing on the pedal and the pedal pushing back on my foot. In the first step, I list the important external forces that will influence the motion, which are the forces shown in Figure 7-1. For my compass, I use x and y instead of up, down, north, and south. Also, I select x parallel to the plane, so the acceleration and velocity are in only one direction. (Notice that the forces are in two-dimensions, but the acceleration and velocity are in a straight line — one-dimensional motion.) In this figure, I ignore some external forces that aren't important, such as the gravitational attraction of the moon on me.

Step No. 3: Calculate the forces

In the third step, you want to calculate the expressions for the forces in Figure 7-1. The purpose of this list is to split the magnitudes and directions apart for the vectors. I have also added the velocity and acceleration to the list and start with them. *Tip:* Place the free-body diagram right beside you when you make the list, so you don't miss anything.

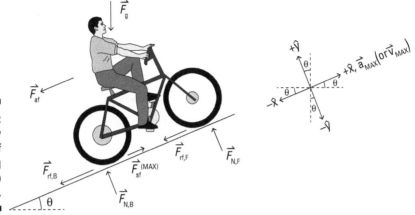

Figure 7-1:
A free-body
diagram of
me riding
my bike up
the hill.

Here are the line entries on the list:

- ✔ **Velocity:** It has a magnitude v, and your bike is moving up the hill in the +x-axis direction.

- ✔ **Acceleration:** It has a magnitude a, and your bike is accelerating up the hill in the +x-axis direction.

- ✔ **Applied force:** The third line is the force pushing your bike up the hill against all the other forces. In this problem, it's the maximum static friction between the tire and the road, and the force is up the hill in the +x-axis direction.

- ✔ **Rolling resistance:** This represents the rolling resistance forces of the wheels against the road. The resistance is trying to stop the bike so the rolling resistance forces are pointing down the hill in the –x-axis direction.

- ✔ **Normal forces:** They're the sixth and seventh lines. These forces are preventing the bike from accelerating into the road. The normal forces are perpendicular to the surface, so they are in the +y-axis direction.

- ✔ **Air resistance force:** This force, which is the eighth line, is always opposite to the direction of the velocity.

- ✔ **Weight:** This represents your weight and your bike's weight, which always points toward the center of the earth. (Don't forget to replace my numbers with your own numbers.)

Here is the list:

$$\vec{v} = +\hat{x}\,v,\text{ the velocity}$$

$$\vec{a}_{max} = +\hat{x}\,a_{max},\text{ the acceleration}$$

$$\vec{F}_{sf}^{(max)} = +\hat{x}\,\mu_s\,F_{N,B},\text{ the static friction on the back wheel}$$

$$\vec{F}_{rf,F} = -\hat{x}\,\mu_r\,F_{N,F},\text{ the rolling resistance on the front wheel}$$

$$\vec{F}_{rf,B} = -\hat{x}\,\mu_r\,F_{N,B},\text{ the rolling resistance on the back wheel}$$

$$\vec{F}_{N,F} = +\hat{y}\,F_{N,F},\text{ the normal on the front wheel}$$

$$\vec{F}_{N,B} = +\hat{y}\,F_{N,B},\text{ the normal on the back wheel}$$

$$\vec{F}_{af} = -\hat{x}\,\frac{D\rho A v^2}{2},\text{ the air resistance acting on me and my bike}$$

$$\vec{F}_{g} = mg\left[-\hat{x}\,\sin(\theta)-\hat{y}\,\cos(\theta)\right],\text{ the weight of me and my bike}$$

Step No. 4: Apply the laws of physics

You have to calculate the forces and acceleration acting on you and your bike, so you should use Newton's laws of motion. Newton's second law of motion, the law of acceleration, states that the acceleration is equal to the net external force divided by the mass, which for this example looks like this (just copy the list in step No. 3):

$$x\text{-axis: }a_{max} = \frac{1}{m}\left[\mu_s\,F_{N,B}-\mu_r\,F_{N,F}-\mu_r\,F_{N,B}-\frac{D\rho A v^2}{2}-mg\sin(\theta)\right]$$

$$y\text{-axis: }0 = \frac{1}{m}\left[F_{N,F}+F_{N,B}-mg\cos(\theta)\right]$$

You have three unknowns: the acceleration and the two normal forces. The first line gives you an answer for the acceleration if you know the normal forces. You need to solve for the normal forces first. You can achieve this step by solving the y-axis equation first. (Remember to use your weight distribution over the tires.) The solution is as follows:

$$F_{N,B} = 2F_{N,F} = \frac{2}{3}mg\cos(\theta)$$

You can write the normal forces as a number because you have your weight and the angle of the hill. I'm not going to because I want to discuss several things about the acceleration and that would be hard with the numbers. You can find the acceleration by substituting these expressions for the normal forces into the x-axis equation:

$$a_{max} = g\left[\frac{2}{3}\mu_s\,\cos(\theta)-\mu_r\,\cos(\theta)-\frac{D\rho A v^2}{2mg}-\sin(\theta)\right]$$

The equation for the maximum acceleration has several interesting pieces of information:

✔ The first term on the right-hand side of the equality is your contribution to the acceleration. The "max" is the best you can do because if you try to add more power, then the back wheel will start to spin and slip. The coefficient of kinetic friction is less than the coefficient of static friction, so if the bike wheel starts to slip, then the bike will have a smaller maximum acceleration.

You can substitute your numbers into the expression to see what your best value is. I obtained with my numbers $2g\mu_s\cos(\theta)/3 = 2(32.2 \text{ ft/s}^2)$ $(0.8)\cos(4.5°)/3 = 17.1$ feet per square second (5.22 meters per square second).

✔ The second term on the right hand side of the equality is the rolling resistance. This depends a lot on what type of surface you're riding on and the type of tires on your bike. Calculate the number for my bike, $-g\mu_r\cos(\theta) = (-32.2 \text{ ft/s}^2) (0.01)\cos(4.5°) = -0.321$ feet per square second (–0.0978 meters per square second).

✔ The third term on the right hand side of the equality is the air resistance, which will have an important impact at high speeds or if you're out riding in a strong wind. In the formula, v is your speed relative to air, so if you have a *tail wind* (a wind blowing on your back), it will actually help you accelerate up the hill. In this problem you can ignore the wind.

✔ The fourth term on the right hand side of the equality represents the work you have to do against gravity. Fortunately, gravity is a conservative force, so this work is being stored as gravitational potential energy, which you'll be able to use when you come back down the hill. You can calculate the contribution to the acceleration. Here I do it for my numbers: $-g \sin(\theta) = (-32.2 \text{ ft/s}^2) \sin(4.5°) = -2.53$ feet per square second (–0.770 meters per square second).

✔ Because you're out for a bike ride with no wind, the maximum acceleration will occur when you're just starting out (speed is zero). You can calculate your maximum acceleration when the speed is zero. Similarly, I can calculate my maximum acceleration with zero speed by adding the numbers calculated together, or from:

$$a_{max} = 32.2\frac{\text{ft}}{\text{s}^2}\left[\frac{2}{3}(0.8)\cos(4.5°) - (0.01)\cos(4.5°) - 0 - \sin(4.5°)\right]$$

The maximum acceleration is 14.3 feet per square second (4.35 meters per square second), but decreases rapidly as the speed increases because of the air resistance.

- ✔ Eventually, the acceleration will be zero when you reach the terminal velocity. This is the fastest you can go. You can calculate your terminal velocity by solving the a_{max} equation with a_{max} = 0. I help you rearrange the formula, so you can calculate your terminal speed. In addition, I find my terminal speed using the same formula in the second line:

$$v_{max} = \sqrt{\frac{2mg}{D\rho A}\left[\frac{2}{3}\mu_s\ \cos(\theta)-\mu_r\ \cos(\theta)-\sin(\theta)\right]}$$

$$= \sqrt{\frac{2(198\ \text{lb})\left[\frac{2}{3}(0.8)\ \cos(4.5)-0.01\ \cos(4.5)-\sin(4.5)\right]}{(0.8)(0.00233\ \text{lb s}^2\ \text{ft}^{-4})(6.46\ \text{ft}^2)}}$$

  How fast can you go up the hill on your bike? My maximum speed up the hill is 121 feet per second (36.8 meters per second), which is equivalent to 82.4 miles per hour (133 kilometers per hour). I can't pedal that fast, so I would definitely need to attach a motor to my bike to achieve this speed.

- ✔ The source of power to move your bike is your muscles. It would be interesting to see if your muscles can produce enough power to reach the maximum speed. Remember, the power output is equal to the average force times the average velocity. To achieve the terminal speed, the minimum power output you would require is

$$P_{output} = \mu_s F_{N,B} v_{max} = \frac{2}{3}\mu_s mg v_{max} \cos(\theta)$$

  The minimum power output I would need is 2(0.8)(198 lb)(121 ft/s) cos(4.5°) = 1.27×10^4 foot pounds per second (1.73×10^4 watts = 23.1 horsepower (hp)). A typical human can put out about ⅓ horsepower over a long period, so I would definitely need a motor in order to achieve a speed of 82.4 miles per hour (133 kilometers per hour). Going downhill with a tailwind would help too.

Racing the horses

Biophysics changed horse racing forever. In the 1890s, jockeys changed their posture from sitting straight up in the saddle to a standing crouch in the stirrups with their heads forward almost touching the horse's neck. This change caused a 7 percent improvement in race times — the biggest single improvement in the history of the sport.

To help you understand what caused this improvement, suppose Wayne is a retired jockey and has put on a couple of extra pounds. You take some measurements of Wayne's body: He has a weight of 97 pounds (431 newtons) or a mass of 3.01 slugs (44.0 kilograms), a frontal cross-sectional area of 3.0 square feet (0.28 square meters) while sitting and a frontal cross-sectional area of 1.5 square feet (0.14 square meters) while in the standing crouch.

You want to look at the change in the rider's posture because it's the only thing that changed and caused a huge improvement in the race times. You can ignore the horse except for providing a force on Wayne. The steps to figuring out this change are as follows:

1. **Determine all forces acting on the biological system and all sources of energy within the system.**

   Three forces are the primary external forces acting on Wayne:

   - Wayne's weight is acting straight downwards and it's located at his center of mass. The center of mass is a special point within the body (or object) where the gravitational force thinks all the mass is located.

   - The horse is running fast enough that air resistance could play a role. This is an approximately horizontal force opposing his motion.

   - The contact force between Wayne and the horse. The horse is holding Wayne up against gravity and pushing his body forward against the air resistance.

2. **Draw a free-body diagram.**

   You need to draw a free-body diagram of Wayne showing all the forces acting on his body. Figure 7-2 shows an example. How does your free-body diagram compare to mine? I include a few extras in my free-body diagram in Figure 7-2. I show Wayne, the forces acting on Wayne's body based on the forces in Step No. 1, a compass, and the trajectory of Wayne's body (actually, the center of mass).

   In Figure 7-2a the acceleration is constantly changing as Wayne bounces up and down, and in Figure 7-2b the acceleration is zero if the horse is running at a constant velocity. Instead of Wayne's acceleration, I include the *trajectory* (the path the object follows through space) of his center of mass in Figure 7-2. The tangent to this line is the direction of the velocity, which is also shown in Figure 7-2.

Visualizing what's going on is important to solving this problem. Figure 7-2 has two parts:

- In Figure 7-2a, Wayne experiences air resistance, contact forces with the horse, and weight. Wayne is continuously changing acceleration as his body is bounced up and down, shown by the dashed line.

- In Figure 7-2b, Wayne experiences essentially the same forces: air resistance, contact forces with the horse, and weight, but only the weight stays the same compared with Figure 7-2a. The other major change with Figure 7-2a is the direction of the velocity doesn't change. If the horse is running in a straight line at maximum velocity, then Wayne's acceleration is zero and the velocity is a constant. Wayne uses his legs as shock absorbers so he doesn't bounce with the horse.

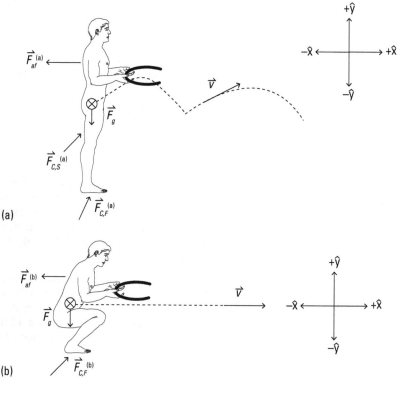

Figure 7-2:
A free-body
diagram of
Wayne as **(a)**
if he were
riding the
horse while
sitting in the
saddle (a)
and while
standing in
the jockey
crouch
position (b). **(b)**

3. **Calculate the external forces by using the free-body diagram as a guide.**

The three external forces for both sitting and crouching are

- $\vec{F}_g = -\hat{y}\, mg$, Wayne's weight.

- $\vec{F}_{af}^{(a)} = -\hat{x}\dfrac{D^{(a)}\rho A^{(a)}v^2}{2}$, the air resistance acting on Wayne while he is sitting.

- $\vec{F}_{af}^{(b)} = -\hat{x}\dfrac{D^{(b)}\rho A^{(b)}v^2}{2}$, the air resistance acting on Wayne while he is crouching.

- $\vec{F}_{C,S}^{(a)} + \vec{F}_{C,F}^{(a)} = \hat{x}\, F_{C,x}^{(a)} + \hat{y}\, F_{C,y}^{(a)}$, the contact force between Wayne and the horse while he is sitting.

- $\vec{F}_{C,F}^{(b)} = \hat{x}\, F_{C,x}^{(b)} + \hat{y}\, F_{C,y}^{(b)}$, the contact force between Wayne and the horse while he is crouching.

4. **Apply the laws of physics.**

Figure 7-2 has two major differences between the two postures: the force of air resistance and Wayne's trajectory. Take a look at each of these separately:

- **Air resistance:** The difference in the magnitude of the air resistance force at the same speed is

$$\Delta F_{af} = \frac{D^{(b)}\rho A^{(b)}v^2}{2} - \frac{D^{(a)}\rho A^{(a)}v^2}{2} = \left[D^{(b)}A^{(b)} - D^{(a)}A^{(a)} \right]\frac{\rho v^2}{2}$$

You can assume the drag coefficient is 0.4 and doesn't change with posture.

You measured Wayne's cross-sectional areas as 1.5 square feet (0.14 square meters) when crouching and 3.0 square feet (0.28 square meters) when sitting upright.

The weight density of air is 0.0749 pounds per cubic foot (11.8 newtons per cubic meter) or the mass density of air is 0.00233 slugs per cubic foot (1.2041 kilograms per cubic meter).

The horse is moving at 59.2 feet per second (18.1 meters per second), which is equivalent to 40.4 miles per hour (65.0 kilometers per hour).

If you substitute these numbers into the equation and do the calculation, you find the change in the force is –2.47 pounds (–11 newtons). The change in the force is negative, so the sitting position has a larger force.

The work done equals the force times the displacement, so the total energy saved by the horse during a one-mile race having Wayne in the crouch position is $\Delta W = \Delta F\, \Delta x = (-2.47 \text{ pounds}) (5{,}280 \text{ feet}) = -13{,}000 \text{ foot pounds} (-1.77 \times 10^4 \text{ joules})$.

- **Wayne's trajectory:** In both cases the horse is pushing Wayne forward, but in the first case the horse is also pushing Wayne upward whenever it pushes off the ground. If you're an inexperienced horse rider and go for a long horse ride, you know what this feels like and walking is a challenge for the next few days. The horse's muscles must do work against gravity lifting Wayne up.

  You need to estimate the amount of work the horse does against gravity. I looked at some typical racehorses and you can assume the horse lifts Wayne's center of mass a distance of 1 foot (30.48 centimeters) 550 times during the one-mile race. The total energy saved by the horse during the one-mile race having Wayne in the crouch position and keeping his center of mass horizontal is $\Delta W = \Delta F \, \Delta x = -mg \, \Delta x = (-97 \text{ pounds}) (1 \text{ foot} \times 550) = -53{,}400$ foot pounds $(-7.23 \times 10^4$ joules).

The one-mile race takes the horse 1.5 minutes to complete at 40 miles per hour (65 kilometers per hour), and Wayne can save the horse 66,400 foot pounds $(9.00 \times 10^4$ joules) of energy during that time. If the horse isn't wasting its energy on the jockey, it use the energy to run faster. This amount of reduced energy consumption is a huge savings of energy, and this calculation shows that both Wayne's trajectory and his air resistance play an important role in the horse's speed improvement with 70 percent of the savings coming from the change in Wayne's trajectory.

Simplifying the dynamics of multiple objects in contact

Usually, the quickest and easiest method to solve a multi-object system is to treat it as a single system. However, in some situations it may be better to treat each object separately. This is something you have to decide depending on the problem. Also, if you need to find specific internal forces such as contact forces or tension in a muscle, then you'll need to split the system apart into the individual objects and apply the steps to each object.

 If you split a biological system apart into the individual objects, then it's easier to figure out what forces are involved. If you're not sure what forces are involved and how they interact with the biological system, then this is the way to go. The problem with this method is you have to apply Newton's laws to each individual object, so there are many mathematical equations. The alternative approach is to combine all the objects together into one. This method has the advantage of having only one object and therefore one set of equations from Newton's laws. The disadvantage is that it's a lot easier to miss a force or get the direction or magnitude of a force wrong.

This section looks at biological systems with multiple objects.

You live on the second story of an apartment and you have your biophysics friends over for your Saturday night biophysics party. Abby from marketing lives across the hallway and has her friends over for their party. Abby ordered a 15.5 US gallon (58.66 liter) keg of beer for her party; but the delivery person wouldn't carry it up the stairs because it weighs 170 pounds (77.1 kilogram mass), so he left it beside the back door. Abby asks you for help. To help, you and your friends think up the following biophysics home experiment.

You lay planks on the stairs so you have a smooth ramp that makes an angle of 40 degrees relative to the horizontal. At the top of the ramp (stairs), Abby sits on a skateboard and a rope is tied to the back of the skateboard. You have a frictionless pulley, which you've fastened to the windowsill. The moment of inertia of the pulley I = 0.884 slug square foot (1.20 kilogram square meter) and the radius of the pulley is 12.2 inches (1.02 feet = 0.310 meters). (Yes, this is a big pulley.) You take the rope from the skateboard throw it over the pulley and out the window. Your friends tie the other end of the rope to the keg. After the rope is tied to the keg, you jump on the skateboard with Abby and give the skateboard a push down the stairs. The combined weight of you, Abby, and the skateboard is 300 pounds (1,330 newtons) or the mass is 9.32 slugs (136 kilogram mass). The rope doesn't slip on the pulley. What's Abby's acceleration?

The steps needed to solve the problem are listed right before the problem, so follow these steps:

1. **Determine all forces acting on the biological system and all sources of energy within the system.**

   You can assume there are no dissipative forces (friction) in the system, so the system has conservation of mechanical energy.

   The forces acting on you, Abby, and the skateboard (no dissipative forces) are gravity, normal force, and the tension in the rope. (I call this force: *tension inside.*)

   The forces acting on the Pulley (no dissipative forces) are

   - Tension in the rope trying to rotate the pulley toward the skateboard. (I call this force: tension inside. This is the same contact force that is acting on you, Abby, and the skateboard.)

   - Tension in the rope trying to rotate it toward the keg. (I call this force: *tension outside.* This is the same contact force as that acting on the keg.)

The forces acting on the keg (no dissipative forces) include gravity and tension in the rope pulling the keg up to the window. (I call this force: tension outside.)

The external forces are gravity, which is conservative, and the normal force, which is perpendicular to the motion and does no work. (It keeps you and Abby from falling through the stairs.) The rest of the forces are the tension in the rope, which you can ignore if you treat everything as a single system. Therefore, you can use conservation of energy to solve this problem.

2. **Draw a free-body diagram.**

Figure 7-3 shows my free-body diagrams. In my figures, I include compasses and the direction of the accelerations. Figure 7-3a is a combination of Figures 7-3b, 7-3c, and 7-3d. Figure 7-3a is the one you need for this problem, if you combine all the objects together to form a single object; but if the problem wanted you to find the tensions, you would need to use figures 7-3b, 7-3c, and 7-3d. I show only Abby in Figure 7-3b, so you need to add yourself to the figure.

To combine Figures 7-3b, 7-3c, and 7-3d to make Figure 7-3a, start with one figure, say Figure 7-3b and begin with the acceleration. It's easiest to start with the acceleration and place it on the object in Figure 7-3a. Now add all the forces from Figure 7-3b to Figure 7-3a; the directions will be fixed relative to the acceleration. You can now add the forces from Figure 7-3c and Figure 7-3d. At this point you'll get more forces than I have in Figure 7-3a because you'll have all the internal forces. For example, from Abby in Figure 7-3b you have the inside tension pointing in the +x-axis direction and then from the pulley in Figure 7-3c you will have the inside tension pointing in the –x-axis direction. The two have the same magnitude but opposite direction so they cancel out in Figure 7-3a (the action-reaction pair).

The rope doesn't slip, so there's a relationship between the pulley's motion and the motion of Abby and the keg — namely, the rope has to move the same amount, must have the same speed, and must have the same acceleration at the three locations: Abby, the pulley, and the keg. Otherwise, the rope will break. Mathematically, the rope (three objects) having the same motion means that the angular variables of the pulley are: $\alpha = a_T/R = a/R$, $\omega = v_T/R = v/R$ and $\Delta\theta = \Delta s/R = \Delta y/R$.

Figure 7-3a is easy to work with in the next step, but going from Figures 7-3b, 7-3c, and 7-3d to Figure 7-3a takes practice. The trick here is to realize that all three objects have the same acceleration (and same speed), so you put them together with all their accelerations pointing in the same direction. For the pulley, you think of the circular motion in the direction of the acceleration as linear motion, $\Delta s = R\,\Delta\theta$.

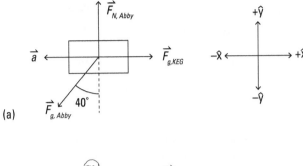

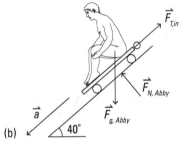

(a)

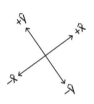

(b)

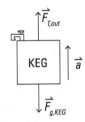

(c)

Figure 7-3:
A free-body
diagram of
the entire
system.

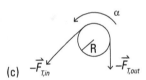

(d)

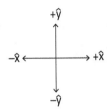

3. Calculate the forces by using the free-body diagram(s) as a guide.

Use Figure 7-3a because it's quicker and the same without the internal forces, shown here:

$$\vec{F}_{g,\text{Abby}} = m_{\text{Abby}}g\left[-\hat{x}\sin(40) - \hat{y}\cos(40)\right], \text{ Abby's weight}$$

$$\vec{F}_{g,\text{KEG}} = m_{\text{KEG}}g\left[+\hat{x}\right], \text{ the KEG's weight}$$

$$\vec{F}_{N,\text{Abby}} = F_N\left[+\hat{y}\right], \text{ the normal force acting on Abby}$$

$$\vec{a} = a\left[-\hat{x}\right], \text{ the acceleration of the system}$$

4. **Apply the laws of physics.**

Use Newton's second law of motion, the law of acceleration, because you want to calculate the acceleration and you know the forces. The mathematical formula of Newton's second law, the law of acceleration is

x-axis: $-M_{\text{TOTAL}}a = -m_{\text{Abby}}g\sin(40) + m_{\text{KEG}}g$

y-axis: $0 = -m_{\text{Abby}}g\cos(40) + F_N$

To calculate Abby's and your acceleration, you need to know the total effective mass of the system. The total mass is Abby's mass, your mass, the skateboard's mass, the keg's mass, and the pulley's effective mass. To calculate the pulley's effective mass, divide the moment of inertia by the radius squared (center of the pulley to the rope). Solving for the acceleration in the x-axis equation:

$$a = \frac{m_{\text{Abby}}g\sin(40) - m_{\text{KEG}}g}{M_{\text{TOTAL}}} = \left(\frac{32.2 \text{ ft}}{s^2}\right)\left[\frac{(300 \text{ lb})\sin(40) - 170 \text{ lb}}{300 \text{ lb} + 170 \text{ lb} + 28.5 \text{ lb ft}^2/(1.02 \text{ lb})^2}\right]$$

Abby's acceleration down the stairs is 1.48 feet per square second (0.450 meters per square second).

Discovering Forces and Torques Involved with Circular Motion

Objects moving in circles are always experiencing a net force and therefore an acceleration. Recall Newton's first law of motion — the law of inertia, which states that objects travel in straight lines at constant velocity unless forced to change their motion. This is important to remember because it can have important consequences such as in sports that involve motion in circular paths.

The motion in a circle can be split into *radial motion* and *tangential motion*. The radial direction is the direction toward (and away from) the center of the circle. For circular motion the distance r from the center is a constant; the radial velocity is zero and the magnitude of the *radial (centripetal) acceleration* is $a_r = r\,\omega^2$, where ω is the angular speed. (As long as you have sufficient force in the radial to produce this acceleration, the object will go in a circle.) The tangential direction is the motion around the center of the circle, which can be completely described by the angle when the radius r is a constant. The motion is described by the magnitude of the angular position θ, the magnitude of the angular velocity ω, and the magnitude of the angular acceleration α.

Understanding the motion of biological systems and how the motion changes involves five steps:

1. **Determine all forces acting on the biological system and all sources of energy within the system.**

2. **Draw free-body diagrams.**

   The free-body diagram includes the object and the all external forces acting on the object. For circular motion, I like to draw two free-body diagrams: one from the top looking down the axis of rotation of the object and one from the side that gives the clearest picture of the forces. It's also helpful to include the acceleration in the diagrams because there usually is both a radial (centripetal) acceleration and a tangential acceleration.

   In addition, I like to include a compass in my free-body diagram so I know my directions. When I do a vector sum of all the forces, it has to point in the same direction as the acceleration; if it doesn't, there's a mistake somewhere.

3. **Calculate the forces in the radial direction by using the free-body diagrams as a guide.**

4. **Calculate the forces and torques in the tangential direction by using the free-body diagrams as a guide.**

5. **Apply the laws of physics to solve the problem.**

This section illustrates how to apply these five steps. You discover why it can be important to consider what's going on in the radial direction, when you look at motion around the circle.

Racing on a circular track — forces and acceleration

This section looks at the forces in the radial direction and the radial (centripetal) acceleration. I consider an example that illustrates an important aspect of biomechanics, which has application in many different sports that require the athlete to travel in a circle at least part of the time. You need sufficient force in the radial direction to make the turn.

Every day after work, Bob thinks he is in a Formula 1 race on the highway. During a rainstorm, the coefficient of static friction between the road and his car tires is 0.3. What's the maximum speed Bob can take a turn if the road is banked at 2.5 degrees? (The path in many summer and winter sports are banked so the normal force is partially applied in the radial direction and helps the athlete make the turn at a faster speed.) In this case, the radius of curvature is 820 feet (250 meters) and the car has a weight of 2,750 pounds (12,200 newtons) or a mass of 85.4 slugs (1,240 kilograms).

The steps involved in solving for the speed are as follows:

1. **Determine all forces acting on the biological system and all sources of energy within the system that you believe are relevant.**

   Bob is traveling around a bend at a constant speed. If he goes any faster he will fly off the road. Three forces are acting on Bob's car:

   - Gravity, which pulls the car straight down

   - Normal force, which pushes the car perpendicular to the road's surface

   - Static friction, which is parallel to the road's surface and keeping the car on the road. The magnitude of the static friction is at its maximum value, which is the coefficient of static friction times the magnitude of the normal force. A small portion of the static friction is in the tangential direction to counter the rolling resistance of the tires on the road and maintain a constant speed. (For convenience, I assume no rolling resistance and hence no static friction in the tangential direction.)

   You can ignore the other dissipative forces.

2. **Draw free-body diagrams for Bob's car.**

   After you draw your diagrams, you can look at my free-body diagrams for Bob's car as shown in Figure 7-4. Figure 7-4a shows the car from above, whereas Figure 7-4b shows the front of the car as it comes toward you. How does your diagram compare with mine?

A common error arises in the choice of the direction of the radial acceleration. It's always toward the center of the circle. For example, if I asked you the direction of your radial (centripetal) acceleration because of the earth's rotation, many people would say it's toward the center of the earth. This is only true for people living at the equator. For everyone else, the circle is the plane of constant latitude.

Another common error is to get the direction of the normal force or the direction of the weight wrong because people get in the habit of placing the normal force opposite to the weight.

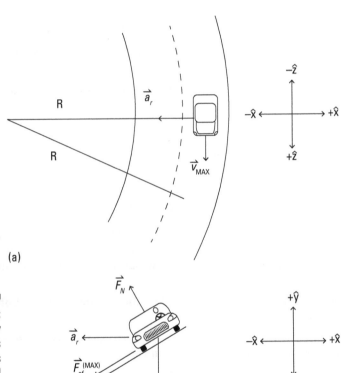

Figure 7-4: Free-body diagrams of Bob's car viewed from above (a) and the front (b).

3. **Calculate the forces in the radial direction by using the free-body diagrams as a guide.**

   The radial direction for Bob's car is parallel to the x-axis according to Figure 7-4b. The forces in the radial direction are

   $$F_{N,x} = -F_N \sin(\theta); \ F_{sf,x}^{(max)} = -\mu_s F_N \cos(\theta); \text{ and } F_{g,x} = 0$$

4. **Calculate the forces and/or torques in the tangential directions by using the free-body diagram as a guide.**

   Space is three-dimensional and there's one radial direction. This means two dimensions are tangential. In the case of Bob's car, I show the two tangential directions in Figures 7-4a and 7-4b. The forces in these directions are

   $$F_{N,y} = +F_N \cos(\theta); \ F_{sf,y}^{(max)} = -\mu_s F_N \sin(\theta); \text{ and } F_{g,y} = -m\,g$$

5. **Apply the laws of biophysics to solve for the normal force acting on Bob's car.**

   You can solve this step by using Newton's laws of motion. Apply Newton's second law, the law of acceleration:

   $$0 = +F_N \cos(\theta) - \mu_s F_N \sin(\theta) - m\,g$$

   You know everything for this formula except the normal force. Solve for the normal force:

   $$F_N = \frac{m\,g}{\cos(\theta) - \mu_s \sin(\theta)} = \frac{2750 \text{ lb}}{\cos(2.5) - 0.3\sin(2.5)}$$

   The normal force acting on the car is 2,790 pounds (12,400 newtons). The normal force is greater than the weight because the normal force is countering the weight and providing part of the acceleration around the corner.

6. **Apply the laws of biophysics to find the maximum speed Bob's car can make the turn.**

   Now, apply Newton's second law, the law of acceleration, to the radial direction and solve for the maximum speed.

   $$-\frac{m\,v_{max}^2}{R} = -m\,a_r = -F_N \sin(\theta) - \mu_s F_N \cos(\theta) + 0$$

   $$v_{max} = \sqrt{\frac{RF_N[\sin(\theta) + \mu_s \cos(\theta)]}{m}} = \sqrt{\frac{(820 \text{ ft})(2790 \text{ lb})[\sin(2.5) + 0.3\cos(2.5)]}{85.4 \text{ slugs}}}$$

   The maximum speed Bob can make the corner at is 95.9 feet/second (29.2 meters per second = 65.4 miles per hour = 105 kilometers per hour). Bob better slow down if he wants to make the turn.

Accelerating around the corner — torques and forces

This section is about biological systems that don't have a net force always pointing in the same direction (linear motion) or a net force that always points toward the center of the circle. This section deals with situations where the net force can be split into two parts: a part toward the center of the circle (radial or centripetal force) and a part tangential (perpendicular to the radial component). I restrict the discussion to objects moving in circles and the object speeding up or slowing down.

An object speeding up or slowing down as it goes in a circle can be analyzed using the net tangential force, or by analyzing the net torque acting on the object. Do not forget the radial (centripetal) force is always there.

When forces cause the object to accelerate or decelerate in the tangential direction while moving in a circle then the net acceleration is a combination of both the radial acceleration and the tangential acceleration. The magnitude of the net acceleration is

$$|\vec{a}| = \sqrt{|\vec{a}_r|^2 + |\vec{a}_T|^2} = \sqrt{\frac{|\vec{v}|^4}{R^2} + |\vec{a}_T|^2} = R\sqrt{|\vec{\omega}|^4 + |\vec{\alpha}|^2}$$

The first term is the magnitude of the net acceleration, the second set of terms is the magnitude of the acceleration in terms of its radial and tangential components, the third set of terms has the radial acceleration replaced by the speed squared divided by the radius of the circle, and the last set of terms expresses the acceleration in terms of the radius of the circle, the angular velocity, and the angular acceleration. The magnitudes of the corresponding net force and net torque from Newton's second law, the law of acceleration, are

$$\text{Net force:} |\vec{F}_{NET}| = \sqrt{|\vec{F}_r|^2 + |\vec{F}_T|^2}$$

$$\text{Radial force:} |\vec{F}_r| = m|\vec{a}_r| = m\frac{|\vec{v}_T|^2}{R} = m\,R|\vec{\omega}|^2$$

$$\text{Tangential force:} |\vec{F}_T| = m|\vec{a}_T| = m\,R|\vec{\alpha}|^2$$

$$\text{Net torque:} |\vec{N}_{NET}| = R|\vec{F}_T| = m\,R|\vec{a}_T| = m\,R^2|\vec{\alpha}|$$

All the expressions are in terms of magnitudes, but they start from vectors, which have direction. The vectors with subscript r point toward the center of the circle whereas vectors with a subscript T point tangentially to the circle. The net torque ($\overline{N}$), the angular acceleration ($\overline{\alpha}$) and the angular velocity ($\overline{\omega}$) all point perpendicular to the plane of rotation formed by the radial and tangential acceleration.

The expressions for the net force, the net torque, and accelerations are the starting point for any problem where the object is speeding up or slowing down as it goes in a circle. If you're involved with a sport such as bike or car racing around a circular track, the forces providing the acceleration are the static friction between the tires and the road, the normal force, and the gravitational force.

Gravity can play a role if there's a vertical change in position such as in long distance road races, mountain bike races, and bike races in the velodromes. Gravity is usually used as a linear force or as a tangential force. It isn't usually used as a radial force except in amusement park rides.

The normal force is usually used as a radial force in banked turns, which allows the racer to use more of the static friction force for linear and tangential motion instead of in the radial direction. In fact, if a banked turn is done at the correct speed, then the amount of static friction needed to make the turn is zero and all of it can be used for tangential acceleration, which has a maximum value equal to the coefficient of static friction times the magnitude of the normal force.

This section is an application of these concepts in combination with Newton's laws and energy-work concepts. To illustrate these concepts I want you to work through an example with me.

Fred is practicing at the local velodrome. He goes into the turn that has a radius of curvature of 26.0 feet (7.92 meters) and is banked at 40.0 degrees. The bike's speed is 27.5 feet per second (8.38 meters per second). Fred and the bike weigh 190 pounds (845 newtons) or a mass of 5.90 slugs (86.1 kilograms). Fred maintains a constant acceleration over the next 70.0 feet (21.3 meters) and doubles his speed to 55.0 feet per second (16.8 meters per second) at which point the static friction has reached its maximum value. His bike starts to go into a skid at this point. What's the coefficient of static friction between Fred's tires and the track?

The steps involved in solving this problem are similar to the other sections of this chapter. Follow these steps:

1. **Determine all forces acting on the biological system and all sources of energy within the system that you believe are relevant.**

You need to realize that Fred is accelerating around a bend. Three forces are acting on Fred and his bike:

- Gravity, which pulls Fred and the bike straight down

- Normal force, which pushes the bike perpendicular to the track's surface

- Static frictional force, which contributes to the radial net force and the tangential net force. The combination of the radial static frictional force ($F_{sf,r}$) and the tangential static frictional force ($F_{sf,T}$) gives the total static frictional force.

- The other dissipative forces such as the rolling resistance and the air resistance will be ignored.

2. Draw free-body diagrams for Fred and his bike.

After you draw your free-body diagrams, look at mine. I draw two free-body diagrams so you can see all the forces and accelerations acting on Fred, as well as their directions. Refer to Figure 7-5.

Figure 7-5a shows Fred from above. It shows the velocity in the tangential direction, the change in the arclength along the track (Δs), the radius of the circle (R), the radial (centripetal) acceleration (a_r), the tangential acceleration (a_T), and the static frictional force split into both a tangential and radial component.

Figure 7-5b shows Fred from behind. It shows the radial (centripetal) acceleration, the radial component of the static frictional force, the force of gravity, and the normal force. Notice I have drawn my compass to be parallel and perpendicular to the surface, which makes it easier to solve for the static friction and the normal force.

3. Calculate the forces and/or torques by using the free-body diagrams as a guide.

The accelerations and forces can be written in terms of the unit vectors, using Figure 7-5 as a guide:

$$\vec{a}_T = a_T\left[-\hat{z}\right]; \ \vec{a}_r = \frac{v^2}{R}\left[-\hat{x}\cos(\theta) + \hat{y}\sin(\theta)\right];$$

$$\vec{F}_{sf,T} = F_{sf,T}\left[-\hat{z}\right]; \ \vec{F}_{sf,r} = F_{sf,r}\left[-\hat{x}\right];$$

$$\vec{F}_g = mg\left[-\hat{x}\sin(\theta) - \hat{y}\cos(\theta)\right]; \text{ and } \vec{F}_N = F_N\left[+\hat{y}\right]$$

4. Apply the laws of biophysics to solve the problem.

Break down everything. You need to find the coefficient of static friction, so you need to find the static friction when at its maximum value, which happens when Fred is moving at his final speed of 55.0 feet per second (16.8 meters per second). In other words, you need to find the magnitude of the normal force and hence the static friction at the end of the acceleration period. In Figure 7-5a, you see the static friction has both a radial component and a tangential component, so that is what you want to find.

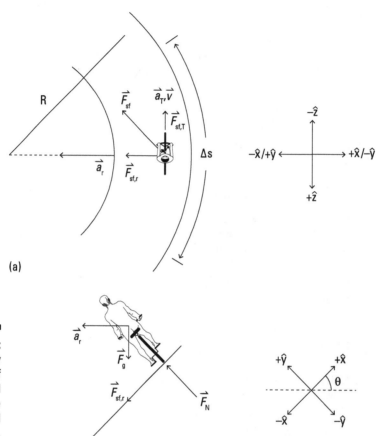

Figure 7-5:
A free-body
diagram of
Fred viewed
from above
(a) and
behind (b).

You can find the tangential static friction by solving the work-energy theorem in the tangential direction. The reason is the problem tells you both the initial and final speeds and you know work equals force times displacement. The work-energy theorem gives

$$F_{sf,T} \Delta s = W_{NET} = \Delta E_K = \frac{mv_{final}^2}{2} - \frac{mv_{initial}^2}{2}$$

$$F_{sf,T}(70 \text{ ft}) = \frac{5.90 \text{ slugs}}{2}\left[\left(55 \text{ ft s}^{-1}\right)^2 - \left(27.5 \text{ ft s}^{-1}\right)^2\right]$$

The tangential static frictional force is 95.6 pounds (425 newtons).

Newton's second law of motion, the law of acceleration, can be used to find the radial static friction because you know the radial acceleration.

Newton's second law of motion, the law of acceleration, for this problem is

$$x\text{-axis: } -m\frac{v_{final}^2}{R}\cos(\theta)=-F_{sf,r}-mg\sin(\theta)$$

$$y\text{-axis: } m\frac{v_{final}^2}{R}\sin(\theta)= F_N - mg\cos(\theta)$$

$$z\text{-axis: } -ma_T =-F_{sf,T}$$

The z-axis equation gives a solution for the tangential acceleration, but you don't need it.

The y-axis equation gives a solution for the normal force:

$$F_N = m\frac{v_{final}^2}{R}\sin(\theta)+ mg\cos(\theta)=\frac{(5.90 \text{ slugs})(55 \text{ ft s}^{-1})^2}{(26 \text{ ft})}\sin(40)+(190 \text{ lb})\cos(40)$$

The normal force is 587 pounds (2,610 newtons).

The x-axis equation gives a solution for the radial component of the static friction:

$$F_{sf,r} = m\frac{v_{final}^2}{R}\cos(\theta)- mg\sin(\theta)=\frac{(5.90 \text{ slugs})(55 \text{ ft s}^{-1})^2}{(26 \text{ ft})}\cos(40)-(190 \text{ lb})\sin(40)$$

The radial component of the static friction is 404 pounds (1800 newtons).

You can now find the coefficient of static friction:

$$\mu_S = \frac{F_{sf}}{F_N}=\frac{\sqrt{F_{sf,T}^2 +F_{sf,r}^2}}{F_N}=\frac{\sqrt{(95.6 \text{ lb})^2 +(404 \text{ lb})^2}}{587 \text{ lb}}$$

The tires on Fred's bike have a coefficient of static friction of 0.707 with the track.

Notice that the tires must share the static friction between the radial direction and the tangential direction. If you think you're going to go in a skid going around the corner and can't slow down in time, then just take your foot off the brake and off the gas and coast around the corner. Doing so can put the entire static friction force in the radial direction and maximize the speed, so that you can make the corner. Furthermore, this problem is considered a very hard problem, but notice it's still the same steps used in every other problem in this chapter. If you can master this problem and as long as you follow the steps of setting up the problem and drawing the free-body diagrams, you should have no trouble.

Chapter 8

Looking at Where Moving Objects Go — Kinematics

. . . .

In This Chapter

▶ Examining linear one-dimensional motion

▶ Focusing on circles and motion

▶ Tackling two-dimensional motion

. . . .

*T*his chapter stands by itself on the subject of biomechanics. Here the main focus is to understand how objects move without worry about what causes the motion. *Kinematics* means mechanics of objects in motion. This branch of biomechanics is important in fields that need to understand the motion of objects.

To completely describe the motion of an object in a biological system, all you need to know is the acceleration, velocity, and position of the object. Understanding biological system's motion and how the motion changes involves a few steps:

1. **Make a table(s) of the quantities of motion at key moments in time.**

   Key moments depend on the problem but can include moments when the forces acting on the system change (and hence the acceleration).

2. **Draw the corresponding graph(s).**

   Velocity versus time graphs are the best, but sometimes you'll need to make acceleration versus time and position versus time graphs.

3. **Solve the problem using the graphs and or formulas.**

This chapter focuses on describing the motion of objects as they move through space by looking at different types of motion, including linear one-dimensional motion, circular motion (and how to treat the two-dimensional circular motion as one-dimensional motion), and noncircular two-dimensional motion.

Grasping One-Dimensional Motion

The easiest way to describe one-dimensional motion of an object is by using a *velocity versus time graph*. Figure 8-1 shows the velocity of my bike while I'm riding on a mountain bike trail over a period of time. The graph conveys all the information you need to know about my bike to completely describe its motion. Some of the information contained in this graph tells you

- ✔ **Velocity:** Also known as *instantaneous velocity.* At any given moment in time you know the value of the velocity.

- ✔ **Acceleration:** Also known as *instantaneous acceleration.* If you calculate the slope at any moment in time, it's the acceleration at that moment.

- ✔ **Average acceleration:** If you calculate the slope between two times, say $t_{initial}$ and t_{final}, then the slope equals the average acceleration between these two times. In Figure 8-1, the curve is a straight line between times $t_{initial}$ and t_{final}, which means the slope is a constant and the instantaneous acceleration is equal to the average acceleration between these times.

- ✔ **Displacement:** The area between the velocity curve and the time axis is equal to the displacement. In Figure 8-1, the displacement of the object from time $t_{initial}$ to time t_{final} is equal to the shaded area. If the shaded region is below the time axis then the displacement is negative.

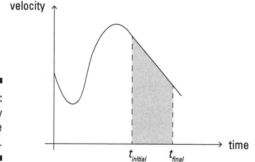

Figure 8-1:
The velocity
of my bike
while riding.

Knowing the displacement, velocity, and acceleration completely describes the motion of the object. But you may be thinking, "I want formulas." The formulas are hidden within Figure 8-1. When there's a straight line in a velocity versus time graph the acceleration is a constant, such as from $t_{initial}$ to t_{final}, then the equations are algebraic.

The formulas when the acceleration is constant are as follows:

source	equation	contains	missing
slope:	$a = \dfrac{\Delta v}{\Delta t} = \dfrac{v_{final} - v_{initial}}{t_{final} - t_{initial}}$	$\Delta t,\ v_{initial},\ v_{final},\ a$	Δx
area:	$\Delta x = \dfrac{\left(v_{final} + v_{initial}\right)}{2}\Delta t$	$\Delta t,\ \Delta x,\ v_{initial},\ v_{final}$	a
work – energy:	$v_{final}^2 - v_{initial}^2 = 2\,a\,\Delta x$	$\Delta x,\ v_{initial},\ v_{final},\ a$	Δt
area + slope:	$\Delta x = v_{final}\Delta t - \dfrac{a}{2}\left(\Delta t\right)^2$	$\Delta t,\ \Delta x,\ v_{final},\ a$	$v_{initial}$
area – slope:	$\Delta x = v_{initial}\Delta t + \dfrac{a}{2}\left(\Delta t\right)^2$	$\Delta t,\ \Delta x,\ v_{initial},\ a$	v_{final}
average velocity:	$\bar{v} = \dfrac{\Delta x}{\Delta t} = \dfrac{\left(v_{final} + v_{initial}\right)}{2}$		

This chart includes a lot of information (just like Figure 8-1 has a lot of information). These formulas show how the physical quantities of motion (velocity and position) change over some time interval. Here are the mathematical symbols explained in plain English: $t_{initial}$ is the initial time you're interested in, t_{final} is the final time, $\Delta t = t_{final} - t_{initial}$ is the elapsed time, $x_{initial} = x(t_{initial})$ is the position at the initial time (also called the *initial position*), $x_{final} = x(t_{final})$ is the position at the final time (also called the *final position*), $\Delta x = x_{final} - x_{initial}$ is the *displacement*, $v_{initial} = v(t_{initial})$ is the velocity at the initial time (also called the *initial velocity*), $v_{final} = v(t_{final})$ is the velocity at the final time (also called the *final velocity*), and a is the acceleration, which is a constant that doesn't change over the time interval.

In addition to the list of formulas, here are a few things worth noting about the chart:

✔ The motion of an object is described by only five physical quantities: elapsed time (Δt), the displacement (Δx), the initial velocity ($v_{initial}$), the final velocity (v_{final}), and the (constant) acceleration (a). Each formula contains only four of the five quantities, so as long as you know three quantities, you can find the other two by using the correct formulas. The third column tells you what quantities are in the formula and the last column in the chart helps you select the correct formula to use by telling you which quantity is missing from the formula.

✔ The first column tells you where the formula came from. The first equation is the formula for the slope of the curve in Figure 8-1. The second equation is the formula for the area under the curve in Figure 8-1.

✔ You can obtain the third equation by multiplying the area equation and the slope equation together. You can also obtain it from the work-energy theorem plus the definition of work plus Newton's second law of motion, the law of acceleration (refer to Chapter 6 for more information).

✔ You can get the fourth and fifth equations by combining the area equation with the slope equation to eliminate one of the velocities by adding (or subtracting) the equations together.

✔ The sixth formula shows the relationship between the average velocity (equals displacement divided by elapsed time), final velocity, and initial velocity when the acceleration is a constant.

Be very careful with the fourth and fifth formulas. They look very similar and are easy to mix up. There is a different sign in front of a, and the velocities are different, so be careful.

Remember, the velocity versus time graph and the equations are equivalent, so use whichever is easier. These sections look at applications of Figure 8-1 and the formulas.

Analyzing sprinters' run — the 100-meter dash

In 2009, Usain Bolt set several sprinting world records. Two of the records were the 100-meter (328 feet) race in 9.58 seconds and the 200-meter (656 feet) race in 19.19 seconds. In addition, in 2009, he ran in the 150-meter (492 feet) race and finished the last 100-meters in 8.70 seconds. Try analyzing his running feats with a few reasonable assumptions:

✔ Bolt's motion is the same for the 100-meter and 200-meter races. Only the distance has changed.

✔ Bolt's maximum speed is the same for all three races.

✔ Bolt's acceleration out of the blocks is the same for the 100-meter and 200-meter races.

✔ Bolt's acceleration out of the blocks is a constant.

You can use the 100-meter and 150-meter race results to calculate his maximum speed, his acceleration out of the blocks, and how long he accelerated for. Finally, test the assumptions by using the 200-meter result.

You can solve the problem by using these steps:

1. **Make a table(s) of the five quantities of motion at key moments in time.**

   You can make a table indicating what the quantities are at each important moment of time. After you have completed your table, you can compare it to mine. My quantities are in Table 8-1.

   One important piece of information I didn't place in the table is that the last 100 meters (328 feet) of the 150-meter race was completed in 8.70 seconds.

Table 8-1   The Quantities of Motion Describing Bolt's Races

Time (second)	Position (ft) (m)	Velocity (ft/s) (m/s)	Acceleration (ft/s²) (m/s²)	Comments
0	0	0	a	Start of the race
t_a	x_a	v_{max}	$a \rightarrow 0$	Maximum speed reached
9.58	328 (100)	v_{max}	0	End of 100-meter race
t_{150}	492 (150)	v_{max}	0	End of 150-meter race
19.19	656 (200)	v_{max}	0	End of 200-meter race

2. **Make the corresponding graph(s).**

   Sketch a velocity versus time graph. Include the assumptions listed in Table 8-1 in your graph. After you've drawn your graph, have a look at mine in Figure 8-2.

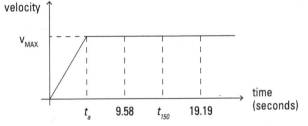

Figure 8-2: The velocity of Bolt from his three 2009 races.

3. Solve the problem by using the graphs and or the formulas.

You have listed all the important information in step No. 1 and you visualize the motion from the graph in step No. 2. The problem wants you to find v_{max}, a, and t_a. You're now ready to solve the problem for these three quantities.

You can use the kinematic equations, but remember the equations are valid only for constant acceleration. This problem has two accelerations, so you have to keep the time interval in the range 0 to t_a (constant acceleration of a), and in the range t_a to 19.19 seconds (constant acceleration of 0) when using the kinematic equations. Alternatively, you can use the information in Figure 8-2, and the formulas for area under the curve and the slope of curve to solve this problem.

I use the graph in Figure 8-2, and the formulas for area under the curve and the slope of the curve to solve this problem.

REMEMBER

If you're going to use the graphical technique, then the area under a velocity versus time graph is equal to the displacement and the slope is equal to the acceleration.

You want to calculate the maximum velocity first, the acceleration time interval (t_a) second, and then you can find the acceleration. The solutions for the three quantities are

- ✓ **The maximum velocity:** Figure 8-2 shows that he is running at a constant velocity after time t_a. Therefore, during the last 328 feet (100 meters) of the 150-meter race he is running at his maximum speed and you can use the area equation, which states displacement equals the (average) velocity times elapsed time. Mathematically, it looks like this:

$$v_{max} = \frac{\Delta x}{\Delta t} = \frac{328 \text{ ft}}{8.70 \text{ s}} = \frac{37.7 \text{ ft}}{s} \left(\frac{11.5 \text{ m}}{s} \right)$$

The maximum speed is fast, which in common terms is 25.7 miles per hour (41.4 kilometers per hour). Bolt would get a speeding ticket running through certain speed zones within cities.

- ✓ **The acceleration time:** From Figure 8-2, you know the area under the curve from the time of 0 seconds to 9.58 seconds is equal to the displacement of 328 feet (100 meters). The area from 0 seconds to t_a is a right-angle triangle and the area from t_a to 9.58 seconds is a rectangle. The solution is (displacement equals area)

$$\Delta x = \frac{v_{max} t_a}{2} + v_{max} \left(9.58 \text{s} - t_a \right)$$

$$t_a = 2 \left[9.58 \text{ s} - \frac{\Delta x}{v_{max}} \right] = 2 \left[9.58 \text{ s} - \frac{328 \text{ ft}}{37.7 \text{ ft/s}} \right] = 1.76 \text{ seconds}$$

The time over which the acceleration is occurring is 1.76 seconds.

✔ **The acceleration:** You can calculate his acceleration from the slope of the curve in Figure 8-2 because you now know v_{max} and t_a. Remember, the slope (acceleration) equals rise (change in the velocity) over run (change in time):

$$a = \frac{v_{max} - 0}{t_a - 0} = \frac{37.7 \text{ ft/s}}{1.76 \text{ s}}$$

The acceleration is 21.4 feet per square second (6.53 meters per square second).

You've now calculated everything about Bolt's 100-meter race. From this information, you can calculate how long it would take him to run the 150-meter race and the 200-meter race:

✔ $t_{Dx} = t_{100} + (\Delta x - 100 \text{ meters})/v_{max}$

✔ $t_{150} = 9.58 \text{ second} + 164 \text{ feet}/(37.7 \text{ feet per second}) = 13.93 \text{ seconds}$

✔ $t_{200} = 9.58 \text{ second} + 328 \text{ feet}/(37.7 \text{ feet per second}) = 18.28 \text{ seconds}$

In reality, his 150-meter time was 14.35 seconds and his 200-meter time was 19.19 seconds, which are 0.42 seconds and 0.91 seconds slower than what we calculated, respectively. His initial 50 meters was slower in the 150-meter race than in the 100 meter race, and it was his last 100 meters in the 200-meter race that was a little slower. Your approximations are reasonable, though, and he deserves a speeding ticket.

Dunking the basketball — people and animals' jumping abilities

If you're studying motion and you don't care about time, the work-energy theorem and the conservation laws are the way to go, but you still follow the steps to get things right. To illustrate this, in this section, you analyze the vertical jumping motion of a human (Harry) dunking a basketball and the philaenus spumarius. The philaenus spumarius (linnaeus) is commonly known as the froghopper or the spittle bug. I have included the froghopper for comparison with humans. You'll find the maximum velocity and the take-off acceleration of both the froghopper and Harry. Strictly speaking, the froghopper jumps at about 58 degrees, so you have to take that into account when calculating the vertical take-off velocity.

Vertical jumping has two accelerations:

✔ Take-off acceleration: When the animal is pushing off the ground

✔ A second acceleration while in the air, which is the acceleration due to gravity, g = 32.2 feet per square second (9.81 meters per square second)

Four quantities are usually of interest:

- ✔ Displacement while pushing off the ground
- ✔ Acceleration while pushing off the ground (take-off acceleration)
- ✔ Take-off velocity, which is the velocity right as the animal leaves the ground
- ✔ Maximum height reached above the ground

Normally, it's easier to experimentally measure the pushing displacement and the maximum height reached, so you would measure those two and then calculate the take-off acceleration and the take-off velocity. Assume you have already made the measurements, so you want to solve for the acceleration and velocity; use the steps listed in these sections.

Step No. 1: Make a table (s)

In this stage, you make a table of the four quantities of motion (time, position, velocity, and acceleration) at key moments in time. This step is important and actually a little more than filling in a table. You figure out what's going on from a biophysical point of view. As you figure it out, you can fill in your table indicating what the quantities are at each important moment of time. Table 8-2 shows you my table for comparison.

To start the table, the animal squats down so the starting position is $-d$ below its normal vertical height. In the case of Harry, $d = 1.75$ feet (0.533 meters), whereas $d = 6.6 \times 10^{-3}$ feet (2.0×10^{-3} meters) for the froghopper.

The animal now pushes off the ground with a take-off acceleration of a_T. Assume that the acceleration is constant because the duration of the take-off is very short (less than 0.001 seconds for the froghopper).

At the take-off time, t_T, the animal has extended its legs and feet, and can no longer push against the ground. The acceleration has changed from a_T to $-g$ (which is the acceleration of gravity), and the animal has reached its maximum vertical velocity. The animal is now moving through the air and will continue to rise until it stops. I call this time t_{max}. At this time the animal will reach a maximum height h. Harry's maximum height was measured to be $h = 4.00$ feet (1.22 meters), whereas $h = 1.93$ feet (0.588 meters) for the froghopper. The animal falls back toward the ground until time t_g, at which point the animal hits the Earth. Table 8-2 summarizes all the physical quantities during this motion.

Table 8-2		The Quantities of Motion Describing an Animal Jumping		
Time (second)	Position (ft) (m)	Velocity (ft/s) (m/s)	Acceleration (ft/s²) (m/s²)	Comments
0	−d	0	a_T	The animal is in the crouch and starts the jump.
t_T	0	v_{max}	$a_T \rightarrow -g$	The animal has reached maximum speed.
t_{max}	h	0	−g	The animal has reached maximum height.
t_g	0	$-v_{max}$	−g	The animal is back on the ground.

Step No. 2: Make the corresponding graph (s)

You have enough information (from the preceding section) to sketch a velocity versus time graph. After you draw your graph, have a look at mine in Figure 8-3.

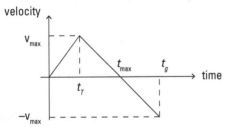

Figure 8-3:
The velocity of the animal while jumping vertically.

The slope of the curve in Figure 8-3 from time 0 to t_T is equal to the take-off acceleration and the slope from time t_T on is equal to −g; the acceleration due to gravity. The area under the curve in Figure 8-3 from time 0 to t_T is equal to the displacement d, the area under the curve from time t_T to t_{max} is equal to the displacement h, and the area under the curve from time t_{max} to t_g is equal to the displacement −h.

The restriction on the slope (acceleration) and the area under the curve (displacement) places further restrictions on the motion. The velocity at impact with the ground must be $-v_{max}$, and $t_g = 2\,t_{max} - t_T$.

Step No. 3: Solve the problem

Use the graphs and the formulas to solve the problem. You don't know the times, so you need to use the work-energy theorem (or conservation of mechanical energy because there are no dissipative forces). When the animal is in the air, the only thing you don't know is the maximum velocity at take-off. Use the information in Table 8-2 from time t_T to t_{max} (and the conservation of energy equation in the preceding section) to solve for v_{max}.

$$v_{\text{max}} = \sqrt{v_{final}^2 - 2a\Delta x} = \sqrt{2gh}$$

You now have a formula to find v_{max}, so you can calculate some values. Remember, $g = 32.2$ feet per second squared (9.81 meters per second squared). Harry's maximum height was 4.00 feet (1.22 meters), so the take-off velocity is $v_{max} = 16.0$ feet per second (4.89 meters per second). On the other hand, the froghopper's maximum height was 1.93 feet (0.588 meters), so his vertical take-off velocity is $v_{max,y} = 11.1$ feet per second (3.40 meters per second). You aren't finished with the froghopper though. Remember it left the ground at an angle of 58 degrees, so the froghopper's maximum speed is $v_{max,y}/\sin(58) = 13.1$ feet per second (4.01 meters per second).

The take-off speed of both Harry and the froghopper are very close to being the same, which illustrates an interesting fact about jumping. The maximum height attained by the animal depends only on the vertical take-off velocity of the animal; it doesn't care how heavy the animal is, how big the animal is, or any other characteristic about the animal.

In addition to the time, which I'm not interested in at the moment, you have only the take-off acceleration to calculate. Use the information in Table 8-2 from time 0 to t_T (and the conservation of energy equation in the preceding section) to solve for a_{max}.

The take-off acceleration is calculated by using the work-energy theorem during the time interval from 0 to t_T.

$$a_T = \frac{v_{final}^2 - v_{initial}^2}{2\Delta x} = \frac{v_{max}^2}{2d}$$

You know v_{max} for both Harry and the froghopper. From the problem, you know the value of d for both Harry and the froghopper, so you can now calculate the value of the take-off acceleration. After doing the calculations, Harry's take-off acceleration is 73.1 feet per square second (22.4 meters per square second), whereas the froghopper's take-off acceleration is 13,000 feet per square second (3,960 meters per square second)!

To put the froghopper's take-off acceleration into perspective, it is 404 times larger than the acceleration due to gravity, whereas an airbag deploying during a car accident has an acceleration of about 60 times the acceleration due to gravity. If Harry jumped with the same take-off acceleration as the froghopper, then Harry would jump to a height of:

$$h = \frac{v_{max}^2}{2g} = \frac{d\,a_T}{g} = 404d$$

Substitute Harry's value of d into the equation, and the height is 707 feet (215 meters), or more than the length of two football fields straight up! That is why the froghopper is ranked one of the best jumpers in the animal kingdom.

Skydiving and non-uniform acceleration

In October 2012, Felix Baumgartner skydived from an altitude of 24.4 miles (39.2 kilometers), and he broke the sound barrier as he fell. During an event like this the displacement is so large that forces acting on the body are changing and so is the acceleration. The mathematical tools needed to solve a body falling from this altitude is beyond the scope of this book, but by using graphs you can still qualitatively analyze what's going on.

This section serves two purposes: analyzing very complex systems by using graphs and correcting a myth some people have about skydiving from watching TV. If I ask people to sketch the velocity versus time curve of a skydiver, many people get it wrong. In biophysics, there can be abrupt changes in the acceleration caused by sudden changes in the forces acting on the biological system. On the other hand, the velocity and position change smoothly, but they can change rapidly.

In cases where the acceleration is changing in time, it's usually better to draw an acceleration versus time graph before drawing a velocity versus time graph. To illustrate the importance of including the acceleration versus time graph, consider the following example. A person jumps out of a plane, falls for awhile and reaches terminal speed, opens the parachute, and then floats the rest of the way down at a new terminal speed. Sketch the acceleration versus time and the velocity versus time graphs for this motion.

You want to draw the acceleration versus time graph and remember the forces acting on the person, which are the gravitational force and air resistance. At first, gravity is the dominant force and then air resistance kicks in and increases until the forces balance and the net force is zero. The parachute is opened and the air resistance becomes greater than gravity. This causes a deceleration and the person slows down until reaching a new terminal velocity. You can now sketch your acceleration versus time curve; I have drawn mine in Figure 8-4. How does your graph compare to mine?

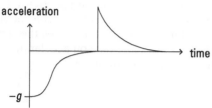

Figure 8-4:
The accel-
eration
versus time
curve of a
skydiver.

You know what the acceleration looks like as a function of time from Figure 8-4. The change in the velocity is equal to the area under the curve, so you can use Figure 8-4 to draw the velocity versus time graph. After you've drawn your velocity versus time graph, have a look at mine in Figure 8-5.

On TV when you see someone skydiving and the person opens her para-chute, she appears to fly upward. In reality, she doesn't shoot upward, but she does decelerate while the cameraperson continues falling at the original terminal speed. Notice from Figure 8-5 that the velocity is always negative, which means the person continues to fall downward and doesn't shoot upward with a positive velocity.

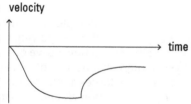

Figure 8-5:
The velocity
versus time
curve of a
skydiver.

The terminal velocity shown in Figure 8-5 is represented mathematically by the following equation:

$$v_{max} = \sqrt{\frac{2mg}{D\rho A}}$$

Here m is the mass of the skydiver, g is the acceleration due to gravity, so mg is the weight of the skydiver, D is the drag coefficient, ρ is the air's mass density, and A is the frontal cross-sectional area. The frontal cross-sectional area of a parachute is a lot larger than your body, so the terminal velocity is smaller when the parachute is open. A typical parachute has a cross-sectional area proportional to the weight of the person, 0.6 to 0.7 pounds per square foot (or 2.9 to 3.0 kilograms per square meter).

You now know how to make the graphs and calculate the terminal velocity, so when you go skydiving, you can also calculate and analyze your fall. You can also calculate how large a parachute you need.

Spinning In Circles

When objects move in circles, they're moving in two dimensions, but the distance from the center of the circle doesn't change. Only the angle is changing in time, which you can think of as one-dimensional motion.

The easiest way to describe the motion of an object is using an *angular velocity versus time graph* similar to what you did for linear motion in Figure 8-1. The figure conveys all the information you need about the object's circular motion. Some of the information contained in this graph tells you the following:

- ✔ **Angular velocity (ω):** At any given moment in time you know the value of the angular velocity directly from the graph.

- ✔ **Tangential speed (v_T):** The angular velocity also gives you the speed at which the object is traveling around the circle, $v_T = R\,\omega$. R is the distance from the center of the circle to the object.

- ✔ **Instantaneous angular acceleration (α):** If you calculate the slope at any moment in time, that value is the angular acceleration.

- ✔ **Tangential acceleration (a_T):** The angular acceleration also gives you the tangential acceleration of the object as it travels around the circle. The magnitude of the tangential acceleration is $a_T = R\,\alpha$. R is the distance from the center of the circle to the object.

- ✔ **Radial acceleration (a_r):** The radial acceleration is related to how fast the object is moving around the circle. The magnitude of the radial (centripetal) acceleration is $a_r = R\,\omega^2$. R is the distance from the center of the circle to the object.

- ✔ **Average angular acceleration:** If you calculate the slope between two times, say $t_{initial}$ and t_{final}, the slope equals the average angular acceleration between these two times.

- ✔ **Angular displacement (Δθ):** The area between the angular velocity curve and the time axis is equal to the angular displacement, which is the change in the object's angle.

- ✔ **Arclength (Δs):** The distance the object moves around the circle is the *arclength*. The arclength is $\Delta s = R\,\Delta\theta$.

Algebraic equations exist if the angular acceleration is constant. The motion of an object is described by five quantities:

- Elapsed time (Δt)
- Angular displacement ($\Delta \theta$)
- Initial angular velocity ($\omega_{initial}$)
- Final angular velocity (ω_{final})
- Constant angular acceleration (α)

Each formula contains four of the five quantities, so as long as you know three quantities, you can find the other two using the correct formulas. The formulas from an angular velocity versus time graph with constant angular acceleration are

source	equation	contains	missing
slope	$\alpha = \dfrac{\Delta \omega}{\Delta t} = \dfrac{\omega_{final} - \omega_{initial}}{t_{final} - t_{initial}}$	Δt, $\omega_{initial}$, ω_{final}, α	$\Delta \theta$
area	$\Delta \theta = \dfrac{\left(\omega_{final} + \omega_{initial} \right)}{2} \Delta t$	Δt, $\Delta \theta$, $\omega_{initial}$, ω_{final}	α
work – energy	$\omega_{final}^2 - \omega_{initial}^2 = 2 \alpha \Delta \theta$	$\Delta \theta$, $\omega_{initial}$, ω_{final}, α	Δt
area + slope	$\Delta \theta = \omega_{final} \Delta t - \dfrac{\alpha}{2} \left(\Delta t \right)^2$	Δt, $\Delta \theta$, ω_{final}, α	$\omega_{initial}$
area – slope	$\Delta \theta = \omega_{initial} \Delta t + \dfrac{\alpha}{2} \left(\Delta t \right)^2$	Δt, $\Delta \theta$, $\omega_{initial}$, α	ω_{final}
average velocity	$\overline{\omega} = \dfrac{\Delta \theta}{\Delta t} = \dfrac{\left(\omega_{final} + \omega_{initial} \right)}{2}$		

The first column tells you where the formula comes from when the angular acceleration is a constant. The first formula is the slope of the curve in angular velocity versus time graph. The second formula is the area under the curve in angular velocity versus time graph. You can obtain the third equation by multiplying the slope equation with the area equation. You can also obtain it by combining the work-energy theorem plus the definition of the work plus Newton's second law of motion, the law of acceleration. You can obtain the fourth and fifth equations by adding and subtracting the slope and area equations.

The fourth and fifth formulas look very similar; it's easy to mix them up, so be careful.

Analyze the discus throwing sport by using the concepts of circular motion. During the discus throw, the athlete spins with the discus held in an outstretched arm. After spinning around in a circle for two revolutions, the discus is released, and it flies through the air. In men's discus, the discus weighs 4.41 pounds (19.6 newtons) corresponding to a mass of 0.137 slugs (2.00 kilograms), and the world-record distance was achieved when it was thrown with a speed of approximately 85.6 feet per second (27.0 meters per second). The discus during the spin was approximately 3.00 feet (0.914 meters) from the center of the circle. Calculate three quantities:

✔ The average angular acceleration

✔ The time of the spin

✔ The work done on the discus by the athlete

For you to calculate these quantities, follow these steps outlined:

1. **Make a table(s) of the five quantities of motion at key moments in time.**

   The table needs to indicate what the quantities are at each important moment in time. (*Hint:* You get this information at the beginning and end of the spin.) Check out Table 8-3 for comparison.

   Don't forget to change the tangential speed to the angular velocity $(\omega_{final}) = v_T/R = 85.6$ feet per second/3.00 feet = 28.5 radians per second.

Table 8-3   The Quantities of Motion Describing a Discus Throw

Time(s)	Angular Position (Radians)	Angular Velocity (Radians/s)	Angular Acceleration (Radians/s²)	Comments
0	0	0	α	Start of the spin
t_{final}	4π	28.5	α	End of the spin

2. **Make the corresponding graph(s).**

   Sketch the angular velocity versus time graph. Your graph should look similar to Figure 8-6.

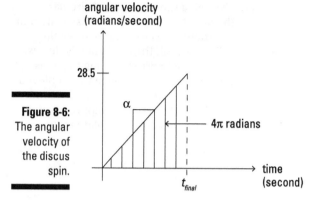

You can solve the problem by using either the formulas or the graph. This time, use the formulas and find the average angular acceleration without knowing the time. From the formulas, look for the one missing time:

$$\text{Work} - \text{energy: } \alpha = \frac{\omega_{final}^2 - \omega_{initial}^2}{2\Delta\theta} = \frac{\left(28.5 \text{ radians/s}\right)^2 - 0}{2\left(4\pi \text{ radians}\right)}$$

The angular acceleration is 32.4 radians per square second. The tangential acceleration (equals the radius times the angular acceleration) is 97.0 feet per square second (29.6 meters per square second).

The next quantity you want to calculate is the time of the spin. You know everything else so you can use the graph or any of the formulas.

If you have a choice, try to avoid using calculated numbers and use those given to you in the problem. I use the formula that doesn't contain the angular acceleration:

$$\text{Area: } \Delta t = \frac{2\Delta\theta}{\left(\omega_{final} + \omega_{initial}\right)} = \frac{2\left(4\pi \text{ radians}\right)}{\left(28.5 \text{ radians/s} + 0\right)}$$

The time of the spin was 0.881 seconds.

You can calculate the amount of work done by the discus thrower on the discus by using the work-energy theorem. First calculate the moment of inertia of the discus (mass times radius squared). You also need to remember the kinetic energy equals half the moment of inertia times the angular velocity squared:

$$I = mR^2 = \left(0.137 \text{ slugs}\right)\left(3 \text{ ft}\right)^2 = 1.23 \text{ slugs ft}^2$$

$$W_{NET} = \frac{I\omega_{final}^2}{2} - \frac{I\omega_{initial}^2}{2} = \frac{\left(1.23 \text{ slugs ft}^2\right)\left(28.5 \text{ radians s}^{-1}\right)^2}{2} - 0$$

The work done by the thrower on the discus is 502 foot-pounds (680 joules).

Moving With Noncircular Two-Dimensional Motion

Motion in each direction for most biophysical systems doesn't depend on the motion in the other directions (air resistance is an exception). You can treat each direction independent of the other. You just have to repeat the first section for each spatial dimension following the steps listed here:

1. **Make table(s) of the quantities of motion at key moments in time.**

   I find making a separate table for each direction helpful. This way I don't mix and match the wrong variables, and it keeps the directions separate. The time columns of your table allow you to connect the motion in each direction.

2. **Make the corresponding graph(s).**

   Make a velocity versus time graph for each direction. If it doesn't make the graph confusing, you can put all the velocity curves on the same graph. Sometimes you need to make acceleration versus time and position versus time graphs, but keep these to a minimum.

   A *trajectory graph* is another graph that's helpful. This graph is a map showing the path (trajectory) of the object moving through space. This is the same as using a *map app* and typing in the starting location and destination.

3. **Solve the problem by using the graphs and formulas.**

   The formulas for constant acceleration are the same as in the first section, but they're now vectors:

$$\vec{a} = \frac{\Delta \vec{v}}{\Delta t} = \frac{\vec{v}_{final} - \vec{v}_{initial}}{t_{final} - t_{initial}}$$

$$\Delta \vec{s} = \frac{\left(\vec{v}_{final} + \vec{v}_{initial}\right)}{2}\Delta t$$

$$v_{final}^2 - v_{initial}^2 = 2\vec{a} \cdot \Delta \vec{s}$$

$$\Delta \vec{s} = \vec{v}_{final}\Delta t - \frac{\vec{a}}{2}\left(\Delta t\right)^2$$

$$\Delta \vec{s} = \vec{v}_{initial}\Delta t + \frac{\vec{a}}{2}\left(\Delta t\right)^2$$

Average velocity: $\overline{\vec{v}} = \frac{\Delta \vec{s}}{\Delta t} = \frac{\left(\vec{v}_{final} + \vec{v}_{initial}\right)}{2}$

Serving in tennis — projectile motion

Projectiles are any objects where the dominant force acting on them is the gravitational force and the other forces such as air resistance can be ignored. Objects that are thrown straight up in the air, dropped, or thrown horizontally are all projectiles. The only difference is the horizontal component of the velocity.

For projectile motion close to the earth, the kinematic equations (equations of motion) simplify to the following set of equations with the y-axis being the vertical direction and the x-axis being the horizontal direction:

$$-g = \frac{\Delta v_y}{\Delta t} = \frac{v_{y,final} - v_{y,initial}}{t_{final} - t_{initial}}$$

$$v_{y,final}^2 - v_{y,initial}^2 = -2g\,\Delta y$$

$$\Delta y = v_{y,final}\Delta t + \frac{g}{2}(\Delta t)^2$$

$$\Delta y = v_{y,initial}\Delta t - \frac{g}{2}(\Delta t)^2$$

$$\Delta x = v_x \Delta t \quad \text{and} \quad \Delta y = \frac{(v_{y,final} + v_{y,initial})}{2}\Delta t$$

Note these equations are only valid for close to the surface of the Earth, and for large vertical altitude changes these equations do not hold because g isn't constant and air resistance will be a factor.

If the acceleration is uniform (constant magnitude and constant direction), select one of the axis directions parallel to the acceleration. This makes the acceleration in the other direction (x-axis above) zero. Instead of having ten equations in two dimensions (15 equations in three dimensions), you have only six equations in two dimensions (seven equations in three dimensions).

Try this example:

Mark is a tennis player, and he can serve the tennis ball at 40 miles per hour (64.4 kilometers per hour). He hits the ball during the serve at a height of 8.0 feet (2.44 meters) above the ground and an angle of 5 degrees above the horizontal. He's standing at the edge of the court, which is 39 feet (11.9 meters) from the net. The net is 3 feet (0.914 meters) high at the center and the service line is 21 feet (6.4 meters) beyond the net. Do you think his ball will clear the net? In other words, you want to find the vertical height of the ball when it's at the net.

To solve the problem, follow these steps:

1. **Make tables of the quantities of motion at key moments in time.**

   Tables 8-4 and 8-5 show my tables. I used the initial speed ($v_{initial}$) and the angle relative to the horizontal to calculate the component form of the initial velocity. In the vertical direction, $v_{initial,y}$ = 58.7 ft/s (17.9 m/s) sin(5°) = 5.11 feet per second (1.56 meters per second) and in the horizontal direction, $v_{initial}$ = 58.7 ft/s (17.9 m/s) cos(5°) = 58.4 feet per second (17.8 meters per second).

Table 8-4   **The Quantities Describing the Vertical Motion of Mark's Serve**

Time(s)	Position (ft) (m)	Velocity (ft/s) (m/s)	Acceleration (ft/s²) (m/s²)	Comments
0	8.0 (2.44)	5.11 (1.56)	$-g$	Mark has just finished hitting the ball.
t_{max}	y_{max}	0	$-g$	The ball has reached its maximum vertical height.
t_{net}	y_{net}	$v_{y,net}$	$-g$	The ball just clears the net — maybe.

Table 8-5   **The Quantities Describing the Horizontal Motion of Mark's Serve**

Time(s)	Position (ft) (m)	Velocity (ft/s) (m/s)	Acceleration (ft/s²) (m/s²)	Comments
0	0	58.4 (17.8)	0	Mark has just finished hitting the ball.
t_{max}	x_{max}	58.4 (17.8)	0	The ball has reached its maximum vertical height.
t_{net}	39 (11.9)	58.4 (17.8)	0	The ball just clears the net.

2. **Draw the graphs.**

   Draw both a trajectory graph and a velocity versus time graph. Figure 8-7 shows my trajectory graph with the ball leaving the racket, reaching a maximum height, and dropping over the net. Figure 8-8 shows my velocity versus time graph with both the horizontal velocity and the vertical velocity shown. Remember, the slope is equal to the acceleration, and the area under the curve equals the displacement.

Figure 8-7: The trajectory of the tennis ball as it travels from Mark to the net.

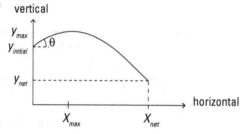

Figure 8-8: The velocity versus time graph of the tennis ball as it travels from Mark to the net.

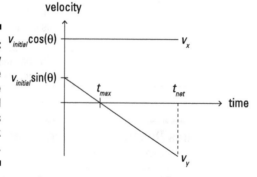

3. **Solve the problem by using the graphs and formulas.**

   The formulas for constant acceleration and independent motion in each direction have time in all the equations except the work-energy theorem. In many projectile motion problems in biomechanics, you aren't interested in the time, such as in this problem. If you eliminate the time from the formulas, they become

$a_x = 0$; and $a_y = -g$

$$-g = \frac{\left(v_{y,final} - v_{y,initial}\right)v_x}{\Delta x}$$

$$v_{y,final}^2 - v_{y,initial}^2 = -2g\,\Delta y$$

$$\Delta y = v_{y,final}\frac{\Delta x}{v_x} + \frac{g}{2}\left(\frac{\Delta x}{v_x}\right)^2$$

$$\Delta y = v_{y,initial}\frac{\Delta x}{v_x} - \frac{g}{2}\left(\frac{\Delta x}{v_x}\right)^2$$

$$\frac{\Delta y}{\Delta x} = \frac{\left(v_{y,final} + v_{y,initial}\right)}{2v_x} = \frac{\bar{v}_y}{v_x}$$

You want to find y_{net}, which is hidden in Δy. Looking at the tables, you notice that you know everything else except $v_{y,net}$, which is your $v_{y,final}$. In the list of equations, every equation has $v_{y,final}$ except the second to last one:

$$\Delta y = v_{y,initial}\frac{\Delta x}{v_x} - \frac{g}{2}\left(\frac{\Delta x}{v_x}\right)^2 = \frac{(5.11 \text{ ft/s})(39 \text{ ft})}{(58.4 \text{ ft/s})} - \frac{32.2 \text{ ft/s}^2}{2}\left(\frac{39 \text{ ft}}{58.4 \text{ ft/s}}\right)^2$$

The vertical height of the ball at the net is $y_{net} = y_{initial} + \Delta y = 8.0$ ft $+ (-3.77$ ft$) = 4.23$ feet (1.29 meters). The net is 3 feet high, so the tennis ball will clear the net by 1.23 feet (0.38 meters).

The angle of the hit in the serve is very important as is the vertical height at which the ball is hit. The higher the hit the lower the initial angle can be, so taller players don't have to lob the ball as much as shorter players. Also, the angle the ball can be served at depends on the speed of the serve; the slower the serve, the larger the initial angle must be. But, the lower the initial angle the shorter the time of flight to the other court and the less time your opponent has to react to the serve. Professional tennis players can serve the ball at speeds up to 160 miles per hour (257 kilometers per hour).

Pouncing on prey — combining jumping and projectile motion

In projectile motion, there are two quantities that are usually of interest: the range (R) the projectile will travel, and the maximum height (H) the projectile will reach. Mathematically, the equations are

$$H = y_{initial} + \frac{v_{y,initial}^2}{2g}$$

$$R = \frac{v_x}{g}\left[v_{y,initial} + \sqrt{v_{y,initial}^2 - 2g\,\Delta y}\,\right]$$

These formulas are very useful if you don't care about the time. Also, you may have noticed that the range formula doesn't look the same as what's usually found in a textbook. This formula allows for a cougar to leap out of a tree onto its prey; whereas the formula in the textbooks usually assumes no overall change in the vertical height — that is $\Delta y = 0$.

If you set $\Delta y = 0$, then the range formula changes to

$$R = \frac{2v_x v_{y,initial}}{g} = \frac{v_{initial}^2 \sin(2\theta)}{g}$$

You can use these formulas to study animal projectile motion.

Part III
Making Your Blood Boil — The Physics of Fluids

Blood Flow

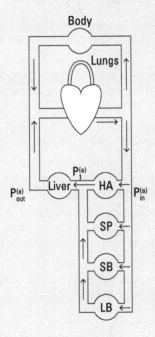

Part III

Making Your Blood Boil — The Physics of Fluids

In this part . . .

✔ Become acquainted with the laws of fluids, some of which have been around since Archimedes shouted eureka, and how they're applied in biophysics.

✔ Discover why some objects float, how sap gets to the top of the tallest trees, how oxygen gets into the blood, and how some bugs can walk on water.

✔ Find about the ins and outs of Newtonian and non-Newtonian fluid, as well as when you need to consider viscosity or not.

✔ Look at different aspects of the human body, such as how the body maintains a constant temperature, how the heart works, and how blood moves in the body.

✔ Grasp how molecules diffuse and are able to cross membranes, as well as the process by which the body metabolizes and eliminates molecules from the body.

✔ Uncover how the human body conserves energy and the contribution of oxygen and food to supply energy to the body and where that energy is used.

Chapter 9

Understanding the Mechanics of Fluids and Cohesive Forces

. . . .

In This Chapter

▶ Comprehending pressure and fluids

▶ Focusing on the Archimedes principle

▶ Conserving energy and the amount of fluid

▶ Becoming more cohesive

. . . .

*M*any situations in biophysics involve biological systems that are fluid and need something to contain them. For instance, take your morning cup of coffee. Could you drink it (and enjoy it) without a cup to contain it? Organisms need fluids for the transportation of nutrients and waste within the system (because solids aren't very good at doing that).

This chapter discusses the forces and energy involved with the mechanics of fluids and their application to biophysics. Here I introduce the concepts of pressure, density, and Pascal's principle; discuss Archimedes' principle and the buoyant force; cover the conservation laws, so you can understand that what doesn't change is just as important as what does change; and examine cohesive forces and some of their applications in biophysics.

Pushing On Fluids — Pressure and Density

The big difference between solids and fluids is solids don't need a container to help the solid maintain its shape when under the influence of forces, whereas fluids need a container to maintain their shape or else they'll deform under the influence of forces. Calculating the force on individual molecules in a fluid and studying the motion of each molecule individually is very difficult especially considering fluids can deform and change shape.

In many biophysical situations, it's easier to discuss how many molecules you have in a given region of space, which is called *density*. Also, it's easier to understand the biophysics if the average force placed on the molecules is considered instead of the force on each individual molecule. For example, when you shake hands with a friend, the force applied to each molecule doesn't matter; what matters is the overall average force over the surface of the hand, which is called *pressure*. Your body is mostly water and biological organisms are mostly fluid, so understanding fluids is important to understanding the biophysics of biological systems.

In this section, you discover the concepts of pressure and density, which make it easier to analyze fluids. I also discuss how heavy air and fluids can weigh and what's involved with your blood pressure.

Squeezing atoms together — density and pressure

In biophysics, density and pressure are two of the most important concepts. This is because biological organisms are mostly fluid and pressure is a measure of a fluid's interaction with its surroundings. When people say "density," they're usually referring to mass density. But when speaking in biophysics, they should say "mass density" because there are many different densities, such as mass density, weight density, energy density, number density, and molar density . . . just to mention a few.

Mass density (ρ_m) is defined as the mass divided by the volume. Density tells you how much of the fluid has been squeezed together in that region of space. Mathematically, the relationship is as follows:

$$\rho_m = \frac{m}{V}$$

Specific gravity is defined as the density of the object divided by the density of water. This is a convenient method of measuring density because it's a dimensionless quantity, which is the same in all systems of units.

A useful property of fluids is the density of fluids in biological systems don't change that much and can usually be treated as a constant. Treating the density as constant makes many problems in biophysics a lot easier to solve. For example, water at 32 degrees Fahrenheit (0 degrees Celsius) has a density of $\rho_m = 1.94$ slugs per cubic feet (1,000 kilograms per cubic meter) while at 212 degrees Fahrenheit (100 degrees Celsius) water has a density of $\rho_m = 1.86$ slugs per cubic feet (958 kilograms per cubic meter), which is only a 4.2 percent change over the entire range of temperatures from freezing to boiling. This makes density a convenient quantity for some applications in biophysics.

In biophysics, most things are fluids with approximately constant density. The next thing you need to know is how fluids interact with their surroundings. *Force* is the quantity used in physics to describe the interaction of an object with its surroundings. Force is a local quantity, but it's more beneficial to discuss forces over a region, especially when dealing with fluids. Pressure is the magnitude of the normal force (the force perpendicular to the surface) divided by the surface area over which the force is applied. Mathematically, the equation looks like this:

$$P^{(a)} = \frac{F}{A}$$

The superscript (a) indicates that this is the absolute pressure.

The forces acting on a fluid can be split into two types:

- ✔ **Body forces:** A force (like gravity) that acts throughout the system.

- ✔ **Surface forces:** Contact forces between the fluid and the surroundings. Contact forces can be split into two types.

  - • **Tangential forces:** The tangential forces are parallel to the surface like friction.

  - • **Normal forces:** The normal force pushes into the fluid. The normal force is related to pressure.

You should be aware that many different units are used for pressure, and the type of unit used depends on the specific field of application. The more common units in everyday use and in scientific research are 1 pound per square inch (psi) = 144 pounds per square foot (lb/ft^2) = 6,895 pascals (Pa) = 0.06805 atmospheres = 0.06895 bars = 51.73 torrs = 51.73 millimeters of mercury (mm Hg).

When pressure is applied to most fluids in biophysics, they won't shrink, and they maintain a constant density. The fluid is said to be *incompressible* in that case. Humans are mostly water and water is approximately incompressible, but it does compress a little under extremely high pressures such as in the deepest depths of Mariana Trench. At the bottom of Mariana Trench, 6.833 miles (10.99 kilometers) below sea level, the density of water increases by only 5 percent with a hydrostatic pressure of 1,071 atmospheres (15,700 pounds per square inch = 1.09×10^8 pascals)! I discuss hydrostatic pressure more in the next section.

Gases on the other hand are *compressible* so they obey Boyle's law. *Boyle's law* states that at constant temperature, the product of the absolute pressure and volume of a compressible gas (fluid) is constant. Mathematically it looks like this:

$$P_1^{(a)}V_1 = P_2^{(a)}V_2$$

Weighing air and fluids — Pascal's principle

Fluids interact with their surrounding through pressure, and it can go both ways. A fluid can exert a pressure on the surroundings and the surroundings can exert a pressure on the fluid. *Pascal's law* states what happens when an external pressure is exerted on an incompressible fluid. The second part of this section flips things around and introduces the *hydrostatic pressure*, which is a measure of how much the weight of the fluid is pushing down.

Pascal's law (principle) states that the pressure applied anywhere in an enclosed incompressible fluid is transmitted equally throughout the fluid. If the surroundings apply a pressure to a fluid, then the fluid will try to squeeze into the fluid on the other side of the pressure. But if the fluid is incompressible, then it can't get any closer, which means the neighboring fluid will try to move away and create a pressure on the fluid on the other side of this fluid. The fluid eventually transfers the pressure throughout the fluid because the molecules within the fluid push on the neighboring fluid with the same pressure.

When a person goes for an eye examination, they test for glaucoma, which is often caused by excessive pressure on the retina and optical nerve. Pascal's law means that the pressure on the retina and optical nerve is the same as within the eye, so by measuring the pressure on the front of the eye they know what the pressure is on the nerve.

Pascal's law deals with the surroundings applying a pressure on the fluid, now I look at the fluid applying a pressure on the surroundings. One of the forces on the earth is gravity, which gives everything weight, including fluids. Also, everything is immersed in fluid be it water or air. Therefore, if you change your vertical height, then you change the amount of fluid above you pushing down on your body.

The *hydrostatic pressure* is the change in pressure between two points in a column of fluid, and it's equal to the density of the fluid times the acceleration due to gravity times the change in vertical height. This is a measure of the change in the weight of the fluid pushing down. Mathematically, this looks like the following equation:

$$\Delta P_{HP}^{(a)} = -\rho\,g\,\Delta y$$

Most people don't think about it, but the column of air stretching from space down to the surface of the earth is very heavy. The *atmospheric pressure* is P_{atm} = 1 atmosphere = 14.7 pounds per square inch = 1.013×10^5 pascals = 1.013 bars = 760 torr.

If you're a swimmer, you'll notice that the pressure squeezing on your body increases as you go deeper in the water. This is represented by the minus sign in the mathematical formula. Your body has the atmospheric pressure applied to it plus the weight of the water above pushing down on your body. A typical swimming pool in the diving section is 20 feet (6.1 meters) deep. At the bottom of the pool the hydrostatic pressure relative to the surface will be

$$\Delta P_{HP}^{(a)} = -\rho\, g\, \Delta y = -\left(1.94\,\frac{\text{slugs}}{\text{ft}^3}\right)\left(32.2\,\frac{\text{ft}}{\text{s}^2}\right)(-20\text{ ft} - 0\text{ ft}) =$$

$$1{,}250\,\frac{\text{pounds}}{\text{square foot}}\left(5.98\times10^4\,\text{pascals}\right)$$

The pressure on your body will be 1.6 times greater than at the surface.

Gauging blood pressure

When working with pressure, it's always good practice to figure out if the data you have is gauge pressure or absolute pressure, and which one you need for your calculations. Here are their definitions:

- ✔ *Absolute pressure* ($P^{(a)}$) is the total pressure exerted on fluid.

- ✔ *Gauge pressure* ($P^{(g)} = P^{(a)} - P_{atm}$) is the relative pressure and it's equal to the absolute pressure minus the atmospheric pressure. The atmospheric pressure is all around you, so you usually think in terms of gauge pressure. For example, when you blow up a balloon, the pressure that expands the balloon is the gauge pressure.

The heart doesn't pump blood at constant pressure, but, instead, it fluctuates between a maximum and a minimum pressure. A *sphygmomanometer* measures that blood pressure. It consists of an inflatable pressure sleeve called a *cuff* that's wrapped around a part of the body, which is at the same vertical elevation as the heart (don't place it around the neck) because the blood pressure changes with vertical elevation. Attached to the cuff is a *manometer* that measures the gauge pressure in the cuff. The pressure in the cuff is increased until it causes the artery to collapse and prevents the blood from flowing through the cuff. The pressure in the cuff is slowly decreased and when the blood starts to squirt through the artery again, a measurement is taken from the manometer. This measurement is called the *systolic pressure,* the maximum pressure. The pressure in the cuff is reduced further, and when the blood is flowing through the artery continuously, a second reading is taken called the *diastolic pressure,* the minimum pressure.

Hospitals measure blood pressure in torr (1 torr = 1 millimeter of mercury (mm Hg)), and a normal reading for humans is 120 over 80 (120 torrs = 2.32 pounds per square inch = 1.60×10^4 pascals and 80 torrs = 1.55 pounds per square inch = 1.07×10^4 pascals). Your blood pressure is considered okay if it's in the range (100 to 140) over (65 to 90) with 120 over 80 being ideal.

Examining Why Things Float

This section is all about density and some of the important consequences of objects having different densities. The different densities allow you to image objects (for example, X-rays and ultrasound), float in hot-air balloons, explore shipwrecks in submarines, and other applications.

In this section, you discover Archimedes' principle and the buoyant force. This is based on switching a fluid with a given density by an object with a different density. Also, I discuss some applications of the buoyant force and how you can calculate your body density. I show you how to calculate the weight capacity of a boat based on the buoyant force acting on it.

Floating in fluid — Archimedes' principle and the buoyant force

Archimedes' principle states any object, wholly or partially immersed in a fluid (or gas), has a force exerted on it by the fluid called the *buoyant force*. The magnitude of the buoyant force is equal to the weight of the displaced fluid and the direction is upwards against gravity. Archimedes also realized that the volume of the displaced fluid equals the volume occupied by the portion of the object submerged in the fluid.

As an example, consider a crown that is supposed to be made of gold. You place the crown in water and measure the volume of displaced water. You then take a bar of gold (equal mass to the crown) and place it in the water and measure the volume of displaced water. You compare the volumes of displaced water, which are equal to the volumes of the objects placed in the water (remember the gold bar and the crown have the same mass):

$$\frac{V_{crown}}{V_{gold}} = \frac{m_{crown}\rho_{gold}}{\rho_{crown}m_{gold}} = \frac{\rho_{gold}}{\rho_{crown}}$$

If the crown was made of pure gold, it would have the same density as the gold bar, and therefore occupy the same volume as the gold bar. Gold has a mass density of 37.6 slugs per cubic foot (19,320 kilograms per cubic meter) and silver has a mass density of 20.4 slugs per cubic foot (10,500 kilograms per cubic meter). If the crown were pure silver (painted gold), the volume of the crown would be 1.84 times larger than the gold bar.

As a supplement, Archimedes proposed that a floating object will displace its own weight of fluid. In the case of floating objects, their density must be less than the density of the fluid because the forces are being balanced (buoyant force = force of gravity (weight)). Mathematically, the relationship is

$$1 = \frac{|\vec{F}_B|}{|\vec{F}_g|} = \frac{m_{fluid}g}{m_{object}g} = \frac{\rho_{fluid}gV_{fluid}}{\rho_{object}gV_{object}}$$

Note the ρ is the mass density and ρg is the weight density.

If an object floats, it means its volume can't fit into the volume of the displaced fluid and the part that won't fit sticks into the air above the fluid. The equation for floating objects can be rewritten as

$$\frac{\rho_{object}}{\rho_{fluid}} = \frac{V_{fluid}}{V_{object}} < 1$$

If this ratio is greater than 1 instead of less than 1, then the object will sink.

The buoyant force is always opposite to the effective weight (the effective weight equals the weight minus the net external force acting on the object) and not the gravitational force. Air produces a buoyant force, which is why hot air balloons and helium balloons float.

Suppose you and a friend are hired to go fishing on some lakes to test the fish for mercury poisoning. On the lake, the two of you have a flat-bottom boat that is 20 feet (6.10 meters) long and 2 feet wide (1.22 meters). The combined weight of the boat and the two of you is 500 pounds (227 kilogram mass); how deep does this boat sink into the water with both of you on board?

You can calculate how far into the water the boat sinks by using this ratio (and the volume (V) equals the height (h) times the area (A)):

$$\rho_{fluid}V_{fluid} = m_{object}$$

$$h_{fluid} = \frac{m_{object}}{\rho_{fluid}A_{fluid}} = \frac{500 \text{ lb}}{\left(62.5 \text{ lb ft}^{-3}\right)(20 \text{ ft})(2 \text{ ft})} = 0.200 \text{ feet}(0.0610 \text{ meters})$$

The boat will sink 2.4 inches (6.1 centimeters) into the water.

Measuring the density of the human body

The preceding section shows that density plays an important role in the interaction of objects with fluids, so here you can calculate the density of your body. I calculate the density of Peter's body because I don't know your dimensions. To calculate the density, you need to know your weight (or mass), and you can calculate your volume (or you can experimentally measure your volume). Peter's weight is 225 pounds (1000 newtons) and his mass is 6.99 slugs (102 kilograms). You can determine your volume by following these steps and splitting the body into sections:

1. **Calculate the volume of a leg by assuming a cylindrical shape.**

   Peter's leg is about 3 feet long (0.914 meter) and an average of 0.5 feet (0.152 meters) across. The volume of Peter's leg is $V_{leg} = \pi (0.25 \text{ ft})^2 (3 \text{ ft})$ = 0.589 cubic feet (0.0167 cubic meters).

2. **Calculate the volume of your torso by assuming a box shape.**

   Peter's torso is about 2 feet (0.61 meters) long, 1.33 feet (0.405 meters) wide, and 1 foot (0.305 meters) deep. The volume of Peter's torso is $V_{torso} = (2 \text{ ft}) (1.33 \text{ ft}) (1.00 \text{ ft}) = 2.66$ cubic feet (0.0753 cubic meters).

3. **Calculate the volume of an arm by assuming a cylindrical shape just like the leg.**

   Peter's arm is about 2 feet long (0.61 meters) and an average of 0.35 feet (0.107 meters) across. The volume of Peter's arm is $V_{arm} = \pi (0.175 \text{ ft})^2 (2 \text{ ft}) = 0.192$ cubic feet (5.45×10^{-3} cubic meters).

4. **Calculate the volume of the head by assuming a spherical shape.**

   Peter's head has an average diameter of 7 inches. The volume of Peter's head is $V_{head} = (4/3)\pi (3.5 \text{ in})^3 = 0.104$ cubic feet (2.94×10^{-3} cubic meters).

   Peter's total volume is $V_{Peter} = 2 (0.589 \text{ ft}^3) + 2.66 \text{ ft}^3 + 2 (0.192 \text{ ft}^3) + 0.104$ $\text{ft}^3 = 4.33$ cubic feet (0.123 cubic meters).

You can now calculate your density by taking your mass divided by the volume. In Peter's case, his weight density is 52.0 pounds per cubic foot (8,130 newtons per cubic meter) and his mass density is 1.61 slugs per cubic foot (829 kilograms per cubic meter). Recall that water has a weight density of 62.5 pounds per cubic foot (9,810 newtons per cubic meter) or a mass density of 1.94 slugs per cubic foot (1,000 kilograms per cubic meter), so only about 83 percent of Peter's body would be underwater while he's floating. Your density should be close to Peter's density.

Solving Conservation Laws

Conservation laws are important in all areas of biophysics. Knowing what doesn't change is as important as knowing what does change. The properties of the material differ, depending on what is conserved and what isn't conserved. For example, blood is an incompressible fluid, so it has a constant density, whereas air is a compressible fluid, so its density can change.

This section covers some of the things in fluids that don't change. It discusses the consequences of the mass of the fluid not changing as the fluid flows through restricted regions. You also look at systems with no dissipative forces; in other words, the mechanical energy doesn't change. You finally see an application of Bernoulli's equation when the fluid isn't moving.

Grasping the continuity equation

A *continuity equation* is a mathematical expression that shows how a property doesn't change even if things are in motion. For example, if I pour water into a sink with the drain open, then the amount of water flowing out the drain equals the amount I pour into the sink. In this section I look at a few properties of fluids that don't change even though things are changing, like pouring water down the drain. To begin, an incompressible fluid has constant density that equals mass divided by volume. The property of constant density gives the first conservation law.

The conservation of mass for an incompressible fluid states that the mass of the fluid can't be destroyed or created, so the amount of fluid within a given volume doesn't change and what flows in equals what flows out. Mathematically, this is represented by *the continuity equation*: ρQ = constant.

ρ is the *mass density* of the fluid, which is a constant for an incompressible fluid, so Q = constant.

Q is called the *flow rate*. $Q = A v$ is the rate at which the total fluid flows past some point. It's equal to the speed (v) of the fluid times the cross-sectional area (A). When the flow rate (Q) is multiplied by the mass density (ρ), it gives the amount of mass of the fluid per unit of time flowing past a specific point.

If you go down to the river and watch the water flow, the water flows slowly where the river is wide and much faster where the river is narrow. This is because the flow rate is a constant and when the cross-sectional area (A) becomes small the speed must become large. The amount of fluid (mass) must flow at the same rate everywhere.

Understanding Bernoulli's equation

A fluid doesn't move unless some pressure (force per unit area) does work on the fluid. Whenever you see a fluid moving, something has done work on the fluid. For example, in the case of water starting on a mountain and flowing down to the ocean, gravity is doing the work.

Remember the work-energy theorem, which states that the net work done on an object is equal to the change in the object's kinetic energy. *Bernoulli's equation* is the work-energy theorem for a fluid with a few reasonable assumptions.

Bernoulli's equation is as follows:

$$\frac{\rho v_{final}^2}{2} - \frac{\rho v_{initial}^2}{2} = -\rho g\left[y_{final} - y_{initial}\right] + P_{initial}^{(a)} - P_{final}^{(a)}$$

To understand the connection with the work-energy theorem, take into account the following points:

- The equation on the left side of the equality is the change in the fluid's kinetic energy per unit volume. The kinetic energy of a fluid is

$$\frac{\rho V_{fluid} v^2}{2}$$

- The first set of terms on the right hand side of the equality are the work done per unit volume by gravity on the fluid. The work done by gravity is

$$W_g = -\rho V_{fluid} g\left[y_{final} - y_{initial}\right]$$

- The last two terms on the right hand side of Bernoulli's equation is the work done per unit volume by the applied pressure (force) acting on the fluid. The work done by external sources is

$$W_A = \left[P_{initial}^{(a)} - P_{final}^{(a)}\right]V_{fluid}$$

Bernoulli's equation has a few assumptions:

- The fluid has a *steady flow* (*steady state*), which means the velocity doesn't suddenly change at any given point in space. The velocity can be different at different points in space but not at the same point at different times.
- The fluid is incompressible so the density is a constant.
- The fluid is *nonviscous* (no *viscosity*), which means there's no friction between the fluid and objects (and barriers) it's in contact with. Refer to Chapter 10 for more on nonviscous fluids.

✔ The fluid is *streamline* and not *turbulent. Streamline flow* is also called *laminar flow*. This type of flow is very orderly and predictable where the fluid follows trajectories. For example, suppose water is flowing down a smooth channel at a slow speed, and I place a stick in the water. The water carries the stick down the channel. I then place a second stick in the water at the same location as the first stick. The two sticks follow the exact same path down the channel.

Turbulent flow (turbulence) is chaotic motion and the path of the flow can't be predicted. Suppose I place my two sticks in a creek that has rocks sticking out of the water and waterfalls as well. The swirling motion and the mixing of the streamlines cause the two sticks to follow different paths even though they start at the same position but at different times.

Applying Bernoulli's equation to static fluids

Bernoulli's equation is very useful in many biophysical situations. One of those situations is taking blood pressure.

Consider this example: Olivia is 6 feet (1.83 meters) tall, and her systolic pressure is 130 torrs (= 2.51 pounds per square inch = 1.73×10^4 pascals). What's the maximum blood pressure in her feet? In order for you to answer this question, you need a few extra pieces of biophysical information:

✔ Olivia's heart is approximately 4.75 feet (1.45 meters) above her feet.

✔ The mass density of whole blood is 6 percent greater than pure water, 2.0554 slugs per cubic foot (1059.5 kilograms per cubic meter). The weight density of whole blood is 66.2 pounds per cubic foot (1.04×10^4 newtons per cubic meter).

✔ The viscosity (friction) of blood in the body is small and can be ignored for this calculation. This means you can use Bernoulli's equation.

✔ The kinetic energy of the blood is approximately the same and small throughout the body.

You can now apply Bernoulli's equation to solve the problem:

$$0 = -\rho g \left[y_{feet} - y_{heart} \right] + P_{heart}^{(a)} - P_{feet}^{(a)}$$

$$P_{feet}^{(a)} = 2.51 \text{ psi} - \left(2.0554 \times 32.2 \frac{\text{lb}}{\text{ft}^3} \right) \left(-4.75 \text{ ft} \left(\frac{1\text{ft}^2}{144 \text{ in}^2} \right) \right)$$

The pressure in the feet of 4.69 pounds per square inch (= 243 torrs = 3.24×10^4 pascals) is almost double the pressure at the heart.

Sticking Together — Cohesive Forces

You may have noticed that pure solids don't make good biological organisms because nothing moves and pure gases don't make good biological organisms because they fly apart. Fluids are the perfect compromise by sticking together but still being able to move around. Fluids are fundamental to all biological organisms and are very important in biophysics. This section looks at the forces involved with fluids. The *cohesive force* keeps the fluid from flying apart. It's the collective attraction between the molecules making up the fluid. The molecules like being part of the collective because of the cohesive force, and this has several consequences. I also tell you about the *adhesion* or the *adhesive force,* which is the same as the cohesive force except between molecules of different types. I talk about these forces and the boundary of the fluid. These are the molecules living on the edge. You also discover the surface in contact with flat surfaces and inside tubes, Laplace's law, and membranes.

Fighting surface tension

If you've ever looked at the surface of a pond, you've noticed bugs sitting on the surface of the water without sinking or even breaking the surface. The molecules stick together because of the cohesive force, so when you get to the molecules at the surface, they have a cohesive force only in one direction. This effective force toward the rest of the fluid produces a tension in the surface called the *surface tension.*

The surface tension (γ) is defined as the amount of work required to change the surface area of a fluid divided by the amount of change in the surface area. Work is equal to the applied force times the displacement, so the surface tension can be defined as the applied force per unit length:

$$\gamma = \frac{W}{\Delta A} = \frac{F_{applied}}{L}$$

In the case of bubbles and membranes with two surfaces, γ needs to have a factor of two in front of it.

Water striders (insects that walk on water) spread their weight over the surface of the water. The bugs are covered in hydrophobic microhairs to keep their bodies dry and not gain weight from water clinging to their bodies. The legs don't break the surface and only cause small indentations in the water as the surface tension holds them up. You notice a water strider has six legs, and the length of the leg in contact with the water is $L = 0.25$ inches (6.4 millimeters) and the width of each leg is in contact with the water is $w = 0.060$ inches (1.5 millimeters). The water is indented to a depth of $d = 0.020$ inches (0.51 millimeters) by the bug. Figure 9-1 shows the indentation looking down the length of the leg so it's easier to see d and w. I haven't shown the bug's leg and foot, but it's what's causing the indentation. Figure 9-1 also shows the surface tension force and the weight of the bug divided by six. I have indicated the radius of curvature, the angle of the arc, and the arclength in Figure 9-1. These quantities are needed to calculate the area of the indentation. Calculate the weight of the water strider.

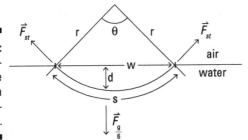

Figure 9-1:
The indenta-
tion of the
water by a
water strid-
er's leg.

You can calculate the weight of the bug by following these steps:

1. **Understand the biophysics of the problem.**

   You want to figure out the bug's weight. The bug is standing on the water and not moving so the net force acting on the bug is zero. Therefore, the surface tension must provide sufficient force to equal the weight of the bug. If you find the force of surface tension, then you'll know the weight of the bug.

   To find the force of surface tension, it equals the surface tension times the change in the surface area. You now know what to find.

2. **Draw a figure.**

   Visualizing the problem can really help. Figure 9-1 shows my figure of the water being bent by the bug's leg. It also shows the forces.

3. **Finding the numbers.**

   You can look up the number for the surface tension. You will find the surface tension of water at room temperature is 0.00499 pounds per foot (0.0728 newtons per meter).

   The lake is originally flat, so the old surface area of the water is a rectangle:

   $$A_{old} = Lw$$

The bug pushes the water downward so the new surface area of the water is part of a cylinder:

$$A_{new} = Ls = Lr\theta$$

TIP

I ignore the areas at the ends of the cylinders because they're very small, but if you want to include them, you have to add an extra $[r^2\,\theta - w\,(r - d)]$.

In the formula for area, you need to use radians and not degrees.

You now have all the numbers you need.

4. **Find the numbers to calculate the change in the area.**

You can calculate the radius of curvature and the angle by using the geometry in Figure 9-1. They are

$$r = \frac{4d^2 + w^2}{8d} = \frac{4(0.020 \text{ in})^2 + (0.060 \text{ in})^2}{8(0.020 \text{ in})}$$

$$\sin\left(\frac{\theta}{2}\right) = \frac{4wd}{w^2 + 4d^2} = \frac{4(0.060 \text{ in})(0.020 \text{ in})}{(0.060 \text{ in})^2 + 4(0.020 \text{ in})^2}$$

From this you can see that the radius of curvature is $r = 0.0325$ inches (0.826 millimeters) and the angle is $\theta = 135$ degrees $= 2.35$ radians.

5. **Substitute everything into the surface tension formula and solve for the weight.**

You have to remember several formulas to get the answer. The first formula is the force of the weight of the bug divided by 6, the second formula is the force times displacement equals the work done, the third formula is the work done on a fluid's surface is equal to the surface tension times the change in the area, and the fourth formula is the change in the area that is the new area minus the old area given in the third step. You can now put all four formulas together to solve the problem:

$$\frac{\left|\vec{F}_g\right|}{6}d = W = \gamma\,\Delta A = \gamma\left[r\theta L - wL\right]$$

$$\left|\vec{F}_g\right| = \frac{6\gamma L}{d}[r\theta - w] = \frac{6(0.00499 \text{ in/ft})(0.25 \text{ in})}{(0.020 \text{ in})(12 \text{ in/ft})}\left[0.0325 \text{ in}(2.35 \text{ radians}) - 0.060 \text{ in}\right]$$

You find that the weight of this water strider is 5.11×10^{-4} pounds (2.27×10^{-3} newtons) or it has a mass of 1.59×10^{-5} slugs (2.32×10^{-4} kilograms).

Making contact with capillarity and contact angles

Have ever looked at the water in your drinking glass and wondered why the surface looks different and is curved? The previous section discusses the cohesive force as the bonding between the molecules within the fluid. These molecules also bind with molecules within other materials such as the drinking glass (container). The *meniscus* is the name of the curvature of the upper surface of a fluid near the edge of the container or surface of an object. This is caused by the surface tension and a competition between the *cohesion* within the fluid and the *adhesion* of the fluid to the container or object.

Again, take a look at the water in your drinking glass (not plastic). You'll notice the surface of the water in contact with the glass is higher than the surface of the water in the middle; it's concave. This means the adhesive force between glass and water is strong, and it pulls the water up the glass walls. Some fluids and materials bend this way whereas others bend the other way. The contact angle determines this distinction.

Contact angle θ is defined as the angle from the container's surface inside the fluid to the meniscus. A *concave meniscus,* like the water in your glass, has an angle between 0 degrees and 90 degrees. The fluid is said to "wet" the glass. On the other hand, a bead of water sitting on a waxed car hood has a *convex meniscus* and has a contact angle between 90 degrees and 180 degrees. The water in this case does "not wet" the car.

You may probably be thinking that this attraction (adhesive force) that pulls water up the glass can be used to our advantage. You're correct. *Capillarity* or *capillary action* is the rise or fall of a fluid in a narrow tube as the tube is stuck into a large container of the fluid.

If you stick a narrow straw into your glass of water, it will rise up the straw higher than the water in the surrounding glass; whereas if you stick your straw into mercury, the meniscus is convex and the mercury inside the tube will be lower than the mercury surrounding the straw on the outside.

The *height* of the *capillary rise* is calculated by the following equation:

$$h = \frac{2\gamma \cos(\theta)}{\rho g R}$$

The parts of the equation stand for the following:

- h is the height of the center of the meniscus.
- γ is the surface tension of the fluid.

✔ θ is the contact angle.

✔ ρ is the density of the fluid.

✔ g is the acceleration due to gravity.

✔ R is the radius of your tube.

You see from this formula fluids that "wet" a surface rise up the capillary because the contact angle is between 0 degrees and 90 degrees, which has a positive cosine, whereas fluids that don't wet a surface drop down a capillary because the contact angle is between 90 degrees and 180 degrees, which has a negative cosine.

To derive this equation, you have to balance the vertical component of the surface tension at the container ($F_{vertical} = F_{applied} \cos(\theta) = \gamma L \cos(\theta) = \gamma 2\pi R \cos(\theta)$) with the weight of the fluid it's holding above the surrounding fluid ($m g = \rho \pi R^2 h g$). L is the circumference of the tube, R is the radius of the tube, m is the mass of the fluid lifted above the flat surface, and h is the height it was lifted.

In the formula you may have noticed that the narrower the tube, the higher the fluid will go up the tube. You can calculate the height the water rises in your drinking glass or you can calculate the capillary rise of sap in your friend's roses instead.

Blocking fluids with Laplace's law

Fluids need to be contained and biological systems do so with membranes. Some membranes are solid barriers (such as the skin of your water balloon), whereas others are semi-permeable, such as cell walls.

Suppose the membrane is solid to the fluid. You can think of the balloon filled with water or air. The pressure of the fluid inside the balloon is greater than the outside pressure and the balloon expands until the force produced by the surface tension in the balloon matches the force produced by the difference in the pressures. This example illustrates Laplace's law.

Laplace's law states the internal pressure inside a membrane is

$$P_{in}^{(a)} = \gamma \left[\frac{1}{r_1} + \frac{1}{r_2} \right] + P_{out}^{(a)}$$

Here, r_1 and r_2 are the principal radii of curvature of the membrane, γ is the surface tension, and P_{out} is the pressure on the outside of the membrane.

If I go back to the balloon, the surface tension is the force in your balloon's skin trying to bring it back to its original shape. Also, you know the pressures within and outside the balloon are the forces exerted by the fluids on the balloon wall divided by the surface area of the balloon. A balloon has only one radius, so $r_1 = r_2 = r$. An example of something with two radii is an egg, which has a short radius and a long radius.

A few special cases are

- ✔ **A spherical membrane or a solid ball of fluid:** $r_1 = r_2 = r$, the radius of the sphere. For the solid ball of fluid, the surface tension and the external pressure are squeezing the drop, and the force countering it is the outward force of the internal molecules, preventing the surface molecules from getting closer to the center.

- ✔ **A soap bubble with two surfaces:** γ needs to have an extra factor of two in front of it.

- ✔ **A capillary tube:** $r_1 = r_2 = r$, is the radius of curvature of the meniscus, which is equal to $R/\cos(\theta)$, the radius of the tube divided by the cosine of the contact angle.

- ✔ **A cylindrical membrane:** $r_1 = r$, the radius of the cylinder. $r_2 = \infty$ because the wall of the cylinder doesn't curve or bend.

Laplace's law is also known as the *Young-Laplace equation*. I call it Laplace's law so there's no confusion with Young's equation and Laplace's equation, which are completely different.

Sneaking oxygen into the body

One of the most interesting devices in humans is the lungs, where the body allows gases to pass between the air and the blood. The role of the lungs is to exchange oxygen molecules with other gas molecules in the blood, mostly carbon dioxide. In this section, I discuss the application of Laplace's law to the lungs, but some general biophysical information is needed:

- ✔ The smallest blood vessels are called capillaries with a diameter of 5 microns. They connect the arterioles to the venules.

- ✔ The capillaries in the lungs are right beside the gas and the only thing preventing the blood from leaking into the lungs or the gas from forming air bubbles in the blood is a membrane two cells thick, plus a "basement membrane" between the two cells that is less than a micron thick. This membrane is permeable to the gas molecules but not to the cells in the blood.

✔ The alveoli are small air sacs (cavities) that can be approximated as spheres. The pressure inside the alveoli is $P_a^{(a)}$.

✔ Surrounding the lungs is the fluid in the pleural cavity, which has a pressure $P_p^{(a)}$.

✔ The membrane inside the alveoli is covered with a liquid containing *surfactant*, which is a lipoprotein.

The body covers your lungs with surfactant because in your lungs, the membrane is the same as a balloon, and it would be easy to either rupture your lungs if you inhaled too much or cause a lung to collapse if you exhaled too much air. The surfactant is nature's answer to this problem. These long chain lipoproteins resist the expansion of the alveoli during inhalation by increasing the surface tension during expansion. During exhalation the alveoli contract, compressing the surfactant together and decreasing the surface tension so the alveoli don't collapse.

A more technical explanation begins with Laplace's law for a sphere, namely

$$r\left(P_a^{(a)} - P_p^{(a)}\right) = 2\gamma$$

From Laplace's law: If r is small then a large pressure difference is needed to overcome the surface tension in the balloon, but for a large radius, r, the pressure difference doesn't need to be as large. (This is assuming the surface tension is a constant.)

The importance of the surfactant can't be overstated, without it animals would be in trouble. You should try this problem assuming no surfactant: Without the surfactant, the membrane has a wall tension of 0.0034 pounds per foot (0.05 newtons per meter). During an exhalation the gauge pressure of the air within the alveoli is –8 pounds per square foot (–383 pascals = –0.056 psi), and the gauge pressure of the pleural cavity is –12 pounds per square foot (–575 pascals = –0.083 psi). The radius of the alveoli is 1.75×10^{-4} feet (5.3×10^{-5} m). What's the percentage of the surface (wall) tension in the alveoli compared to the wall tension without the surfactant?

You can use Laplace's law to find the wall (surface) tension and then the percentage:

$$\gamma = \frac{r\left(P_{in}^{(a)} - P_{in}^{(a)}\right)}{2} = \frac{1.75 \times 10^{-4} \text{ ft}\left(-8 \text{ lb/ft}^2 - \left(-12 \text{ lb/ft}^2\right)\right)}{2}$$

I obtain a value of $\gamma = 3.5 \times 10^{-4}$ pounds per foot (5.09×10^{-3} newtons per meter), which is only 10.2 percent of the membrane's wall tension. The reduction in the wall tension by 90 percent is because of the surfactant, which prevents the lung from collapsing during an exhalation.

Looking into negative pressure in water columns

How do trees get so tall? The top of the tree must receive water and nutrients through the xylem, which is achieved through *negative pressure*. A *compressive force (compression)* produces a positive pressure on a fluid as it tries to squeeze the fluid together, but a *tensile force (tension)* produces a negative pressure on a fluid as it tries to pull the fluid apart.

To understand how to produce a negative pressure, you can perform a biophysics experiment. Consider a capillary with water in it. Stick a plunger (piston) down the top of the capillary and remove all the gas so the water is in contact with every surface. Now slowly pull the plunger (piston) upward and measure the force. The adhesive force between the water and the plunger pulls the water upward while the cohesive force tries to prevent the water from being ripped apart. You should be able to measure a downward force on the plunger as you try to stretch the water. Negative pressures as large as –635,000 pounds per square foot (–3.04 × 10⁷ pascals) have been measured for water! This is an absolutely huge number.

The *Pseudotsuga menziesii* (Douglas Fir) and *Sequoia sempervirens* (Redwood) trees along the Pacific Coast of North America can reach heights of 325 feet (100 meters) to 350 feet (110 meters). What negative pressure is needed for a 400-foot (122-meter) tree?

To solve this problem, use Bernoulli's equation:

$$\frac{\rho v_{leaf}^2}{2} - \frac{\rho v_{ground}^2}{2} = -\rho g\left[y_{leaf} - y_{ground} \right] + P_{ground}^{(a)} - P_{leaf}^{(a)}$$

I can help you figure out what the values are for each of the quantities.

✔ The sap in the xylem is mostly water so you can use the density of water. The mass density of water is 1.94 slugs per cubic foot (1,000 kilograms per cubic meter).

✔ You can assume the sap isn't moving so the speed is zero.

✔ *g* is the acceleration due to gravity of 32.2 feet per square second (9.81 meters per square second).

✔ The vertical change in height is 400 feet (122 meter).

✔ You can assume the pressure at ground level is equal to the atmospheric pressure, 2,117 pounds per square foot (1.013 × 10⁵ pascals).

You are now ready to solve the problem. I rearrange the formula to isolate for the pressure:

$$P_{leaf}^{(a)} = -\rho g\left[y_{leaf} - y_{ground} \right] + P_{ground}^{(a)} = -\left(62.5\,\text{lb/ft}^3 \right)\left[400\,\text{ft} \right] + 2{,}117\,\text{lb/ft}^2$$

The negative pressure at the leaf is –22,900 pounds per square foot (-1.10×10^6 pascals), which is only 3.6 percent of the maximum negative pressure measured experimentally for water in the piston. This is a reasonable method for getting sap to the top of the tallest trees on the Earth.

You're probably thinking that negative pressure is good in a lab, but how does a tree produce negative pressure? The negative pressure is caused by the process called *transpiration*, which is basically the tree sweating. The xylems travel from the roots up into the tree's leaves, and the sap diffuses amongst the cells in the leaves. The leaves have stomata (pores), which the guard cells close at night and open during the day. When the stomata are open, the water leaks from the leaves and evaporates creating a negative pressure within the leaves. Current estimates place the negative pressure within the leaves at the same order of magnitude as what you calculated for the 400-foot tall tree.

Chapter 10

Going with the Fluid Flow — Fluid Dynamics

· ·

In This Chapter

▶ Distinguishing different fluids

▶ Staying in the air

▶ Fighting the viscosity

▶ Spinning in a centrifuge

· ·

*F*luid dynamics is the study of flowing fluids. All liquids and gases are considered fluids, so fluid dynamics plays an important role in many areas of biophysics. An understanding of fluid dynamics is particularly important in the study of blood flow in the body and in airflow in the lungs, the flight of birds, weather patterns in meteorology, and environmental science.

All fluids have internal friction between the molecules and also friction with the boundaries containing the fluid. *Friction* is the force that resists motion and arises from the electromagnetic force between the molecules of the fluid and between the molecules of the fluid with the atoms and molecules forming the container. The frictional force within a fluid is called *viscosity*. (Think of maple syrup, which has a high viscosity compared to water, which has a low viscosity.) This chapter focuses on two situations: nonviscous fluids and viscous fluids.

Ignoring Friction Nonviscous Fluids

Nonviscous fluids (also known as *inviscid* fluids) have no friction. This section is about fluids where external forces are more dominant than the viscosity, so the fluid can be treated as nonviscous. Many different situations in biophysics allow you to ignore the viscosity of the fluid; such as air flow when the boundaries are not important. Gases for the most part can be treated as nonviscous fluids. The bottom line: The following sections examine biological systems where you can't ignore friction.

Conserving energy with Bernoulli's equation

Bernoulli's equation is the work-energy theorem (conservation of energy) for fluids under a few reasonable assumptions. Here I apply Bernoulli's equation to *dynamic* (moving) *nonviscous* (frictionless) fluids. The *work-energy theorem* states that the change in the kinetic energy is equal to the net work done by external forces acting on the system. In addition to Bernoulli's equation, the fluid is assumed to have no sources or sinks, so the total mass of the fluid doesn't change, which is called *conservation of mass*.

These two important laws can be represented mathematically. The conservation of mass equation depends on if the fluid is *compressible* (like air) or *incompressible* (like water under reasonable pressures). The three important equations for nonviscous fluids are as follows. Check out Chapter 9 for more information.

- **Bernoulli's equation:** It states that the kinetic energy of the fluid plus the gravitational potential energy plus the pressure acting on the fluid is a constant. This equation reads as follows:

$$\frac{\rho v_{final}^2}{2} + \rho g y_{final} + P_{final}^{(a)} = \frac{\rho v_{initial}^2}{2} + \rho g y_{initial} + P_{initial}^{(a)}$$

  It shows that the pressure drops the faster the fluid flows or the higher its vertical elevation. The initial pressure minus the final pressure is the work done per unit volume by applied forces on the fluid. The change in the vertical height is the work done by gravity on the fluid per unit volume, where ρ is the mass density of the fluid, g is the acceleration due to gravity constant = 32.2 feet per second squared = 9.81 meters per second squared, and y is the vertical position of the fluid. The difference of the terms with the speed squared is the change in the kinetic energy per unit volume, where v is the speed of the fluid.

- **Conservation of the total flow rate:** This is the conservation of mass for an incompressible fluid.

$$Q_{final} = v_{final} A_{final} = v_{initial} A_{initial} = Q_{initial}$$

  The conservation of the total flow rate, Q, is a consequence of the fluid being incompressible combined with conservation of mass, ρ = constant, so the amount of fluid flowing into a region must equal the amount of fluid flowing out of a region.

- **Continuity equation:** It generalizes the conservation of total flow to conservation of mass flow when the fluid is compressible (ρ isn't a constant):

$$\rho_{final} v_{final} A_{final} = \rho_{initial} v_{initial} A_{initial}$$

You can apply these equations to solve many problems in biophysics. For example, a person in environmental science can find these equations useful in planning an irrigation system. Go ahead and work this problem.

Suppose you're renting a room in a house with several other renters. You're getting tired of the shower wars and most of the other renters are night owl students, so you decide to have your shower at 5 a.m. when no one else is using water. Your shower is on the top floor, which is 30 feet (9.14 meters) above where the water comes into the house. Now that you have the water to yourself, what is the speed of the water coming out of the showerhead?

To figure out the problem, follow these steps:

1. **Determine the appropriate physical law(s) to use.**

   Here you need Bernoulli's equation because you want to find the speed of the water. The water has to rise vertically up through the house. Hint: The water has to do work against gravity to go up, which can remind you of the work-energy theorem (or conservation of energy).

   You need the conservation of the total flow rate: Bernoulli's equation has two different speeds, so you need a relationship between the speeds. Water is approximately incompressible at atmospheric pressures, so you need the conservation of the total flow rate.

2. **Find the numbers needed in the conservation of the total flow rate.**

   A typical water pipe coming into a house has a diameter of 1 inch (2.54 centimeter).

   > The diameter of a typical pipe in the showerhead is 0.5 inch (1.27 centimeters).

   > The pipes are cylindrical, so the cross-sectional area is $A = \pi r^2$.

3. **Calculate the ratio of speed of the water coming into the house divided by the speed at the showerhead:**

   You can now use the conservation of the total flow rate:

   $$\frac{v_{in}}{v_{shower}} = \frac{A_{shower}}{A_{in}} = \frac{\pi r_{shower}^2}{\pi r_{in}^2} = \frac{(0.5 \text{ in})^2}{(1.0 \text{ in})^2} = 0.25$$

4. **Find the numbers needed for Bernoulli's equation.**

   > **Pressure (in):** The typical pressure of water entering a house is 40 psi (40 pounds per square inch, which equals 5,760 pounds per square foot or 2.76×10^5 pascals).

Pressure (shower): The water coming out of the shower will be at atmospheric pressure 14.7 pounds per square inch = 2,117 pounds per square foot (1.013×10^5 pascals).

The change in the vertical height of the fluid (water) is 30 feet (9.14 meters).

The mass density of the fluid (water) is ρ = 1.94 slugs per cubic foot (1,000 kilograms per cubic meter).

5. **Solve Bernoulli's equation for the speed.**

   You have all the numbers except the speed of the water coming out of the shower, so you can solve by substituting the numbers into Bernoulli's equation:

   $$\frac{\rho}{2}\left(v_{shower}^2 - v_{in}^2\right) = \left[\rho g\left(y_{in} - y_{shower}\right) + P_{in}^{(a)} - P_{shower}^{(a)}\right]$$

   $$v_{shower} = \sqrt{\frac{2\left[1.94 \text{ slug ft}^{-3}\left(32.2 \text{ ft s}^{-2}\right)(-30 \text{ ft}) + (5760 - 2117)\text{lb ft}^{-2}\right]}{1.94 \text{ slug ft}^{-3}\left(1 - (0.25)^2\right)}}$$

   $$= 44.1 \frac{\text{feet}}{\text{second}} \left(13.4 \frac{\text{meters}}{\text{second}}\right)$$

   You're very happy that you have hot water and more than a trickle of water coming out of the shower.

Flowing air — wind, birds, planes, and baseball

Bernoulli's equation shows that the pressure of a gas decreases as the speed increases. Biophysicists in some fields such as environmental science need to keep this under consideration. This property can cause dangers such as during hurricanes and tornados because of the risk of windows being blown out. Suppose you have a barrier with an area A separating the flowing fluid (gas or liquid). The force applied to the barrier by the fluid is

$$\left|\vec{F}\right| = \left(P_{in} - P_{out}\right)A = \frac{\rho A}{2}\left[v_{out}^2 - v_{in}^2\right]$$

During your Saturday night biophysics party, you decide to test Bernoulli's equation. You take two pieces of paper and hold one sheet in each and hand. You then hold them up to your mouth and blow between the two sheets of paper. The result is that the two pieces of paper move together. Bernoulli's equation tells you that the faster the fluid (liquid or gas) is moving, the lower its pressure. In the case of the two pieces of paper, the air is moving faster between the sheets of paper and has a lower pressure, so the higher air pressure on the outside pushes the paper together. You can estimate how fast you're blowing the air by making a few reasonable assumptions. Go ahead and try during your next biophysics party.

Viscous fluids have an effect in sports. Bernoulli's equation isn't the whole story in many ball sports when the spin of the ball is important. An important effect is known as the *magnus effect*. All normal fluids including air are *viscous* fluids. (Remember, viscosity means the fluid has frictional forces.) Normally, air can be treated as a nonviscous fluid but when it's in contact with a solid object, it forms a small *boundary layer* (a layer of air that is stationary relative to the object). When the object is spinning, the air in contact with the object wants to spin with the object, which causes the speed of the air on the opposite sides of the ball to change, creating a difference in the pressures on the ball and hence a force to the side.

For example, in baseball the pitcher can put a spin on the ball. The faster the spin, the more it will curve. The side of the ball spinning toward the batter is moving through the air faster than the other side and tries to drag the air backwards causing a higher air pressure on that side of the ball, which causes the ball to curve in the opposite direction. A riser is a fastball with backspin. The top of the baseball is spinning away from the batter, causing the air to be dragged over the top of the ball, which reduces the acceleration in the vertical direction so the ball doesn't drop as fast as you expect. (No, it doesn't actually move upwards.)

Be careful when applying Bernoulli's equation to fluid (liquid or gas) flows. You can run into problems if the kinetic energy of the fluid isn't obvious. Bernoulli's equation doesn't work for planes and birds or at least it isn't the complete story. To understand, look at Figure 10-1, which shows the air flowing around a wing. This figure also shows the four forces acting on the plane and its wings: the thrust, the lift, the drag, and weight. The *angle of attack* in this figure is θ. (When a plane flies through the air, the wing doesn't line up with the airflow, but it makes an angle called the angle of attack.) Normally the angle of attack is small to reduce the force of drag. A clue that Bernoulli's equation isn't the whole story comes from the fact that it doesn't matter if the thick part of a wing is at the front, center, or back of the wing, only that the speed v_{top} is different to v_{bottom}. In reality though, you need the thick part of the wing at the front to create the lift necessary to fly.

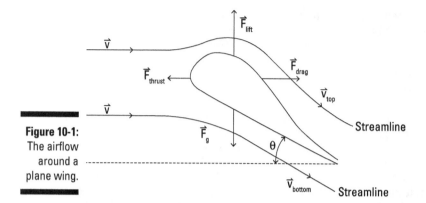

You are probably thinking, why does Bernoulli's equation not work? The problem is in the assumptions. I have implicitly assumed in Figure 10-1 that the speed of the air changes as soon as it's in contact with the wing. Why can't $\vec{v} = \vec{v}_{top} = \vec{v}_{bottom}$? There are no giant lips blowing the air above the wing nor is the wing spinning like a baseball creating different speeds.

To understand how a plane or bird stays in the air, remember Newton's third law of motion — the law of action and reaction. If an object creates a force on a second object, then the second object creates a force on the first object that is equal in magnitude but in the opposite direction. The air striking the wing is forced downwards and there is an overall change in the momentum of the air downwards; the reaction force on the wing is pushing the wing upwards and creating an overall lift to the wing. It's the changing of the air's momentum that generates the lift force on the wing.

Regulating temperature in warm-blooded animals — conservation of heat energy

Heat is a form of energy, and it isn't usually thought of as a fluid. However, heat does flow from hot regions to cold regions, and warm-blooded animals are hot. The human body is always creating heat, so it's continuously getting rid of heat in order to maintain a constant temperature. To maintain a constant temperature, the body obeys conservation of heat energy; the heat flowing from the body equals the heat being created by the body.

Before I can introduce the ways in which the body expels heat, let me explain what all these terms mean:

✔ **Heat energy:** It's one of the five fundamental types of energy. It's associated with the disordered motion of the molecules and atoms within an object. Alternatively, heat is the amount of energy entering (or leaving) an object when it's brought into contact with another object that is at a different temperature. Common sets of units used for energy when dealing with heat energy and food energy are 1 Calorie (food calorie = Cal) = 1,000 calorie (energy calorie = cal) = 3,086 foot pound (ft lb) = 4,184 joules (J). Note two different types of calories exist.

✔ **Temperature:** It's a measure of the heat energy per particle. The temperature of an object is proportional to the average kinetic energy of the atoms and molecules within the object.

✔ **Temperature and heat:** An object's heat energy and temperature are connected through the formula: $\Delta Q_{TOTAL} = m\ c\ \Delta T$. In this formula, ΔQ_{TOTAL} is the change in an object's heat energy, m is the mass of the object, c is the specific heat capacity of the object, and ΔT is the change in temperature. T is measured in kelvin (0 kelvin = –459.67 degrees Fahrenheit = –273.15 degrees Celsius), which is the absolute temperature scale. 0 kelvin means there is no heat energy, and the specific heat capacity is a property of the material. Different types of material require a different amount of heat energy to change the temperature by 1 kelvin. Specific heat capacity is a measure of how hard it is to heat a material.

The mass times the specific heat capacity is equal to the *heat capacity,* which you may sometimes see in the books. The heat capacity depends on both the size of the object and the material the object is made of.

✔ **Different temperature scales:** They are as follows:

X kelvin = Y degrees Celsius + 273.15

Y degrees Celsius = X kelvin – 273.15

X kelvin = 5 (Z degrees Fahrenheit)/9 + 255.37

Z degrees Fahrenheit = 9(X kelvin)/5 – 459.67

Y degrees Celsius = 5 (Z degrees Fahrenheit)/9 – 17.78

Z degrees Fahrenheit = 9(Y degrees Celsius)/5 + 32

✔ **Conservation of heat energy within a warm-blooded animal:** Warm-blooded animals maintain a constant body temperature, which means $\Delta T = 0$. The heat-temperature relation above means that the change in the total heat energy within the body is $\Delta Q_{TOTAL} = 0$. Therefore, to maintain a constant temperature in the body, the body relies on the *conservation of heat energy* (constant total heat energy), which means the rate that heat is lost must equal the rate that heat is produced. The mathematical representation of conservation of heat energy is:

$$H_m = H_{HC} + H_{con} + H_{rad} + H_l + H_s$$

Here H_m is the rate heat is generated by the body through metabolism of the food, H_{HC} is the rate heat is lost by the body through heat conduction, H_{con} is the rate heat is lost by the body through convection, H_{rad} is the rate heat is lost by the body through electromagnetic radiation, H_l is the rate heat is lost by the body through evaporation of moisture in the lungs, and H_s is the rate heat is lost by the body through sweating.

The brain's hypothalamus is the body's thermostat and maintains a constant temperature by measuring the blood's temperature. Flesh has a low thermal conductivity and is a good insulator, so the body controls the amount of blood flowing to the skin's surface to help regulate the rate of heat lost. The limbs and skin are usually several degrees cooler than the core, depending on the air temperature.

Chapter 11 discusses in detail the metabolism of food within the human body and the fact half of the initial energy absorbed is converted directly into heat. Eventually, almost all the energy becomes heat energy with very little energy used in doing work on the surroundings. Therefore, in the following sections I take a closer look at the three of the five methods how the body can lose heat.

Flowing from hot to cold — heat conduction

Heat conduction is the rate at which heat flows from hot regions to cold regions. The formula is as follows:

$$H_{HC} = \frac{\Delta Q_T}{\Delta t} = k_T A \frac{T_{hot} - T_{cold}}{L}$$

H_{HC} is the heat flow rate by conduction and is a measure of how fast the heat is flowing from the hot region to the cold region, k_T is the thermal conductivity, A is the cross-sectional area that the heat is flowing through. The *length* L is the distance from the hot region to the cold region. (**Note:** ΔQ_T is the change in the heat energy and Δt is the elapsed time.)

For example, think of a window in your home on a hot summer day. The outside air is very hot and the air inside your home is cool. A is the size of the window and L is the thickness of your window. Air is an excellent thermal insulator, which is why high quality windows are double or triple paned. The air pockets between the panes slow the flow of heat. Skin is also a good thermal insulator, whereas metals are very bad thermal insulators. The thermal conductivity of air is $k_T = 0.0054$ pound per (second kelvin) (0.024 watts per (meter kelvin)) and for muscle and fat it is $k_T = 0.045$ pound per (second kelvin) (0.2 watts per (meter kelvin)).

Creating currents with heat convection

H_{con} is the heat flow rate by convection and is a measure of how fast a hot surface in contact with a cool fluid will cool due to convection of the fluid. The formula associated with convection is

$$H_{con} = q_T \, A \, (T_{hot} - T_{cold})$$

As the fluid close to the surface heats, it moves away from the hot surface and is replaced by cooler fluid. This cooler fluid heats, and it moves away repeating the cycle. q_T is the convective heat transfer constant, and A is the area of the surface. The T_{cold} is the fluid's temperature far from the hot surface, which has a temperature of T_{hot}.

Many apartment heaters work this way. The radiator heats the air in one part of a room. The air then rises to the ceiling and flows to the other side of the room, down the far wall, and back to the radiator across the floor. The convective heat transfer constant for a naked human is $q_T = 0.49$ pound per (foot second kelvin) (7.1 watts per (square meter kelvin)).

Glowing with heat through electromagnetic radiation — Stefan's law

Stefan's law is the rate at which heat energy is converted into electromagnetic radiation, which is then emitted from the body. The mathematical formula for Stefan's law is

$$H_{rad} = H_{out} - H_{in} = e_R \sigma_S A \left(T_{surface}^4 - T_{outside}^4 \right)$$

H_{rad} is the heat flow rate by electromagnetic radiation being emitted by the body minus the heat flow rate into the body by absorption of electromagnetic radiation. H_{out} is the heat flow rate out of the body and is a measure of how fast heat is leaving the body through electromagnetic radiation. H_{in} is the heat flow rate into the body and is a measure of how fast heat is entering the body through electromagnetic radiation. e_R is the emissivity of the body and is a measure of how much of that type of radiation can enter or leave the body. σ_S is Stefan's constant. (*Stefan's constant* is the number $\sigma_S = 3.89 \times 10^{-9}$ pound per [second foot kelvin to the fourth power] $= 5.67 \times 10^{-8}$ watts per [square meter kelvin to the fourth power].) A is the surface area of the body that the radiation is coming out of. $T_{surface}$ is the surface temperature of the body and $T_{outside}$ is the temperature of the surroundings.

Objects at a nonzero kelvin temperature emit and absorb electromagnetic radiation. The heat energy is being converted into electromagnetic radiation within the body, and electromagnetic radiation absorbed by the body is being converted into heat energy. This is how the sun heats your body. The human body is too cold to glow in visible light, so if you were to stand in a pitch dark cave, no one would be able to see you; however, if people could see in the infrared range of the electromagnetic spectrum, then you would look like a very bright light bulb glowing in the dark. Chapter 16 examines electromagnetic radiation.

e_R is the emissivity of the object and is a measure of how much of the radiation can escape or be absorbed by the object through its surface. For human skin the emissivity varies from $e_R = 0.65$ (lightest skin) to $e_R = 0.82$ (darkest skin) for visible light, but the emissivity is almost 1 for all human skin in the infrared. A perfect absorber (and emitter) has an emissivity of 1 while a perfect barrier that allows no radiation through has an emissivity of 0.

Wien's displacement law gives the wavelength (and frequency) of the electromagnetic radiation with the maximum intensity. If you look at the distribution of the radiation as a function of wavelength, it forms a curve with a peak at a specific wavelength (λ) and frequency (f). The Wien displacement law gives the wavelength where the peak is located. The mathematical formula for Wien's displacement law is

$$\lambda = \frac{c}{f} = \frac{B_W}{T}$$

λ is the wavelength and f is the frequency of the electromagnetic radiation at the peak. The speed of light (electromagnetic radiation) is $c = 9.84 \times 10^8$ feet per second $= 3.00 \times 10^8$ meters per second. The Wien displacement constant is $B_W = 9.51 \times 10^{-3}$ feet kelvin $= 2.898 \times 10^{-3}$ meter kelvin.

The human body has a temperature of 310.2 kelvin (98.6 degrees Fahrenheit = 37.0 degrees Celsius). The peak intensity of the electromagnetic radiation produced by the human body is at a frequency of 3.21×10^{13} hertz, which is infrared radiation. This is why thermal cameras work in the infrared range. Humans glow in the infrared while hotter objects glow in visible light.

Applying the heat formulas to biophysics

Now you can use the information in the preceding sections to solve this problem.

Todd weighs 154 pounds (70 kilogram mass), has a surface area of 18.8 square feet (1.75 square meters), and at rest generates heat through metabolism at a rate of 59 foot pounds per second (80 watts). Todd has just returned from a jog and is lying on the floor in only his jogging shorts.

(Assume one-quarter of his skin is touching the floor and three-quarters of his skin is in contact with the air, ignoring his shorts.) Calculate each of his heat loss flow rates assuming his metabolic rate is currently 885 foot pounds per second (1,200 watts), his outer skin temperature is 95 degrees Fahrenheit (35 degrees Celsius, or 308 kelvin), his breathing is back to normal, and the room air and floor are at 70 degrees Fahrenheit (21 degrees Celsius, or 294 kelvin).

In the problem and in the previous section I give you numbers and formulas to calculate the heat flow rates, except for sweating. However, for sweating, you can use conservation of heat energy after you know the rest. To solve this problem, follow these steps:

1. **Calculate H_m, the rate the body generates heat.**

   The metabolic rate of heat generation is H_m = 885 foot pound per second (1,200 watts).

2. **Calculate H_{HC}, the rate the body loses heat through heat conduction.**

   You are told to assume that three-quarters of his skin is in contact with air (k_T for air is 0.0054 pound per (second kelvin) = 0.024 watts per (meter kelvin)) and a quarter of his skin is in contact with the stone floor. (k_T for stone is 0.18 pound per (second kelvin) = 0.8 watts per (meter kelvin).)

   You can make the reasonable assumption that the effective length between the temperatures is L = 0.5 inch (= 0.0417 feet = 0.0127 meters) for both the floor and the air. The problem tells you the temperatures and Todd's body surface area.

   $$H_{HC} = k_{floor}\left(\frac{A}{4}\right)\frac{T_{skin} - T_{floor}}{L} + k_{air}\left(\frac{3A}{4}\right)\frac{T_{skin} - T_{air}}{L}$$

   $$= \left[\left(0.18\frac{lb}{s\,K}\right) + 3\left(0.0054\frac{lb}{s\,K}\right)\right]\frac{\left(18.8ft^2\right)\left(308K - 294K\right)}{4\left(0.0417ft\right)}$$

   $$= 284\frac{ft\,lb}{s} + 25.6\frac{ft\,lb}{s} = 310\frac{foot\ pounds}{second}\left(420\ watts\right)$$

   The heat flow from conduction is 310 foot pounds per second (420 watts) with 92 percent of the heat flowing into the floor even though three-quarters of the skin is in contact with the air. Air is a very poor thermal conductor. If Todd were standing, the heat flow rate would be only 34 foot pounds per second (46 watts).

3. **Calculate the convective heat flow rate, H_{con}, the rate the body loses heat through convection.**

You need the convective heat transfer constant. Look up this value in a reference source or measure it experimentally, because it depends on the person's shape and orientation (and the clothing worn). For example, the convective heat transfer constant is large for a naked person lying on their back: the convective heat transfer for Todd is 0.49 pound per (foot second Kelvin) (7.1 watts per (square meter Kelvin)).

$$H_{con} = q_T A \left(T_{skin} - T_{air} \right) = \left(0.49 \frac{lb}{ft\ s\ K} \right) \left(18.8\ ft^2 \right) \left(308K - 294\ K \right) =$$

$$129 \frac{\text{foot-pounds}}{\text{second}} (174\ \text{watts})$$

4. **Calculate the radiation heat flow rate, H_{rad}, the rate the body loses heat through electromagnetic radiation.**

   For humans, almost all the electromagnetic radiation emitted is in the infrared spectrum. The emissivity of the human body at these frequencies is approximately 1. This problem tells you Todd's surface area and the temperatures. You know Stefan's constant from the previous section.

   $$H_{rad} = e_R \sigma_s A \left(T_{skin}^4 - T_{air}^4 \right) =$$

   $$(1) \left(3.89x10^{-9} \frac{lb}{ft\ sK^4} \right) \left(18.8\ ft^2 \right) \left[\left(308\ K \right)^4 - \left(294\ K \right)^4 \right] =$$

   $$112 \frac{\text{foot-pounds}}{\text{second}} (151\ \text{watts})$$

5. **Estimate the evaporation in the lungs' heat flow rate, H_l, the rate the body loses heat through evaporation of moisture in the lungs.**

   The heat flow rate for evaporation in the lungs varies from person to person and depends on how heavily he or she is breathing. An average person breathing normally loses heat energy at a rate of approximately 8 foot pounds per second (10 watts). You can assume Todd is average, and the problem tells you that he's breathing normally.

6. **Calculate the sweat heat flow rate, H_s, the rate the body loses heat through sweating.**

   You know that the body is maintaining a constant temperature, so it's obeying conservation of heat energy. Therefore, the formula for conservation of heat energy is the equation for the heat flow rate for evaporation of sweat, which gives you:

   $$H_s = H_m - H_{HC} - H_{con} - H_{rad} - H_l =$$

   $$885 \frac{ft\ lb}{s} - 310 \frac{ft\ lb}{s} - 129 \frac{ft\ lb}{s} - 112 \frac{ft\ lb}{s} - 8 \frac{ft\ lb}{s}$$

   Todd is losing heat at a rate of 326 foot-pounds per second (442 watts) through sweat.

Fighting the Drag — Viscous Flow

All fluids (liquids and gases) have *viscosity* (frictional forces) except *super-fluids* (which have an exactly zero viscosity, such as liquid helium when it's close to zero kelvin). Therefore, all dynamical fluids in biophysics are viscous flows. Fluids play an important role in most area of biophysics.

The following sections take a closer look at the properties and applications of viscous fluid flows. These sections examine what is (and isn't) a fluid, discuss different fluid flows and their interaction with boundaries, and apply the discussion on blood flow in the human body.

Stressing out with viscous fluids

The internal frictional forces within a fluid are called *viscosity,* which is a measure of the resistance of a fluid to stress. A fluid with a large (stress) viscosity will have a small strain from a given stress compared to a fluid with a small (stress) viscosity. For example, water has low (stress) viscosity whereas honey has high (stress) viscosity.

All fluids have (stress) viscosity except superfluids. I refer to these fluids as viscous fluids. (Note that some people only refer to fluids with a viscosity greater than water as viscous fluids.) I put the word "stress" in parentheses because it's a common practice to call the stress viscosity just the viscosity. Stress viscosity is also known as the *dynamic (stress) viscosity* or the *shear viscosity.* I drop the stress because it should be clear if I'm talking about stress viscosity or the viscosity between the fluid and other objects such as a container.

Allow me to define and clear up a few terms related to stress. Stresses are applied forces that deform a material. Three types of stress are

- ✔ *Tensile stress* is a force that stretches the material.
- ✔ *Compressive stress* is a force that compresses the material.
- ✔ *Shear stress* is a tangential force that tries to rip the material. (Think of rubbing your hand across sandpaper.)

For each of the three stresses a corresponding *strain,* which is the amount of deformation that occurs to the material when a stress is applied. Chapter 6 discusses the stresses and strains when applied to solids.

Classifying viscous fluids — Newtonian and non-Newtonian fluids

A stress causes a strain in a fluid, and how that fluid responds to the stress and deforms (strain) tells you a lot about that fluid's properties. The behavior of fluids under stress allows you to place the fluids into one of two categories:

- ✔ **Newtonian fluid:** The *Newtonian fluids* have the property that the stresses are directly proportional to the strain rates. In other words, if a graph of the stress as a function of the strain rate produces a linear curve with the slope equal to the viscosity then the fluid is Newtonian. Water and air (gases) are good examples of Newtonian fluids having a small constant viscosity.

- ✔ **Non-Newtonian fluid:** If the curve is nonlinear, then the fluid is non-Newtonian. The strain rate is a measure of how the velocity of the fluid changes in space, so you can think of it as the *rate of deformation* (change in the velocity divided by the displacement). Meanwhile some non-Newtonian fluids include the following:

    - **Shear thinning fluids:** The viscosity of the fluid decreases with an increasing rate of shear stress. An example is ketchup and synovial fluid.

    - **Bingham plastic:** The plastic is solid at low stresses and becomes a viscous fluid with large stresses. A plastic squeeze bottle is a good example.

    - **Thixotropic materials:** They become less viscous over time when shaken, agitated, or otherwise stressed. Yogurt is an example. You have to stir for a while before it's in a liquid form.

    - **Shear thickening fluids:** The viscosity of the fluid increases with the rate of shear strain. These fluids are also known as *dilatant fluids*. The combination of cornstarch and water is the classic example.

    - **Rheopectic materials:** The material becomes more viscous over time when shaken, agitated, or otherwise stressed. An example is kids' slime.

    - **Magnetorheological fluids and ferrofluids:** The viscosity of these fluids increases in the presence of a magnetic field. The magnetorheological fluids have magnetic particles that have a diameter around a micron whereas the ferrofluids use magnetic nanoparticles. The fluid is poured into the cavity and then a permanent magnetic is turned on and the fluid becomes a solid, sealing the cavity.

Flowing slowly at the edge — laminar flow and Poiseuille's law

Streamline flow is when the fluid is flowing in an organized manner with no disorganized motion or sideways mixing. If you look at the smoke rising from a fire very close to the flames, you'll notice the streams of smokes are rising side by side without any swirling. This initial flow of the smoke is a streamline flow.

Laminar flow is a layered viscous fluid where each layer is a streamline flow with a different speed. Strictly speaking, the layers aren't distinct and the speed changes continuously across the layers. For example, consider a straight section of a slow moving river where the bottom of the river isn't a factor on the flow of the surface water. All the water will be flowing in the same direction (streamline flow) and the water in the middle of the river will be moving the fastest. The closer you go to the shore, the slower the speed of the water.

The velocity of the fluid in contact with a boundary is traveling at the same speed as the boundary. The velocity changes in magnitude as you move away from the boundary in a perpendicular direction. In the case of my river example, the water touching the shore isn't moving.

You probably want to see how this works in the human body. The synovial fluid, which is a non-Newtonian shear thinning fluid, fills the synovial cavity in many joints in your body, such as the knee joint. Synovial fluid forms laminar flow when your bones are moving. The articular cartilage covers the ends of the bones in the synovial joint, which are kept separated by the synovial fluid. The synovial fluid against each articular cartilage is moving with the same speed as the cartilage, so as you move from one cartilage to the other, the fluid's velocity changes continuously from one speed to the other speed, forming speed layers, hence the name laminar flow. Figure 10-2 shows the synovial fluid's velocity in a synovial joint.

Figure 10-2:
Laminar flow in a synovial joint.

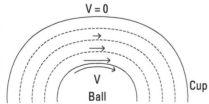

An important application in biophysics is the flow of air in the alveoli or blood flow within the body. These applications are described by *Poiseuille's law*, which states how the pressure of an incompressible, viscous fluid will drop in a laminar flow through a long horizontal pipe (of length L) that has a constant radius R. R is assumed to be much smaller than L. The mathematical expression of Poiseuille's law for the pressure drop of an incompressible viscous fluid in a pipe (tube) is

$$\frac{\pi R^4 \left(P_{inital}^{(a)} - P_{final}^{(a)} \right)}{8\eta L} = Q$$

Here, R is the radius of the pipe, L is the length of the pipe from the location of the initial pressure to the location of the final pressure, η is the fluid's viscosity, and Q is the total fluid flow rate.

The total fluid flow rate, Q, is related to the velocity of the fluid, where v_{avg} is the average speed of the fluid from the center to the edge against the pipe wall, and v_{center} is the maximum speed of the fluid, which at the center of the cylindrical pipe:

$$Q = \pi R^2 v_{avg} = \frac{\pi R^2 v_{center}}{2}$$

Poiseuille's law shows the pressure is dropping as the fluid moves through the pipe, which means the fluid is losing energy due to viscosity (frictional forces). In Chapter 4 I talked about the relationship between power, work and force. The power lost to the system is equal to the force times the average velocity. The force is related to the pressure and the average velocity is related to the total fluid flow rate. The *power loss* within the fluid is given by the formula:

$$P_{lost} = \left(P_{inital}^{(a)} - P_{final}^{(a)} \right) Q$$

The two pressures and the total fluid flow rate on the right-hand side are the same quantities as in Poiseuille's law.

You can use the preceding formulas to solve this type of problem you may encounter in your biophysics course.

Venus has a partially clogged artery from plaque buildup. The flow rate of the blood is 3.5×10^{-5} cubic feet per second (9.9×10^{-7} cubic meters per second). The plaque buildup is 0.08 feet (2.4 centimeters) long and has closed the artery to a radius of 6.5×10^{-4} feet (2.0×10^{-4} m). What is the pressure drop across the plaque and the corresponding power loss?

To solve this problem, follow these steps:

1. **Determine what physical principles or laws you need.**

   The problem wants you to find the drop in the pressure across the clogged artery, so you need Poiseuille's law.

   The problem also wants you to calculate the power loss. You have the formula relating the power loss to the flow rate and pressure change.

2. **Find the numbers needed for Poiseuille's law.**

   The problem says that the total flow rate is $Q = 3.5 \times 10^{-5}$ cubic feet per second (9.9×10^{-7} cubic meters per second).

   You're told the length of the partially clogged artery is $L = 0.08$ feet (2.4 centimeters) and the radius of the opening is $R = 6.5 \times 10^{-4}$ feet (2.0×10^{-4} m).

   You know everything in Poiseuille's law to solve the problem except the viscosity of the blood. You have to search for this number (use a medical resource book or website for this information.) or perform an experiment. To save you some time, I give you the number. Whole blood at 98.6 Fahrenheit (37 degrees Celsius) has a viscosity of $\eta = 4.353 \times 10^{-5}$ pound second per square foot (2.084×10^{-3} pascals second).

3. **Solve for the change in the pressure by solving Poiseuille's law.**

   $$\Delta P^{(a)} = P^{(a)}_{initial} - P^{(a)}_{final} = \frac{8\eta L Q}{\pi R^4} = \frac{8\left(4.353 \times 10^{-5} \text{lb s ft}^2\right)(0.08 \text{ ft})\left(3.5 \times 10^{-5} \text{ft}^3 \text{ s}^{-1}\right)}{\pi\left(6.5 \times 10^{-4} \text{ft}\right)^4}$$

   The change in the pressure across the length of the plaque is 1.74×10^3 pounds per square foot (8.33×10^4 pascals).

4. **Solve for the power loss along the partially clogged artery.**

   The formula states that the power lost is equal to the pressure drop time the total fluid flow rate: $P_{lost} = \Delta P^{(a)} Q = (1.74 \times 10^3 \text{ lb ft}^{-2}) (3.5 \times 10^{-5} \text{ ft}^3 \text{ s}^{-1}) = 0.0609$ foot pound per second (0.0825 watts). This lost energy is being converted into heat energy.

Flowing of the blood and flow resistance

You can view the blood's circulatory system in the same way as an electrical DC circuit with a battery being the heart, the current being the total flow rate, and the resistance being the flow resistance.

This flow resistance is a measure of the work the fluid has to do against the viscosity. Use the following formula for measuring flow resistance:

$$R_f = \frac{\Delta P^{(a)}}{Q} = \frac{8\eta L}{\pi R^4}$$

Poiseuille's law is the blood version ($\Delta P^{(a)} = Q\,R_f$) of *Ohm's law* ($V = I\,R$), and the equation in the previous section for power lost ($P_{lost} = \Delta P^{(a)}\,Q$) is just the blood version for the *Joule losses*, namely $P = V\,I$. (I discuss circuits and Ohm's law in Chapter 16.) If you're more familiar with electrical circuits and adding resistance together, then you can understand how this law works because the same rules apply here for fluids.

In addition, when considering multiple paths for the blood to flow through, you need to consider the following:

- **Flow resistances in parallel:** When the blood has several parallel paths to choose from, then

    - The conservation law states that the pressure at the beginning and at the end must be the same, independent of which path is taken by the blood.

    - The rule states that the sum of the blood flow rates through all the paths must equal the total blood flow rate Q. (You can't create or destroy blood. Other parts of the body are doing that.)

    The combination of the conservation law and the rule gives the following formula for the flow resistances of blood through several parallel paths:

    $$\frac{1}{R_{f,total}} = \frac{1}{R_{f,1}} + \frac{1}{R_{f,2}} + \frac{1}{R_{f,3}} + \dots$$

    The left side of the formula is the total effective flow resistance of all the paths combined together. The right side has all the individual flow resistances for each separate path.

- **Flow resistances in series:** When the blood has only one path to flow through then:

    - The conservation law states that the total blood flow rate is the same through each organ (only one route for the blood to go).

    - The rule states that the sum of the pressure drops across each organ must equal the total pressure drop across all of them combined.

    The combination of the conservation law and the rule gives the following formula for the flow resistances:

    $$R_{f,total} = R_{f,1} + R_{f,2} + R_{f,3} + \dots$$

Different channels in parallel have the same pressure difference across them and different flow resistances in series have the same flow rate through them. The two rules are: the amount of fluid (blood) flowing in must equal the amount of fluid (blood) flowing out.

Use these formulas to solve this problem.

In the human body, some of the arteries branch off and go to the hepatic artery, spleen, small bowel, and large bowel. The four combine and then flow through the liver before traveling through a main vein to the vena cava and back into the heart. Assume in Walen's body the blood's flow rate through the liver is $Q_{liver} = 8.8 \times 10^{-4}$ cubic feet per second (2.5×10^{-5} cubic meter per second). What is the change in the blood pressure and the blood flow rate through all five paths just mentioned?

To solve this problem, follow along with these steps:

1. **Draw a diagram of the blood flow from the heart through the organs and back to the heart.**

   A figure can help visualize what is going on here. Figure 10-3 shows the figure I drew for this problem. In the figure, HA is the hepatic artery, SP is the spleen, SB is the small bowel, and LB is the large bowel.

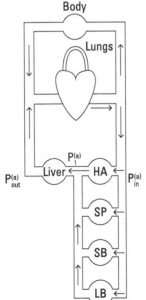

Figure 10-3:
Blood flow
through part
of Walen's
body.

2. **Find the five flow resistances.**

   Here I help you and look up the numbers for the five flow resistances:

   - The HA branches into the arterioles, then the capillaries, and then the venules with an overall flow resistance of $R_{f,HA} = 1.3 \times 10^6$ pound second per fifth power of foot (2.2×10^9 pascal seconds per cubic meter).

   - The arterioles, capillaries, and venules for the SP have an overall flow resistance of $R_{f,SP} = 7.7 \times 10^5$ pound second per fifth power of foot (1.3×10^9 pascal seconds per cubic meter).

   - The arterioles, capillaries, and venules for the SB have an overall flow resistance of $R_{f,SB} = 7.1 \times 10^5$ pound second per fifth power of foot (1.2×10^9 pascal seconds per cubic meter).

   - The arterioles, capillaries, and venules for the LB have an overall flow resistance of $R_{f,LB} = 2.2 \times 10^6$ pound second per fifth power of foot (3.7×10^9 pascal seconds per cubic meter).

   - The liver has an overall flow resistance of $R_{f,liver} = 2.4 \times 10^4$ pound second per fifth power of foot (4.1×10^7 pascal seconds per cubic meter).

3. **Find the pressure drop across the liver.**

   For the liver you know the flow rate and the flow resistance, so all you need to find for the liver is the pressure drop using Poiseuille's law:

   $$\Delta P^{(a)}_{liver} = P^{(a)}_1 - P^{(a)}_{out} = Q_{liver} R_{fl,\,iver} = \left(8.8 \times 10^{-4} ft^3 s\right)\left(2.4 \times 10^4 lb\ s\ ft^{-5}\right) =$$

   $$21.1 \frac{pounds}{square\ foot}\left(1.01 \times 10^3 pascals\right)$$

4. **Find the effective flow resistance across HA, SP, SB, and LB.**

   You know the four individual flow resistances and the rule for blood flows in parallel:

   $$\frac{1}{R_{f,eff}} = \frac{1}{R_{f,HA}} + \frac{1}{R_{f,SP}} + \frac{1}{R_{f,SB}} + \frac{1}{R_{f,LB}} =$$

   $$\frac{1\ ft^5}{1.3 \times 10^6 lb\ s} + \frac{1\ ft^5}{7.7 \times 10^5 lb\ s} + \frac{1\ ft^5}{7.1 \times 10^5 lb\ s} + \frac{1\ ft^5}{2.2 \times 10^6 lb\ s}$$

   The effective flow resistance is 2.54×10^5 pound second per fifth power of foot (4.3×10^8 pascal seconds per cubic meter).

5. **Find the pressure drop across HA, SP, SB, and LB.**

   Figure 11-3 shows the blood flow is parallel. The conservation law is $Q_{HA} + Q_{SP} + Q_{SB} + Q_{LB} = Q_{liver} = 8.8 \times 10^{-4}$ cubic feet per second (2.5×10^{-5} cubic meter per second). This law states the pressure drop across all four is the same, which means that the pressure drop equals the total blood flow rate (Q_{liver}) times the effective flow resistance.

$$\Delta P_{HA}^{(a)} = \Delta P_{SP}^{(a)} = \Delta P_{SB}^{(a)} = \Delta P_{LB}^{(a)} = \Delta P_{eff}^{(a)} = P_{in}^{(a)} - P_1^{(a)} = Q_{liver}R_{f,eff}$$

$$= \left(8.8 \times 10^{-4} \frac{ft^3}{s}\right)\left(2.54 \times 10^5 \frac{lb\ s}{ft^5}\right) = 224 \frac{pounds}{square\ foot}\left(1.07 \times 10^4 pascals\right)$$

6. **Find the flow rate through HA, SP, SB, and LB.**

   You now know the pressure drop and the flow resistance, so calculate the flow rates:

   $$Q_{HA} = \frac{\Delta P_{eff}^{(a)}}{R_{f,HA}} = \frac{224\ lb\ ft^{-2}}{1.3 \times 10^6 lb\ s\ ft^{-5}} = 1.72 \times 10^{-4} \frac{cubic\ feet}{second}\left(4.88 \times 10^{-6} \frac{cubic\ meters}{second}\right)$$

   $$Q_{SP} = \frac{\Delta P_{eff}^{(a)}}{R_{f,SP}} = \frac{224\ lb\ ft^{-2}}{7.7 \times 10^5 lb\ s\ ft^{-5}} = 2.91 \times 10^{-4} \frac{cubic\ feet}{second}\left(8.24 \times 10^{-6} \frac{cubic\ meters}{second}\right)$$

   $$Q_{SB} = \frac{\Delta P_{eff}^{(a)}}{R_{f,SB}} = \frac{224\ lb\ ft^{-2}}{7.1 \times 10^5 lb\ s\ ft^{-5}} = 3.15 \times 10^{-4} \frac{cubic\ feet}{second}\left(8.93 \times 10^{-6} \frac{cubic\ meters}{second}\right)$$

   $$Q_{LB} = \frac{\Delta P_{eff}^{(a)}}{R_{f,LB}} = \frac{224\ lb\ ft^{-2}}{2.2 \times 10^6 lb\ s\ ft^{-5}} = 1.02 \times 10^{-4} \frac{cubic\ feet}{second}\left(2.88 \times 10^{-6} \frac{cubic\ meters}{second}\right)$$

   Notice that $Q_{HA} + Q_{SP} + Q_{SB} + Q_{LB} = 8.80 \times 10^{-4}$ cubic feet per second (25×10^{-6} cubic meters per second) $= Q_{liver}$, which is what was expected. Also, the total drop in pressure is $\Delta P^{(a)}_{eff} + \Delta P^{(a)}_{liver} = P^{(a)}_{in} - P^{(a)}_{out} = 245$ pounds per square foot (1.17×10^4 pascals).

Pumping of the heart — making the blood move

The heart is one of the most marvelous machines (it's also a pump). It's designed to work nonstop for decades and pump more than 145,000 cubic feet (4,100 cubic meters) of blood every year. You can use concepts in this chapter and Chapter 9 to estimate how much energy you need to keep your heart pumping each day.

Xi has volunteered for me to calculate her heart's energy consumption. The heart needs to do work against three primary forces (frictional forces within the heart, applied force pushing the blood, and wall [surface] tension in the heart muscle). The rest of the forces can be ignored for now, because their contribution is very small.

To figure out her heart's energy consumption, follow these steps to determine how much work needs to be done in one day for these three forces.

1. **Find the energy dissipated as heat in a day.**

   Unfortunately, the problem did not give me any numbers so I will have to look up the numbers. The rate of heat production for an average human's heart while resting is 1.50 foot pound per second (2.03 watts). (At rest Xi's entire body rate of heat production is about 59 foot pound per second [80 watts] of heat. See Chapter 11 for more information.)

   I estimate Xi's rate of heat production in the heart is 2.25 foot pound per second (3.05 watts) during her eight-hour work shift.

   You can assume your heart produces the same amount of heat. Also, in Chapter 4, I discuss the relationship between power, work, and time. (Work equals power times elapsed time.) I now know the power and the time in this case is 16 hours of rest and 8 hours of work. The work done producing heat for 1 day is

   $$W_{heat} = P_{heat}\Delta t = \left[\left(1.50\frac{\text{ft lb}}{\text{s}}\right)(16 \text{ hr}) + \left(2.25\frac{\text{ft lb}}{\text{s}}\right)(8 \text{ hr})\right]\left(\frac{3600 \text{ s}}{1 \text{ hr}}\right)$$

   The energy lost to heat in a day is 1.51×10^5 foot pound (2.05×10^5 joules = 49.0 Calories).

2. **Find the energy used pumping blood through the body all day.**

   Unfortunately, the problem doesn't give you any numbers again, so you have to search for some numbers. Before you search for numbers, you need to know what numbers you need:

   Work equals the power times the elapsed time, so you need to find the power. The elapsed time is 16 hours of rest and 8 hours of work. The previous two sections explain that the power is equal to the flow rate times the pressure drop.

   The flow rate at rest, the flow rate at work, the pressure change at rest, and the pressure change at work are the four numbers you need to calculate the power and hence the work:

   Xi's heart pumps blood at a flow rate of 3.5×10^{-3} cubic feet per second (9.9×10^{-4} cubic meters per second) while resting.

   Xi's heart pumps blood at a flow rate of 7.0×10^{-3} cubic feet per second (2.0×10^{-3} cubic meters per second) during her work.

   Xi's blood's average pressure in the arteries near the aorta is 270 pounds per square foot (1.29×10^4 pascals) and 22 pounds per square foot (1.05×10^3 pascals) in the veins near the vena cava while resting.

While at work, Xi's blood's average pressure in the arteries near the aorta is 300 pounds per square foot (1.44×10^4 pascals) and 22 pounds per square foot (1.05×10^3 pascals) in the veins near the vena cava.

You now have all your numbers, so you can first calculate the power and then calculate the work done by the heart pumping blood through the body against the viscosity all day:

$$P_{pump} = Q\left(P_{artery}^{(a)} - P_{vein}^{(a)}\right) = \begin{cases} \left(3.5 \times 10^{-3}\,\frac{ft^3}{s}\right)\left(270\,\frac{lb}{ft^2} - 22\,\frac{lb}{ft^2}\right), \text{ at rest} \\[2mm] \left(7 \times 10^{-3}\,\frac{ft^3}{s}\right)\left(300\,\frac{lb}{ft^2} - 22\,\frac{lb}{ft^2}\right), \text{ at work} \end{cases}$$

$$W_{pump} = P_{pump}\Delta t = \left[\left(0.868\,\frac{ft\,lb}{s}\right)(16\text{ hr}) + \left(1.95\,\frac{ft\,lb}{s}\right)(8\text{ hr})\right]\left(\frac{3600\text{ s}}{1\text{ hr}}\right)$$

The work done in a day to pump blood through the body is 1.06×10^5 foot pounds (1.44×10^5 joules = 34.4 Calories).

3. **Find the formula for the work done keeping the heart muscle tense all day.**

   You have to figure out what formula you need to calculate the work done. To do so, you need to look at the physics of the problem:

   Xi's heart is a pump that squeezes and then expands continuously, pushing the blood through the body. The heart muscles form a wall around the chambers, which can't expand too far or contract too much, so the surface (wall) tension in the wall (muscle) must be continuously maintained. If you think of a balloon and you fill the balloon with air or water, then the balloon is very tense. When you open the balloon's end, the surface tension in the balloon squeezes the air or water out of the balloon.

   This calculation needs surface tension and Laplace's law, which I discuss in Chapter 9. If I assume the left ventricle is a sphere with a contracted radius r_{in} and an expanded radius r_{out}, then the work done during a half beat is

   $$W_{tension} = \gamma\,\Delta A = \frac{\pi\,\Delta P^{(g)}}{3}\left(r_{out}^3 - r_{in}^3\right)$$

   $W_{tension}$ is the work done by changing the radius of the sphere and changing the pressure, γ is the surface tension, ΔA is the change in the surface area, $\Delta P^{(g)}$ is the change in the blood's gauge pressure (remember, the gauge pressure equals the absolute pressure minus the atmospheric pressure), and r is the radius of the sphere.

The previous equation requires a few reasonable assumptions combined with a little bit of knowledge about calculus in order to derive it. A small amount of work is $dW = F\,dr$ (refer to Chapter 3). In this problem, I assume the left ventricle is a sphere and from Laplace's law (check out Chapter 9), the force due to the surface tension in a sphere is $F = 2\pi\,r\,\gamma = \pi\,r^2\,P^{(g)}$, where r is the radius of the sphere and $P^{(g)}$ is the gauge pressure. You can now assume the gauge pressure is approximately constant at the low value during the expansion of the heart and approximately a constant at the high value during contraction. To find the work done (W) during a beat, you need to integrate the force (F) over the displacement (dr) for a full beat. The work for a full beat with half a heart appears in the previous equation. The total work done by the whole heart during a full beat is twice this value. Therefore, you can think of the preceding equation as the work done by the whole heart for half a beat.

4. **Find the numbers to calculate the work done keeping the heart muscles tense all day.**

   You need to find the change in the blood pressure within the heart. You also need to find the maximum and minimum radii of Xi's left ventricle. In addition, calculate the work done for one day and not for one beat.

   > Xi's change in the blood's gauge pressure is 130 torr (362 pounds per square foot $= 1.73 \times 10^4$ pascals) during a half beat.

   > The heart is beating at 70 beats per minutes.

   > The left ventricle has a contracted radius of 2 inches (0.167 feet = 0.0508 meters).

   The left ventricle expansion causes the radius to change by 10 percent.

5. **Solve for the work done by keeping the heart muscles tense all day.**

   Substitution of the numbers into the equation gives

   $$W_{tension} = \frac{\pi\left(362 \text{ lb ft}^{-2}\right)}{3}\left[\left(0.183\mathit{ft}\right)^3 - \left(0.167 \text{ ft}\right)^3\right] = 0.581 \text{ foot-pounds}\left(0.788 \text{ joules}\right)$$

   The preceding shows the work done for half a beat, so you need to convert it to the amount of work done in a day: $W_T = W_{tension} \times 140$ half beats per minute $\times 1440$ minutes in a day $= 1.17 \times 10^5$ foot pound (1.59×10^5 joules $= 37.9$ Calories).

Xi's heart needs a total of 49.0+34.4+37.9 = 121 Calories (1.21×10^5 energy calories) to operate each day.

Turning in the turbulence — turbulent flow

A viscous fluid flowing at slow speeds has streamlines producing laminar flow. However, when the speed becomes too large, then the streamlines start to mix and the fluid motion becomes chaotic. This chaotic fluid motion is call *turbulent flow.* Turbulence is very complex and well beyond the scope of this book. The work done on turbulence is either done in an experimental lab or with computer simulations.

Reynolds' number is defined as the inertial forces acting on the fluid divided by the viscous forces. Reynolds' number and its specific mathematical formula will change, depending on the system under consideration. The criterion for turbulence is that when the Reynolds' number is small, the viscous fluid will have laminar flow. When the Reynolds' number is large, the viscous fluid will have turbulent flow. The specific numbers vary depending on the system under consideration.

Remember, you need your fluid to have laminar flow to apply the laws and information in this chapter, so check the Reynolds' number before doing any calculations.

Chapter 11

Breaking through to the Other Side — Transport, Membranes, and Porous Material

*T*he ability of living organisms to absorb materials through membranes and transport it to cells is an important area of biophysics. This ability allows organisms to absorb materials and use the energy to continue living. Have you every wondered how your body uses food for energy to keep your body functioning? Or how fast a drug is eliminated from your body? This chapter answers these questions and more.

Here I take a closer look at diffusion, specifically focusing on the flow of fluids through other materials, such as other fluids, animals, plants, and porous materials. This chapter describes Fick's law on how fluids flow through materials, examines how the body absorbs food and metabolizes it into energy for the body, and deals with what goes in must come out and how fast it leaves.

Examining the Ins and Outs of Diffusion

Diffusion is the process of particulates, gas or fluid, to slowly spread out from regions of high concentration to regions of low concentration. Many different examples of diffusion exist in biophysics. For example, when you eat your dinner, your body absorbs the food and the blood diffuses the nutrients throughout the body, or when you add some salt to water, it dissolves and diffuses throughout the water.

These sections define the diffusion coefficient and Fick's law. They also explain osmosis, what osmotic pressure is, and where it can arise in biophysics.

Defining the diffusion coefficient

The *diffusion coefficient* is a measure of how long it would take molecules or particulates to travel a distance through a gas or fluid. The larger the diffusion coefficient means the faster the molecules will diffuse. Mathematically, the diffusion coefficient is

$$D = \frac{\Delta s^2}{2\Delta t}$$

In the formula, Δs^2 is the average of the squared displacement and Δt is the elapsed time needed for this displacement. With this information, try this problem.

You step into an empty elevator and smell the strong fragrance (odor) of perfume from a previous passenger. A high concentration of perfume molecules is still present in the air and is causing your eyes to water. The molecules didn't disappear as soon as the person left the elevator, but they linger there to cause you pain. You leave the elevator and use the stairs instead. How long should you wait before returning to the elevator?

The fan in the elevator is broken, so you have to block the elevator door in the open position and let diffusion do its job (or have victims run in and out of the elevator to stir the air). To solve this problem you can use the diffusion coefficient equation. You need to estimate or find the two variables in the equation:

To solve this problem, follow these steps:

1. **Estimate the average of the squared displacement.**

   The gas needs to diffuse 1 square foot (0.0929 square meters).

2. **Look up the diffusion coefficient for the perfume molecules in room temperature air.**

   I give you this information for this problem. The value of the diffusion coefficient for the perfume molecules in air is 2.25×10^{-5} square feet per second (2.09×10^{-6} square meters per second).

3. Rearrange the equation for the unknown (time) and solve.

$$\Delta t = \frac{\Delta s^2}{2D} = \frac{(1)^2}{2(2.25 \times 10^{-5})} = 2.22 \times 10^4 \text{ seconds}$$

You should use the stairs because 2.22×10^4 seconds = 6.2 hours is a long wait for the elevator. Actually, you don't have to wait that long because you won't be able to smell the perfume after the concentration has dropped below some threshold level, which depends on the individual.

Flowing through materials — Fick's law

Fick's (first) law states that the diffusion coefficient times the change in the concentration from one region to another region, divided by the distance between the two regions is equal to the negative of the flux. (The *flux* is equal to the rate particles passing through a cross-sectional area divided by the cross-sectional area.)

Fick's law is useful for studying the flow of particulates and molecules across membranes, such as oxygen from the lungs into the blood, the diffusion within the red blood cells, and the diffusion from the red blood cells into the muscle.

Mathematically Fick's law looks like this:

$$D \frac{\Delta C}{\Delta x} = -J_{flux}$$

In this formula, D is the diffusion coefficient, ΔC is the change in the concentration from region 1 to region 2, and Δx is the displacement from region 1 to region 2. J_{flux} is the flux across the cross-sectional area, usually a membrane. The minus sign reminds you that the molecules or particulates are flowing from the region with high concentration to a region with low concentration.

REMEMBER

Another way to consider flux: The flux times the cross-sectional area is equal to the rate at which particles flow through the area. To help you understand, think about water pouring out of a tap and you holding an empty glass. If you hold the glass sideways, then none of the water will enter the cup. If you now hold the cup upright with the lip halfway into the stream, then your cup will slowly fill. Finally, if you stick the cup right underneath the water stream, then all the water goes into the cup, and you'll fill it up rapidly. In all three cases, the water flow hasn't changed and the cup hasn't changed, but the flow of water into the cup has changed because the cross-sectional area of the water over the cup's opening has changed. In the first case the cross-sectional area is zero, in the second case it's half the area of the water stream, and in the third case the cross-sectional area equals the cross-sectional area of the water stream. Mathematically, flux is expressed as

$$J_{flux} = \frac{1}{A} \frac{\Delta N}{\Delta t}$$

J_{flux} is the flux, ΔN is the number of particles going through area A in the elapsed time Δt.

You can use the preceding information to solve this problem:

Estimate the time it takes an oxygen molecule to diffuse across the membrane from an alveolus into a capillary within your body. Also, estimate the flux of oxygen molecules from an alveolus into your blood and the rate of the number of oxygen molecules flowing from an alveolus into your blood.

To solve this problem, follow these steps:

1. **Draw a diagram.**

   It's helpful to visualize what is going on; however, I forego a diagram at this time, but go ahead and draw one.

2. **Figure out what the problem wants.**

   This step is usually the hard part. This problem wants you to find the time of diffusion, the flux across the membrane, and the rate of flow. Lucky you; these three quantities appear in the previous three formulas.

3. **Find the numbers needed to solve for the time.**

   You need some numbers to solve this problem. The problem usually supplies the numbers, but not always. The diffusion coefficient equation gives you the time, if you can find the diffusion coefficient for the membrane and the thickness of the membrane. The numbers for this problem are as follows:

   > Look up the number (in the appropriate reference book or online) and you'll find that the thickness of the membrane is 2.46×10^{-6} feet (7.50×10^{-7} meters). These membranes can be up to one-third times smaller.

   > I assume the diffusion coefficient for an oxygen molecule through the membrane is the same as diffusion in water, which is 1.08×10^{-8} square feet per second (1×10^{-9} square meters per second).

4. **Solve for the time using the diffusion coefficient equation.**

You obtain the time of diffusion by rearranging the formula for the diffusion coefficient:

$$\Delta t = \frac{\Delta s^2}{2D} = \frac{\left(2.46 \times 10^{-6} \text{ ft}\right)^2}{2\left(1 \times 10^{-8} \text{ ft}^2 \text{ s}^{-1}\right)} = 0.303 \text{ milliseconds}$$

This formula shows you in this problem that an oxygen molecule takes about a third of a millisecond to diffuse across the membrane.

5. **Find the numbers to solve for the flux across the membrane.**

The flux of oxygen molecules into the blood needs according to Fick's law: the diffusion coefficient (see step No. 3), the thickness of the membrane (see step No. 3), and the change in the concentration of oxygen molecules across the membrane. You need to find the two concentrations:

> I assume the concentration of oxygen in the blood to be zero. The average concentration of oxygen molecules in the air within the alveolus is the mass divided by the volume, which I assume to be average air with 21 percent of the air being oxygen.

> A volume of 35.31 cubic foot (1 cubic meter) of air has a mass of 0.0822 slugs (1.20 kilograms). The corresponding mass of oxygen molecules is 0.0173 slugs (0.252 kilograms). Therefore, the corresponding mass concentration is 0.0173 slugs/(35.31 ft³) = 0.000490 slugs per cubic foot (0.252 kilograms per cubic meter).

You aren't finished. Working in moles, which is the number of molecules, is more convenient to work with. Luckily, you know the molar mass of oxygen (the mass of 1 mole of oxygen), so you can change units. The molar mass of oxygen molecule is 0.00219 slugs per mole (0.032 kilograms per mole). The molar concentration is 0.000490/0.00219 = 0.223 moles per cubic foot (7.88 moles per cubic meter).

 1 mole of any substance contains 6.022×10^{23} particles. The number is called Avogadro's number. The concentration of oxygen molecules in the air can be expressed as 1.34×10^{23} molecules per cubic foot (4.74×10^{24} molecules per cubic meter) instead of in terms of moles.

6. **Solve for the flux using the Fick's first law equation.**

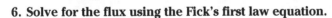

$$J_{flux} = -D\frac{\Delta C}{\Delta x} = -\left(1 \times 10^{-8} \text{ ft}^2\text{s}^{-1}\right)\left[\frac{0 \text{ mol ft}^{-3} - 0.223 \text{ mol ft}^{-3}}{2.46 \times 10^{-6} \text{ ft}}\right]$$

The flux is 9.07×10^{-4} moles per (square foot second) (0.0105 moles per (square meter second)), which corresponds to a flux of 5.46×10^{20} oxygen molecules per (square foot second) (6.32×10^{21} oxygen molecules per (square meter second)).

7. **Find the numbers needed to solve for the flow rate using the flux equation.**

   The rate that oxygen molecules are flowing into the blood from the alveolus needs the flux (see step No. 6) and the cross-sectional area. You need to find the cross-sectional area.

   I will assume the alveolus is completely covered in membrane and is approximately spherical in shape, so the surface area can be approximated as $4 \pi R^2$.

   Finally, the average radius of the alveolus is 3.3×10^{-4} feet (1.0×10^{-4} meters).

8. **Solve for the flow rate using the flux equation.**

   The rate that oxygen molecules are flowing into the blood from the alveolus is

   $$\frac{\Delta N}{\Delta t} = A\, J_{flux} = 4\pi R^2 J_{flux} = 4\pi \left(3.3 \times 10^{-4}\ \text{ft}\right)^2 \left(9.1 \times 10^{-4}\ \frac{\text{mol}}{\text{ft}^2 \text{s}}\right)$$

   The rate oxygen molecules are flowing into the blood from a typical alveolus is 1.2×10^{-9} moles per second = 7.5×10^{14} oxygen molecules per second.

The advanced forms of Fick's first and second laws are differential equations describing the motion of the molecules. The first law states that the current density of the particles is proportional to the gradient of the concentration, where the proportionality constant is the diffusion coefficient. The version given above is for the average flux. The second law is the continuity equation, which is the law of conservation of mass. You can't create or destroy mass. The mathematical form of Fick's first law is

Fick's 1st law: $D \dfrac{dC}{dx} = -J_{flux}$

Restricting what passes through the barrier — osmosis and osmosis pressure

Osmosis is the movement of solvent (usually water) from a *hypotonic solution* (a solution with a lower solute concentration) to a *hypertonic solution* (a solution with a higher solute concentration) across a semipermeable membrane. The membrane is permeable to the solvent and is impermeable to the solute. Osmosis and the osmotic pressure are very important concepts in biological systems involving the membranes within animals and plants. For example, the walls of cells are membranes that are permeable to small molecules such as water, oxygen, and carbon dioxide while being impermeable to large molecules and ions.

As the solvent flows from the hypotonic solution, it increases the solute's concentration. As the solvent flows into the hypertonic solution, it decreases the solute's concentration. The solvent will attempt to continue flowing across the membrane until the concentrations are balanced or it's countered with an opposing force.

The pressure built up in the hypertonic solution that stops the motion of the solvent across the membrane and maintains equilibrium is called the *osmotic pressure*. In the case of no solute in the hypotonic solution and a very *dilute concentration* of solute in the hypertonic solution, the mathematical formula for the *osmotic pressure* is

$$\Pi = CRT$$

In this formula, C is the concentration of the solute in the hypertonic solution, R equals 6.1328 foot pound per (kelvin mole) (8.3145 joules per (kelvin mole)) and is the gas constant, and T is the temperature in kelvin.

Use this information from this section to solve this problem:

Consider the albumin protein group in the blood plasma. What is the osmotic pressure with a mass concentration of 8.7×10^{-8} slugs per cubic foot (4.5×10^{-5} kilograms per cubic meter)?

To solve this problem, follow along in these steps:

1. **Determine the equation needed to calculate the osmotic pressure.**

   The problem doesn't specify, but you can assume no solute (albumin protein) in the hypotonic solution and a very *dilute concentration* of solute in the hypertonic solution (blood plasma). You can then use the mathematical formula given to find the osmotic pressure.

2. **Find the respective numbers.**

   The formula requires three pieces of information: C, R, and T: R is a constant and you know the value of it. R equals 6.1328 foot-pound per (kelvin mole) (8.3145 joules per (kelvin mole))

   The body's temperature (T) is 310 kelvin (98.6 degrees Fahrenheit = 37 degrees Celsius).

   You are given the mass concentration, 8.7×10^{-8} slugs per cubic foot (4.5×10^{-5} kilograms per cubic meter). Unfortunately, the formula needs the molar concentration.

   To convert units you need the molar mass. You can find the molar mass by either looking it up in a reference source, or by looking up the chemical composition of the albumin protein and then calculating the molar mass. I will help you with this. The average molar mass for the albumin protein group is 4.7 slugs per mole (69 kilograms per mole). Therefore, the molar concentration is $8.7 \times 10^{-8}/(4.7) = 1.85 \times 10^{-8}$ moles per cubic foot $(4.5 \times 10^{-5}/(69) = 6.52 \times 10^{-7}$ moles per cubic meter).

3. **Solve for the osmotic pressure.**

You now know everything and can calculate the osmotic pressure.

$$\Pi = \left(1.85 \times 10^{-8} \frac{\text{mol}}{\text{ft}^3}\right)\left(6.1328 \frac{\text{ft lb}}{\text{mol K}}\right)(310 \text{ K}) =$$

$$3.51 \times 10^{-5} \frac{\text{lb}}{\text{ft}^2}\left(1.68 \times 10^{-3} \text{pascals}\right)$$

Understanding Human Metabolism

Metabolism refers to the chemical reactions that occur within living organisms. It's the study of energy and work within living organisms. Living organisms need energy to live, grow, change, and do work.

For example, you eat some food and the body digests the food by breaking it down into its constituents. The blood absorbs them, and some of the chemicals travel to the liver. Enzymes in the liver act as catalysts, changing the chemical compound into a new chemical compound, which then is transported to a different part of the body before entering a cell. It then has a chemical reaction within the cell. The path the chemical takes through the body is referred to as a *metabolic pathway*. Each step along this pathway is a *metabolic process*.

These sections look at metabolism in the human body on the large scale, including energy consumption and the efficiency of the body to do mechanical work.

Eating — balancing your energy

Energy is what makes things go. The most important law in biophysics is *conservation of total energy,* which states you can't create or destroy energy, only change it from one form to another. The first law of thermodynamics is another way of stating the conservation of energy law.

The *first law of thermodynamics* states that the change in the internal energy of a biological organism is equal to the change in the heat energy minus the work done by the organism on its surroundings. Mathematically it looks like this:

$$\Delta E_{\text{int}} = \Delta Q - \Delta W_{done}$$

Humans are constantly losing heat to their surrounding and doing work. (Breathing is important after all.) The consumption of food and the intake of oxygen replenish the energy so there is an overall balance and conservation of energy. This process never stops for living organisms, so it's usual to consider the power instead. Recall power is equal to the work divided by the elapsed time.

Many different sets of units are used in analyzing the energy and power consumption of organisms. The relationship between these different set of units are

1 (food) Calorie (Cal) = 1 kilocalorie (kcal) = 1,000 (energy) calories (cal) = 4,184.5 joules (J) = 3,086 foot pound (ft lb) = 3.970 British thermal unit (BTU).

1 kilocalorie per day (kcal/d) = 4,184.5 joules per day (J/d) = 0.04843 watts (W) = 0.03572 foot pound per second (ft lb/s) = 6.495×10^{-5} horsepower (hp).

Basal metabolic rate (BMR) is the bare minimum power required to keep the body functioning while awake and resting (no stimuli — watching paint dry is okay). This energy is needed to keep the body at 98.6 degrees Fahrenheit (37 degrees Celsius) and keep all the vital organs operating. The most accurate method of determining your BMR is by measuring how much oxygen your body consumes. The metabolism of the food requires oxygen; the human body will produce 1 kilocalorie of energy from food with every 7.31×10^{-3} cubic feet (0.207 liters = 2.07×10^{-4} cubic meters) of oxygen absorbed. You can see how problems may arise if there is insufficient oxygen present to produce the energy needed for your body to function properly, such as can happen at the top of tall mountains.

You can obtain a quick estimate of your BMR by using the Harris-Benedict equation:

$$\text{male: } P = \frac{\Delta E_{int}}{\Delta t} = \left[\frac{6.074M}{1 \text{ pound}} + \frac{12.19H}{1 \text{ inch}} - \frac{5.677A}{1 \text{ year}} + 88 \right] \frac{\text{kilocalorie}}{\text{day}}$$

$$= \left[\frac{56.07M}{1 \text{ kilogram}} + \frac{2008H}{1 \text{ meter}} - \frac{23.76A}{1 \text{ year}} + 370 \right] \frac{\text{kilojoules}}{\text{day}}$$

$$\text{female: } P = \frac{\Delta E_{int}}{\Delta t} = \left[\frac{4.192M}{1 \text{ pound}} + \frac{7.869H}{1 \text{ inch}} - \frac{4.330A}{1 \text{ year}} + 448 \right] \frac{\text{kilocalorie}}{\text{day}}$$

$$= \left[\frac{38.70M}{1 \text{ kilogram}} + \frac{1296H}{1 \text{ meter}} - \frac{18.12A}{1 \text{ year}} + 1873 \right] \frac{\text{kilojoules}}{\text{day}}$$

The BMR is slightly different for males and females, which explains the two formulas. In addition, I have written the equation for each gender in two lines: the first line is in United States customary units and the second line is in SI units. In the equations: *M* is the person's weight (or mass), *H* is the person's height in inches (or meters), and *A* is the person's age in years. At the end of each line are the units for your final answer.

Al is lying on the couch watching the Saturday morning cartoons in a vegetative state. What is his BMR assuming Al is 5 foot, 8 inches (68 inches = 1.73 meters), 180 pounds (81.6 kilogram mass), and 21 years old?

I have already converted Al's numbers to the correct units within the brackets, so you can substitute these numbers into the previous equation to find Al's BMR like this:

$$P = \frac{\Delta E_{int}}{\Delta t} = \left[\frac{6.074(180\ lb)}{1\ lb} + \frac{12.19(68\ in)}{1\ in} - \frac{5.677(21y)}{1\ y} + 88 \right] \frac{kcal}{d}$$

$$= \left[\frac{56.07(81.6\ kg)}{1\ kg} + \frac{2008(1.73\ m)}{1\ m} - \frac{23.76(21y)}{1\ y} + 370 \right] \frac{kJ}{d}$$

Al is using 1,890 kilocalories per day (7,920 kilojoules per day = 91.6 watts) of power while watching the Saturday morning cartoons.

These equations are averages; every person will have a slightly different BMR and different metabolic rates during the same activity. For example, you can take two males (or females) with the same weight, height, age, and physical fitness, and they'll have different basal metabolic rates.

Al decides to go for a one-hour jog. In addition to his BMR, Al is using 93.6 kilocalories per (day-pound) (which equals 10 watts per kilogram) of power during his jog. Remember Al is 5 foot, 8 inches (which equals 68 inches or 1.73 meters), 180 pounds (81.6 kilogram mass), and 21 years old. What is his energy consumption during this hour? How much carbohydrates or protein or fat does Al need to eat to replenish his body's energy reserves?

To solve this problem, follow these steps:

1. **Set up the problem and decide which information you need to solve the problem.**

   The Harris-Benedict equation was already used to find Al's BMR. You know that power output is additive. You know the relationship between power, work (energy), and time, which allows you to find the energy consumption. Finally, you need to find out how much energy Al's body can obtain from different foods.

2. **Find Al's power output during the jog.**

   - Al's BMR is 1,890 kilocalories per day (7,920 kilojoules per day = 91.6 watts).

   - Al's excess power use per pound (kilogram) is 93.6 kilocalories per (day-pound) (10 watts per kilogram). You need to multiply this number by Al's weight (or mass) to find his total power output: $93.6 \times 180 = 16,850$ kilocalories per day ($10 \times 81.6 = 816$ watts).

- Take the sum of these two to find his total power output: 18,740 kilocalories per day (908 watts).

3. **Find Al's energy requirements (output) during the jog.**

You now know Al's total power output and the elapsed time (1 hour = 3600 seconds = 1/24 days). The energy output is 18,740 divided by 24 = 781 kilocalories ($908 \times 3,600 = 3,270$ kilojoules).

4. **Calculate Al's energy intake.**

- Carbohydrates, on average, provide 1,850 kilocalories per pound (17,100 kilojoules per kilogram) of energy. Al would need to absorb 781 divided by 1850, which equals 0.422 pounds (0.191 kilograms) of maple syrup to replace his energy.

- Proteins, on average, provide 1,900 kilocalories per pound (17,500 kilojoules per kilogram) of energy. Al would need to absorb 781 divided by 1,900, which equals 0.411 pounds (0.186 kilograms) of peanut butter.

- Fats, on average, provide 4,200 kilocalories per pound (38,750 kilojoules per kilogram) of energy. Al would need to absorb 781 divided by 4,200, which equals 0.186 pounds (0.0843 kilograms) of fast food.

Searching for efficiency of food energy

Determining the amount of food you should eat so the power in equals the power out is difficult and varies from one person to another. However, consider these general properties when determining the efficiency of the food:

✔ Consider the ability of the body to absorb the food. For example, if you take vitamins or other supplements, then the amount absorbed by your body will depend on the pills' casing. The chalky pills dissolve very rapidly, and the body absorbs the majority of the drug, whereas the easy swallowing plastic pills pass through the body quickly before the body absorbs the drug. Usually, most foods aren't as extreme, but the amount of nutrients absorbed will depend on the individual person's body.

✔ The food is broken down into its nutrients, which are absorbed, and some of the nutrients need enzymes to metabolize. The amount of enzyme changes from person to person and from one ethnic group of people to another. For example, lactase is the enzyme that metabolizes lactose into glucose and galactose. In some regions of Africa, almost the entire adult population has a deficiency of lactase and they're lactose intolerant, whereas in northern Europe the opposite situation is true, allowing the northern European adults the ability to consume milk products.

The nutrients, such as sugar interact with oxygen within the cells to produce materials such as adenosine-5'-triphosphate (ATP). The energy involved in this process is split almost in two with about half going into heat and the other half stored in the ATP. The ATP is the workhorse of biological systems. Whenever energy needs to be transported within a cell or work needs to be done, the ATP does the job.

✔ Only 50 percent of the energy is now available for doing work. When doing more than lying in a vegetative state, the body must do mechanical work, but it also must do internal work, such as maintaining correct posture, breathing harder, making faces, biting your tongue, and so on.

You can quantify the *efficiency* of doing mechanical work with this formula:

$$\eta = \frac{100 \, P_{work}}{P_{int} - P_{BMR}}$$

In this formula, P_{work} is the power needed to do the mechanical work, P_{int} is the total power output while doing the mechanical work, and P_{BMR} is the basal metabolic rate.

You want to estimate Al's efficiency during his one-hour jog. You'll need some numbers in order to calculate the efficiency:

I made some measurements of Al, and found that Al is 5 foot, 8 inches (68 inches = 1.73 meters), 180 pounds (81.6 kilogram mass), and 21 years old.

I was able to calculate Al's BMR to be 1,890 kilocalories per day (7,920 kilojoules per day = 91.6 watts). (See the previous section.)

I measured Al used an extra 93.6 kilocalorie per (day-pound) (which equals 10 watts per kilogram) of power during his jog. This means Al's total power output during his 1-hour jog is 18,740 kilocalories per day (908 watts). (See the previous section.)

Al jogs at an average speed of 5 miles per hour (8.05 kilometers per hour, which is 7.33 feet per second or 2.23 meters per second).

To calculate Al's mechanical power output: I will assume the coefficient of static friction between Al's shoes and the ground is 0.8 and he is using only 10% of his maximum static friction while he is jogging. In addition, I noticed that Al's center of mass remains approximately horizontal while he is jogging so the work done against gravity zero. Therefore, the mechanical power output is

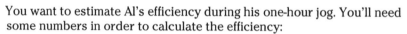

$$P_{work} = \frac{\Delta W}{\Delta t} = \frac{0.1 \mu_s mg \, \Delta x}{\Delta t} = 0.1 \mu_s mg v_{avg} = 0.1(0.8)(180 \text{ lb})(7.33 \text{ ft s}^{-1})$$

The mechanical power output by Al is P_{work} = 106 foot pound per second = 143 watts = 2,950 kilocalories per day.

Solve the problem following these steps:

1. **Determine the equation needed and understand the problem to calculate Al's efficiency.**

   I give the formula you need to calculate the efficiency just before the problem. To calculate the efficiency you need to find — Al's BMR — (which I have given in the problem), Al's total power output (which I have given in the problem), and the mechanical power output (which I have also given in the problem). Therefore, you know everything to find Al's efficiency.

2. **List the numbers given.**

   Making a list eases finding the numbers when you need them:

   Al's BMR is P_{BMR} = 1,890 kilocalories per day (91.6 watts).

   Al's total power output is P_{int} = 18,740 kilocalories per day (908 watts).

   Al's mechanical power output is P_{work} = 106 foot pound per second = 143 watts = 2,950 kilocalories per day.

3. **Calculate the efficiency.**

   You know all three powers, so you can calculate the efficiency by substituting the numbers into the earlier formula. Note you have to be careful you use the powers with the same set of units. I give you all three powers in kilocalories per day and in watts. Feel free to use either set in your calculation. For fun, I convert the first two powers to units of foot-pound per second. Al's efficiency of converting energy into mechanical work while jogging is as follows:

$$\eta = \frac{100 P_{work}}{P_{int} - P_{BMR}} = \frac{100\left(106 \text{ ft lb s}^{-1}\right)}{669 \text{ ft lb s}^{-1} - 67.5 \text{ ft lb s}^{-1}} = 17.6\%$$

Eliminating Product from the Body

The body is constantly recycling — in with the new and out with the old. For example, on average the human body has about 0.5 pounds (0.23 kilograms) of ATP at any given moment, but during a 24-hour period, the body will use its weight in ATP. These sections deal with *enzyme kinetics,* which is the study of the reaction rate of an enzyme in metabolizing a material (a chemical or drug). Here I introduce the concept of the decay constant and half-life and look at when the dosages are large and saturation occurs, which means the body can't eliminate the compound fast enough and drug overdoses can occur.

Keeping doses low — classical kinetics

The simplest model used for modeling the elimination of material from the body is known as classical kinetics. The model assumes that after a given amount of time, called the *half-time,* half the material is eliminated. This half-time is assumed to be a constant.

A *random decay process* can be modeled by the two formulae (which are equivalent):

$$N_{final} = N_{initial}\exp\left[-\lambda\,\Delta t\right] \text{ or } \ln\left[\frac{N_{final}}{N_{initial}}\right] = -\lambda\,\Delta t$$

In the two equations $N_{initial}$ is the initial number of molecules (particles) within the body, and after an elapsed time Δt, the body has N_{final} molecules (particles) left. λ is called the *decay constant,* which tells you how fast the molecules (particles) are disappearing from the body. The two equations are the same equation, but written slightly differently. The equation on the left is convenient for finding $N_{initial}$ or N_{final}; whereas the equation on the right is convenient for λ or Δt.

Meanwhile a *half-life* is how long you have to wait before half the molecules (particles) have disappeared. It's related to the decay constant as this formula shows:

$$T_{1/2} = \frac{\ln(2)}{\lambda}$$

Use the preceding material in this section to solve this problem.

If you don't eat or drink, then half of the carbon in your body will be gone after 35 days. Suppose you fast with no liquids for 24 hours, what percentage of carbon will your body lose?

1. **Determine the equation and understand the problem so you can find the ratio of N_{final} divided by $N_{initial}$.**

   The percentage lost is another way of saying the problem wants you to calculate the ration of N_{final} divided by $N_{initial}$. This means you need to use the left formula, and you have to find the numbers for the decay constant and the elapsed time.

2. **Find the numbers for the decay constant and the elapsed time.**

   The problem states half the carbon in your body will be gone in 35 days, this is the half-life: $T_{1/2}$ = 35 days

   The problem states you are fasting for 1 day, so the elapsed time is 1 day.

3. **Calculate the decay constant from the half-life.**

   The decay constant can be calculated from the half-life: $\lambda = \ln(2)/T_{1/2} = $ 0.0198 per day

4. **Calculate the ratio of the numbers from the decay process formula and the percentage lost.**

   You know enough to use the random decay process formula to calculate the ratio of N_{final} divided by $N_{initial}$ and hence find the percentage lost of carbon in your body:

   $$\frac{N_{final}}{N_{initial}} = \exp\left[-\lambda \Delta t\right] = \exp\left[-(0.0198/d)1d\right] = 0.980$$

   $$100\left[1 - \frac{N_{final}}{N_{initial}}\right] = 100[1 - 0.980] = 1.96\%$$

   Your body will lose 1.96% of its carbon after one day.

You can apply this decay process to the enzyme kinetics. The rate at which a type of molecule is being metabolized by enzymes is:

$$\frac{1}{n_p}\frac{\Delta C_p}{\Delta t} = -\frac{\Delta C_s}{\Delta t} = \lambda_s C_s$$

In this formula, C_s is the concentration of the molecules being metabolized by the enzymes. A molecule bound to an enzyme is called a *substrate*, hence the subscript s. λ_s is the decay constant of the molecules being metabolized by the enzyme. ΔC_s divided by the Δt is the rate at which the concentration of the substrate is being changed by the enzyme. The negative sign indicates the substrate is being eliminated by the body. The term on the left-hand side is the rate at which the product is being produced by the metabolizing of the substrate. The n_p is present because there isn't necessarily a one-to-one correspondence between the amount of product produced and the amount of substrate being eliminated. For example, if you take a single glucose molecule and metabolize it, the molecule can produce up to forty ATPs. (It takes two ATPs to start the process so the maximum yield is 38 ATPs.)

If the substrate's decay constant doesn't change in time, then this equation takes on the following form (using calculus):

$$C_{s,final} = C_{s,initial} \exp\left[-\lambda_s \Delta t\right]$$

$$C_{p,final} = n_p C_{s,initial}\left[1 - \exp\left[-\lambda_s \Delta t\right]\right]$$

Remember $\exp[x] = e^x = (2.718282...)^x$.

Indulging too much — Michaelis-Menten kinetics

One problem with the previous set of equations for classical kinetics is the assumption that your body has an unlimited supply of enzymes to metabolize the molecules. The larger the concentration of molecules means the faster the rate at which they're metabolized. The *Michaelis-Menten kinetics* is a model that takes into account the limited amount of enzymes and enforces a finite maximum rate at which the molecules are metabolized, which the following formula demonstrates:

$$\frac{1}{n_p}\frac{\Delta C_p}{\Delta t} = -\frac{\Delta C_s}{\Delta t} = \frac{v_{max}C_s}{K_m + C_s} \rightarrow \begin{cases} \dfrac{v_{max}C_s}{K_m}, & C_s \ll K_m \\[2mm] \dfrac{v_{max}}{2}, & C_s = K_m \\[2mm] v_{max}, & C_s \gg K_m \end{cases}$$

The formula is left of the arrow. The right side of the arrow shows what happens to the right side of the formula in special limiting cases. In this formula, v_{max} is the maximum rate at which the enzyme can metabolize the molecules and K_m is the substrate concentration at which the rate of metabolization is half the maximum value.

The last part of the formula after the arrow shows what happens to the Michaelis-Menten equation in limiting cases. In the case of very low concentrations, the metabolic rate looks the same as in the previous section, but with two new parameters: v_{max} and K_m instead of a single decay constant. In the case of very large concentration the metabolic rate in the Michaelis-Menten equation is a constant as shown by the bottom expression in the bracket. Figure 11-1 shows the concentration rate of the product as a function of the concentration of the substrate for the Michaelis-Menten model (solid line) and compares it to the classical kinetics result (dashed line).

Figure 11-1:
The rate of production of the product as a function of the concentration of the substrate.

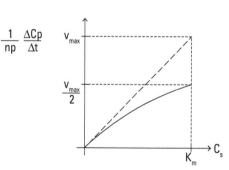

In this figure, the dashed line is the classical model from the previous subsection and the solid line is the Michaelis-Menten model. If v_{max} and K_m are constants, then you can write a relationship between the concentration and time as:

$$\frac{C_{s,final}}{C_{s,initial}} \exp\left[\frac{C_{s,final} - C_{s,initial}}{K_m}\right] = \exp\left[\frac{-v_{max}\Delta t}{K_m}\right]$$

$$C_{p,final} = n_p\left[C_{s,initial} - C_{s,final}\right]$$

The elimination of alcohol that has been consumed obeys the *Michaelis-Menten kinetics* rather than the classical kinetics in the previous section. Approximately 5 percent of the alcohol consumed will be excreted through urine, feces, breath, and sweat, and the rest is eliminated through hepatic metabolism by the alcohol dehydrogenase enzymes. These enzymes reach saturation very rapidly even by normal alcohol consumption, so the rate of conversion of the ethanol into acetaldehyde, which is then metabolized into acetic acid by the acetaldehyde dehydrogenase enzyme, occurs at a constant rate of v_{max}. Unfortunately, the value of v_{max} varies vastly amongst humans, even when you take into consideration the person's ethnic background, gender, and age.

Use the preceding information in this section to solve this problem.

Barney went to the bar with some friends and had a few too many drinks. Barney has his own kit and tested his blood-alcohol level before driving. The test stated his blood-alcohol level was 0.12 percent (by volume). How long does Barney have to wait for his blood level concentration to drop to 0.05 percent (by volume)? Luckily, Barney has recently conducted some tests and determined the parameters describing his body's ability to metabolize alcohol are: v_{max} = 3.5 millimoles per (liter hour) and K_m = 2.0 millimoles per liter.

To figure out this problem, follow these steps:

1. **Determine what equation you need and understanding the problem.**

    You know the amount of alcohol in a person's body obeys the Michaelis-Menten kinetics, so use this formula for this problem. You're asked to find the time given the initial and final concentrations of the alcohol in the blood. You're also given v_{max} and K_m.

2. **Find the numbers for the concentrations and the parameters.**

    You're told the concentrations in a standard form of measurement, but that information doesn't help, so the units need to be changed. I can help you change the units:

- 0.12 percent (by volume) = 26.1 millimoles per liter = 0.428 millimoles per cubic inch

- 0.05 percent (by volume) = 10.9 millimoles per liter = 0.179 millimoles per cubic inch

- Barney said his body's maximum rate at which the enzyme can metabolize the alcohol is v_{max} = 3.5 millimoles per (liter hour)

- Also, Barney said his body's other parameter is K_m = 2.0 millimoles per liter.

3. **Substitute the numbers into the formula and solve.**

I help by rearranging the Michaelis-Menten kinetics equation using the properties of logarithms and isolate for time:

$$\Delta t = \frac{K_m \ln\left[\frac{C_{s,f}}{C_{s,i}}\right] + C_{s,f} - C_{s,i}}{-v_{max}} = \frac{(2 \text{ mmol/L}) \ln\left[\frac{10.9}{26.1}\right] + 10.9 \text{ mmol/L} - 26.1 \text{ mmol/L}}{-3.5 \text{ mmol}/(\text{L h})}$$

Barney's blood-alcohol level will be down to a legal level in 4.84 hours. Barney should call a taxi.

Part IV
Playing the Music Too Loud — Sound and Waves

Important formulas to remember in this part

When working with sound and waves in biophysics, these formulas are a few that play an important role. Master them to help you solve different problems you may encounter:

✔ Diffusion coefficient: $D = \dfrac{\Delta s^2}{2\Delta t}$

✔ Fick's law: $D\dfrac{\Delta C}{\Delta x} = -J_{flux}$

✔ Flux: $J_{flux} = \dfrac{1}{A}\dfrac{\Delta N}{\Delta t}$

✔ An object's position in harmonic motion: $x = x_{eq} + A\sin\left(2\pi f t + \delta\right)$

✔ An object's velocity in harmonic motion: $v = 2\pi f A\cos\left(2\pi f t + \delta\right)$

✔ An object's acceleration in harmonic motion: $\dfrac{F_{NET}}{m} = a = -\left(2\pi f\right)^2 A\sin\left(2\pi f t + \delta\right)$

Visit www.dummies.com/extras/biophysics for more informative Dummies content online about biophysics.

In this part . . .

✔ Discover what harmonic motion is and the physical properties of harmonic oscillators and why they're important in biophysics.

✔ Grasp how sound waves are produced and what the different physical properties of sound waves are and how they combine to form interesting sounds.

✔ Know how to identify the different types of waves and all their interesting properties and see how waves interact with other waves and with objects, both living and inanimate.

✔ Explore the physics of musical instruments, including how they produce specific notes that are pleasant to hear instead of a collection of tuning forks.

✔ See how the Doppler Effect works when the source of the wave and the listener are different and understand how nocturnal animals use the Doppler Effect.

✔ Examine the applications of sound, such as the different ways that animals use echolocation or how ultrasound imaging works.

✔ Delve into the importance of your ears and uncover the functions and properties of the inner ear, middle ear, and outer ear.

Chapter 12

Examining the Physics of Waves and Sound

. . . .

In This Chapter

▶ Tackling harmonic motion

▶ Describing waves and their properties

▶ Bouncing waves off walls

. . . .

*W*aves and harmonic motion (such as a swinging pendulum) play an important role in biophysics. Waves and particles are how objects interact with biological systems. For example, suppose Charles wants to torture you; he can either throw water balloons (particles) at you or he can sing (waves). Many aspects of biophysics depend on the properties of waves. You need to understand waves before you can understand hearing, sound, ultrasound imaging, or echolocation.

So what is a wave? A *wave* is a physical disturbance that transfers energy through space, whereas *harmonic motion* is a motion that repeats itself over some finite period. The definition for harmonic motion seems easy to grasp, but the wave definition doesn't seem that helpful; however, it will make sense if you read on.

This chapter explains what waves and harmonic motion are and some of their properties. This chapter also looks at what happens to a wave when it hits a boundary. Understanding this information can help you grasp how sound travels through walls (and keeps your neighbors awake), how light propagates through windows, and how light bounces off a mirror.

Comprehending Harmonic Motion

Harmonic motion is a repeating motion, such as a beating heart or the first few bounces after bungee jumping off a bridge and then the swinging motion afterwards. Harmonic motion is also the swinging motion of gymnasts on the high bar or the swing of the arm or leg while walking. These examples are just a few of the situations where harmonic motion plays a role. Many areas of biophysics use harmonic motion, so this information not only can help in your biophysics course, but also in your future endeavors. The following sections look at how you can describe repeating motions and apply harmonic motion in biophysics.

Explaining harmonic motion in action

The easiest way to describe harmonic motion is to realize that an object in harmonic motion has an equilibrium position (x_{eq}) that the object oscillates around. In some situations, knowing the position of an object undergoing harmonic motion is important. For example, two trapeze artists need to know the position of each other when doing stunts. The mathematical description of the object's *position* is as follows:

$$x = x_{eq} + A \sin\left(2\pi f t + \delta\right)$$

In this formula, x is the position of the object at time t, x_{eq} is the equilibrium position of the object if it weren't bouncing back and forth, A is the amplitude of the oscillations (how far the object will move from the equilibrium position), f is the frequency of the oscillation (refer to the next section for more on frequency), and δ is the phase shift. The most common phase shifts are $\delta = 0$ radians and $\delta = \pi/2$ radians (90 degrees).

Meanwhile the *velocity* is a measure of how fast the object's position is changing. In the case of harmonic motion, the velocity of the object is represented by this formula:

$$v = 2\pi f A \cos\left(2\pi f t + \delta\right)$$

In this formula, v is the velocity of the object at time t, and the rest are the same as in the expression for position. Notice that the maximum speed of the object is $v_{max} = 2\pi f A$ and it occurs when the object is at its equilibrium position.

Similarly, the acceleration is a measure of how fast the object's velocity is changing. According to Newton's second law of motion, the law of acceleration (refer to Chapter 6 for more information about this law), the acceleration equals the net force divided by the mass. In the case of harmonic motion, this formula represents the *acceleration* of the object:

$$\frac{F_{NET}}{m} = a = -\left(2\pi f\right)^2 A \sin\left(2\pi f t + \delta\right)$$

Here F_{NET} is the net force acting on the object, m is the mass of the object, a is the acceleration of the object at time t, and the rest are the same as in the expression for the position and velocity. Notice that the maximum acceleration of the object is $a_{max} = 2 \pi f v_{max} = (2 \pi f)^2 A$ and it occurs when the object is farthest from its equilibrium position. This makes sense that the farther from equilibrium the greater the force trying to bring the object back to equilibrium.

 REMEMBER

The force, acceleration, velocity, and position are all vectors and should have arrows above them in these three equations. (Refer to Chapter 3 for more information about vectors.) Hence the amplitude, A, is also a vector. I've written the physical quantities as scalars because in most biophysical harmonic motion situations, they're either one-dimensional motion or circular motion. In addition, circular motion can be consider one-dimensional motion if studied in terms of the angle.

Weighing a virus: Applying Hooke's law and harmonic motion

Suppose you have a spring that obeys Hooke's law with a weight on the end. *Hooke's law* states that the force is proportional to the displacement from equilibrium. In other words, if you pull the weight away from equilibrium and let go, then the weight will bounce back and forth around the equilibrium with harmonic motion. Mathematically, *Hooke's law* is as follows:

$$F_{Hooke} = -k_H\left(x - x_{eq}\right) = -k_H A \sin\left(2\pi f t + \delta\right)$$

F_{Hooke} is the force acting on the harmonic oscillator, k_H is Hooke's constant, x is the position of the mass at time t, x_{eq} is the equilibrium position of the mass where there is no force acting on the mass, A is the amplitude of the oscillations in the position, f is the frequency, and δ is the phase shift.

The formula for Hooke's law can be compared to the force equation for harmonic motion in the previous section. The comparison shows that f (the frequency) is fixed to a specific value: $(2\pi f)^2 = k_H/m$ by setting $F_{net} = F_{Hooke}$.

The time it takes the weight to make one full oscillation is called the *period* (*T*), and it's related to the frequency (*f*):

$$T = \frac{1}{f} = 2\pi\sqrt{\frac{m}{k_H}}$$

You can use Hooke's law to weigh a virus. In 2012, scientists were able to weigh a single hemoglobin cell for the first time. They used a device called a nanoelectromechanical resonator, which was capable of measuring the weight of nanoparticles and human antibody molecules. This resonator used the same principles as strumming a guitar string to produce a specific note and then pressing down on the string to change the note played. (Chapter 13 takes a closer look at the acoustic guitar.) Here I discuss a different technique used to measure the weight of a virus.

A *cantilever* vibrates with a certain frequency. If you place a mass on the end of the cantilever, it vibrates with a different frequency. By measuring the change in the frequency, you can calculate how much mass was added to the cantilever. Because a virus is so small, you need a cantilever that is also small.

To help you visualize, picture a swimming pool and a diving board. The board bounces at a different frequency when no one is standing on it compared to when a diver is standing on the end of the diving board. The diving board is the cantilever and the diver the virus.

You're given a virus and asked to calculate the mass and weight of the virus. In addition, you're told the weight of the cantilever is 4.5×10^{-19} pounds (2.00×10^{-19} newtons) or the mass is 1.40×10^{-20} slugs (2.04×10^{-19} kilograms). You go to the lab and perform the experiment to collect the data: You start the cantilever oscillating and measure a frequency of 2.0×10^{15} Hertz without the virus. You then stop the cantilever and place the virus on it and then start it vibrating again. This time you measure the frequency of the oscillations to be 3.0×10^{14} Hertz.

To solve this problem, stick to these steps:

1. **Understand the physics in the problem.**

   Think of the cantilever as a spring tied to the ceiling with a mass attached to the bottom of the spring. The spring is allowed to bounce up and down. This is a harmonic oscillator, and the relationship between the frequency, mass, and Hooke's constant is given in the last formula. You need to use this formula twice: once with the mass being just the mass of the cantilever and once with the mass equal to the mass of the cantilever plus the mass of the virus.

   In addition, Hooke's constant doesn't change when the virus is added to the cantilever.

2. **Find the numbers before the virus is placed on the cantilever.**

   The formula has three variables: k_H, m, and f.

   You're given the weight of the cantilever as 4.5×10^{-19} pounds (2.00×10^{-19} newtons) and the mass of 1.40×10^{-20} slugs (2.04×10^{-19} kg).

   The frequency of oscillation without the virus is 2.0×10^{15} Hertz.

3. **Solve for Hooke's constant.**

   $$k_H = 4\pi^2 f_{without}^2 m_{cantilever} = 4\pi^2 \left(2.0 \times 10^{15}\, s^{-1}\right)^2 \left(1.40 \times 10^{-20}\, \text{slugs}\right) =$$

   $$2.21 \times 10^{12}\ \frac{\text{slugs}}{\text{second}^2} \left(3.22 \times 10^{13}\ \frac{\text{kg}}{\text{second}^2}\right)$$

4. **Find the numbers after the virus is placed on the cantilever.**

   You just found Hooke's constant to be 2.21×10^{12} slugs per second squared (3.22×10^{13} kilograms per second squared).

   You measured the frequency of oscillation with the virus to be 3.0×10^{14} Hertz.

5. **Solve for the weight of the virus.**

   Rearrange the formula for Hooke's constant like this:

   $$m_{virus} = \frac{k_H}{4\pi^2 f_{with}^2} - m_{cantilever} = \frac{2.21 \times 10^{12}\, \text{slugs}\, s^{-2}}{4\pi^2 \left(3.0 \times 10^{14}\, s^{-1}\right)^2} - 1.40 \times 10^{-20}\, \text{slugs} =$$

   $$6.08 \times 10^{-19}\, \text{slugs} \left(8.86 \times 10^{-18}\, \text{kg}\right)$$

   To find the corresponding weight, remember that weight equals mass times g, where g = 32.2 feet per second squared (9.81 meters per second squared) is the acceleration due to gravity constant. The weight of the virus is 1.96×10^{-17} pounds (8.69×10^{-17} newtons).

Swinging in a swing: Applying gravity and harmonic motion

Another type of harmonic motion is the pendulum or swinging motion, such as your arms and legs swinging back and forth relative to the body as you walk. Although the swinging motion is a two-dimensional motion, you can describe the motion using the one-dimensional equations from the previous section. The trick is to describe the motion by using the angular variables. You can do so by switching the linear variables to angular variables in the formulas for harmonic motion. The formulas for harmonic motion become:

✔ The angular position for swinging motion is
$$\theta = \theta_{eq} + A \sin\left(2\pi f t + \delta\right)$$

✔ The angular velocity for swinging motion is
$$\omega = 2\pi f A \cos\left(2\pi f t + \delta\right)$$

✔ The angular acceleration for swinging motion is
$$\alpha = -\left(2\pi f\right)^2 A \sin\left(2\pi f t + \delta\right)$$

The rest of this section focuses on swings and pendulums that are moving under the influence of gravity. Normally, the gravitational force is a constant in most biophysical applications. (An astronaut going into space is an exception.) However, if an object is doing harmonic motion, then the net force needs to look like the force in Hooke's law. Therefore, if an object is swinging and it looks like harmonic motion, there must be more forces than gravity acting on it. Figure 12-1 shows a free-body diagram of an object swinging, assuming conservation of energy (no dissipative forces). This figure shows there are actually two forces acting on the person and the swing: the tension in the rod and the force of gravity. I assume the maximum angle made relative to the vertical direction is small.

Figure 12-1:
The free-
body
diagram
for a swing
under the
influence of
gravity.

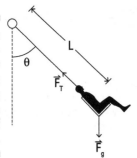

In order to solve this, use Newton's second law, the law of acceleration, which states that the tension in the rod will hold the object in a circular path around the center, and the tangential acceleration is equal to the tangential component of the net force divided by the mass. (Refer to Chapter 4 for more information about this law.) Applying Newton's second law to Figure 12-1, I obtain the first equation in the following formula for the angular acceleration. (Note that if the angle, θ, is small then $\sin(\theta)$ is approximately equal to θ in radian units not degrees.) The second equation is Hooke's law from the previous section converted to angular variables.

$$\text{tangential: } \alpha = \frac{a_T}{L} = \frac{F_{\tan}}{Lm} = \frac{-\left|\vec{F}_g\right|\sin(\theta)}{Lm} \approx \frac{-g\theta}{L}$$

$$\text{Hooke's law: } \alpha = \frac{N_{Hooke}}{I} = -\frac{k_H}{I}\left(\theta - \theta_{eq}\right)$$

Here α is the angular acceleration, a_T is the tangential component of the acceleration (see Chapter 4), L is the length of the rod, F_{tan} is the tangential component of the net force, m is the mass, F_g is the weight, θ is the angular position, N_{Hooke} is the torque from Hooke's law, I is the moment of inertia ($I = m L^2$ in this case), k_H is Hooke's constant, and θ_{eq} is the equilibrium angular position ($\theta_{eq} = 0$ radians in this case).

You can see that the tangential component of Newton's second law is in agreement with Hooke's law if the angular position is always small. Also, Hooke's constant is: $k_H = g\, I/L = g\, m\, L$.

If you use the equations for the angular acceleration and angular position, then for harmonic motion $\alpha = -(2 \pi f)^2 (\theta - \theta_{eq})$. You obtain the *frequency* by comparing the tangential equation with this expression for the angular acceleration:

$$f = \frac{1}{2\pi}\sqrt{\frac{g}{L}}$$

I have two warnings: The first warning is that this formula is an approximation for when the angular position is small. The second warning is that in reality, the pole will have a mass, which may play an important role. The formula for the frequency assumes the pole has a small mass compared to the weight of an object located at L. If the pole has significant mass, then the pole and object will give a combined moment of inertia I, and a center of mass located at L_{cm} to give you the following:

$$f = \frac{1}{2\pi}\sqrt{\frac{mgL_{cm}}{I}}$$

These mathematical equations are useful if you want to know the physical quantities as a function of time. If you don't care about the specific time dependence, then you should use conservation of energy:

$$E_{mech} = \frac{mL^2}{2}\left[\omega_{final}^2 + 8(\pi f)^2\left[1 - \cos(\theta_{final})\right]\right] =$$
$$\frac{mL^2}{2}\left[\omega_{initial}^2 + 8(\pi f)^2\left[1 - \cos(\theta_{initial})\right]\right]$$

In this formula, E_{mech} is the mechanical energy, the middle set of terms between the two equalities is the final kinetic energy plus the final potential energy, and the terms to the right of the equalities is the initial kinetic energy plus the initial potential energy. m is the mass, L is the length of the pole (radius of the circle), ω_{final} is the final angular velocity, f is the frequency, θ_{final} is the final angular position, $\omega_{initial}$ is the initial angular velocity, and $\theta_{initial}$ is the initial angular position.

You have seen a lot of theory so far with no examples of how to apply these concepts. Use these formulas to solve a problem.

Assume you and a friend are swinging back and forth on a swing for two. Assume the length of the poles holding the swing up is 35 feet long (10.7 meters) and the total weight of the seat and the two of you is 500 pounds (2,224 newtons) or a total mass of 15.5 slugs (227 kilograms). Assume the swing starts from rest at an angle of 5.0 degrees from the vertical. What is the mechanical energy of the system? How long does it take the swing to go through one complete cycle? What is the maximum speed at which you're moving?

To solve this problem, follow along with these steps:

1. **Draw a figure.**

   Figure 12-1 shows my figure. Remember, a figure helps to visualize what is going on and helps you solve the problem.

2. **List the numbers given.**

   The problem gives you a lot of numbers, and it's easy to lose numbers and information when so much is given within the problem, which can be just as bad as when the problem doesn't give you enough numbers and you have to go looking for information. The numbers given are

   The weight is $m\,g$ = 500 pounds (2,224 newtons). The mass is m = 15.5 slugs (227 kilograms). Remember g = 32.2 feet per second squared (9.81 meters per second squared).

   The length is L = 35 feet (10.7 meters).

   The initial angular velocity is ω_{inital} = 0 radians per second.

   The initial angular position is θ_{initial} = 5.0 degrees (0.0873 radians). Remember 180 degrees = π radians.

3. **Understand what physics and formulas are relevant.**

   The problem asked you to find the mechanical energy. The mechanical energy is the last formula given, so you need the mass (m) (given), the length (L) of the pole (given), the frequency (f) (not given), the angular velocity (ω_{inital}) (given), and the angular position (θ_{inital}) (given). Find the frequency first.

   The problem asked you find the time it will take to make one complete swing. That is the period (T) of the harmonic motion, which is equal to one divided by the frequency (f) (not given). Find the frequency first.

   The problem asked you to find the maximum speed of the swingers. (For circular motion, $v = R\,\omega$ from Chapter 4.) In this problem, $v_{\text{max}} = L\,\omega_{\text{max}}$. Using the harmonic motion equations, you need the length (L) of the pole (given), the amplitude (A) (not given), and the frequency (f) (not given). Find the frequency and amplitude first.

4. Solve for the frequency.

All three questions need the frequency, so you need to solve for that first. Use the formula for frequency. The frequency is

$$f = \frac{1}{2\pi}\sqrt{\frac{g}{L}} = \frac{1}{2\pi}\sqrt{\frac{32.2 \text{ ft} -\text{s}^2}{35 \text{ ft}}} = 0.959 \text{ hertz}$$

5. Solve for the mechanical energy.

You now have the frequency and the rest of the numbers you need. In addition, use the formula for the mechanical energy, so you have everything. In this case, the mechanical energy is

$$E_{mech} = \frac{mL^2}{2}\left[\omega_{initial}^2 + 8(\pi f)^2\left[1-\cos(\theta_{initial})\right]\right] =$$
$$\frac{(15.5 \text{ slugs})(35 \text{ ft})^2}{2}\left[8\pi^2(0.959s^{-1})^2\left[1-\cos(5°)\right]\right]$$

Solving this formula, you can see that the two swingers are swinging back and forth and have a constant mechanical energy of 2.63×10^3 foot pound (3.56×10^3 joules).

6. Solve for the period.

You now have the frequency and you know the relationship between the period and the frequency. The time of one full swing is the period:

$$T = \frac{1}{f} = \frac{1}{0.959 \text{ s}^{-1}} = 1.04 \text{ seconds}$$

This is a fast swing, but they're swinging over a maximum angle of only 10 degrees.

7. Solve for the amplitude.

The amplitude (A) of the harmonic oscillations is equal to the angular position when the angular velocity is 0 radians per second. Therefore, $A = \theta_{initial} = 5.0$ degrees = 0.0873 radians.

8. Solve for the maximum speed.

You'll reach a maximum speed at the bottom of the swing, where the angular position is zero:

$$v_{max} = L\omega_{max} = L2\pi f A = 2\pi(35 \text{ ft})(0.959 \text{ s}^{-1})(0.0873 \text{ radians})$$

Your maximum speed is 18.4 feet per second (5.61 meters per second = 12.5 miles per hour = 20.2 kilometers per hour). They could probably use seat beats, especially if they're focused on the sights and not the swinging.

Comprehending Waves and Their Properties

Objects in the universe have particle properties and wave properties. For instance, the water waves lapping on the shore are a good example of waves and their properties. Waves are a disturbance that transfers energy and allows objects to do work on other objects. This section helps you understand how this energy propagates through space and some wave properties.

A *disturbance* is anything that will cause things in that area of space to not be in their relaxed (lowest energy) condition. For example, a water wave will cause the water to move up the shore higher than it normally would. Or if you tap the table, it causes the molecules and atoms to move from the resting place.

These sections characterize some different types of waves and their properties, including what happens when waves hit each other.

Dealing with all types of waves

Some waves are bad, such as your roommate singing out of key, whereas other waves are good, such as the person across the room that you want to meet waving at you. In biophysics, you'll encounter all kinds of waves. Here I focus on just the physical waves (no hand waving here). An understanding of waves is important in all areas of life (and jobs) from music (recordings), to singing (microphones), to sight (glasses), to hearing (hearing aids), to all areas of scientific and engineering research.

The following are some types of waves you may encounter:

- **Longitudinal waves:** They have a disturbance parallel to the direction of motion.

  - **Mechanical vibrations:** They're an oscillation back and forth of the material in the direction of wave propagation. Examples include tapping the table and pushing the atoms and molecules to oscillate in the direction of the wave's propagation into the table; a coiled spring where the spring is made to oscillate in the direction of the coils; and elastic materials such as a bungee cord that is tied around your ankles when you jump off a bridge.

- **Pressure waves in a gas:** The compression zone has a high density of air molecules and a rarefaction has a low density of air molecules. A high pressure zone occurs in front of a compression zone and behind a rarefaction zone. A low pressure zone occurs behind a compression zone and in front of a rarefaction zone. Sound waves are longitudinal pressure waves propagating through air, so sound doesn't travel through space (vacuum), contrary to what sci-fi movies tell you.

- **Shock waves:** A star exploding is a blast of material moving in the direction of the wave. An earthquake generates seismic waves and some of these waves are longitudinal.

✔ **Transverse waves:** They have the disturbance perpendicular to the direction of motion. Here are three examples of transverse waves:

- **Vibrating string:** Even though the strings in musical instruments are producing longitudinal waves (sound), the string has a transverse wave moving through the string. Examples include guitar strings, piano strings, harp strings, and violin strings.

- **Surface water waves:** The wave is propagating toward shore while the disturbance is lifting the water vertically upwards.

- **Electromagnetic radiation in free space:** The different types of electromagnetic radiation include from low energy photons to high energy photons: radio waves, microwaves, radar, far infrared, infrared radiation, visible light, ultraviolet radiation, X-rays, and gamma rays.

✔ **Single pulse waves:** They have a single disturbance that propagates through space. They're usually a single antinode. (An *antinode* is the location in space where the disturbance is a maximum.) Some examples include the following:

- **Rogue waves:** They're single crest ocean surface waves that have an amplitude more than double the background waves.

- **Sonic booms:** These are the initial shock waves created by an object moving through air faster than the speed of sound.

✔ **Continuous periodic waves:** They have a continuous sequence of disturbances over a long period of time propagating through space. Some examples include the following:

- **Surface water waves:** Surface water waves lap onto the shore.

- **Sunlight:** All types of electromagnetic radiation from the sun are continuously propagating through space toward the earth.

Grasping physical properties of waves

Physical properties are characteristics of the phenomenon or object that you expect it to have. For example, if I let go of my pen, it will fall toward the ground, whereas if I let go of a helium balloon, I will expect it to float up into the air. Waves have characteristics, which explain how they'll behave in a certain manner. Biophysics is about understanding these properties so you can manipulate waves and use them to your advantage. Here are some of those physical properties.

Medium

Waves require a medium (no psychics here) for the disturbance to travel through, with the exception of electromagnetic waves. A *medium* is the substance that is being disturbed by the wave, and the wave needs the medium in order to exist. For example, fluid waves need a fluid to travel through (water waves stop at the beach). In addition, the medium doesn't propagate with the wave. For example with surface water waves, a huge wall of water would build up on the shore if the water moved with the waves. Instead, the water makes a circular motion, going up and forward with the crest and then down and backwards with the trough. This motion creates undertows that can be very dangerous.

Amplitude

The *amplitude (A)* is the maximum magnitude of the disturbance. Different types of waves will have different types of amplitudes. The amplitude does not have to correspond to an actual displacement of the matter. For example, the amplitude has units of feet (meters) in the case of surface water waves; whereas it has units of pounds per square foot (pascals) in the case of sound.

- *Nodes* are places in the wave where the disturbance is zero. If you looked at a node, you wouldn't be able to tell a wave was present.
- *Antinodes* are the locations in space where the disturbance is a maximum, which means the disturbance has a value of $+A$ or $-A$. (A is the magnitude of the maximum disturbance.)
- *Crests* are antinodes of transverse waves with a disturbance of $+A$.
- *Troughs* are antinodes of transverse waves with a disturbance of $-A$.

Physical measurements of the wave

These are physical properties of waves related to space and time.

- The *wavelength* (λ) is the distance between two successive crests (or compressions) or twice the distance between two successive nodes. The wavelength can be very short (gamma rays) or very long (radio waves).

✔ The *period* (*T*) is the time you have to wait between two successive crests (or compressions) or twice the time between two successive nodes. Suppose you place a float or cork on the water (that can't move horizontally) and you watch it go up and down as the waves pass. When the float is at its highest position, start timing while it drops and when it reaches its highest position again stop timing. That time span is the period.

✔ The *frequency* (*f* = 1/*T*) is the inverse of the period, and it measures how many crests (or compressions) pass a point every second. The units are *Hertz* (Hz), which equals cycles per second. The cycle is added because the period measures the time it takes a full wave to go by, whereas frequency measures how many waves (or cycles) will pass through every second.

✔ The *wavenumber* (*k* = 2π/λ) is proportional to the inverse wavelength. It's a measure of the number of waves that will fit in a given length, hence the name wavenumber. More specifically, the wavenumber (*k*) times some length divided by 2π gives the number of waves that will fit in that length. The units of the wavenumber are radians per foot (radians per meter), but the radians are usually dropped because radians are dimensionless.

✔ The *angular frequency* is ω = 2π *f*. The angular frequency (ω) times a period of time divided by 2 π will tell you how many waves you'll see in that time. The units are radians per second.

Speed

Waves have an acceleration of zero, which means they move with a constant speed. The speed is a measure of how fast the disturbance is changing its position. (I talk in detail about the relationship between position, velocity, and acceleration in Chapter 3.) This property of waves means that the wavelength and the frequency aren't independent. The relationship between the *speed* of the wave and the space-time properties is *v* = λ *f* = ω/*k*. Note that some books use *c* for the speed of the wave because it is a constant.

The speed of a wave depends on the type of wave and the material it's moving through. Check out these examples:

✔ **Light in air:** *v* = 9.84 × 10^8 feet per second (3.00 × 10^8 meters per second = 6.71 × 10^8 miles per hour = 1.08 × 10^9 kilometers per hour)

✔ **Light in glass:** *v* = 6.56 × 10^8 feet per second (2.00 × 10^8 meters per second)

✔ **Sound in air:** *v* = 1,130 feet per second (344 meters per second = 761 miles per hour = 1,220 kilometers per hour)

✔ **Sound in metal:** *v* = 16,500 feet per second (5,000 meters per second)

Two important properties of waves to remember: First, different waves of the same type have the same speed in the same medium. Humans can hear sound waves with frequencies from 20 Hertz up to 20,000 Hertz. All sound waves in air have a speed of 1,130 feet per second (344 meters per second) no matter what the frequency. Second, a wave traveling from one medium into another medium doesn't change frequency. The speed and wavelength change from one medium to the other.

Going the math route with waves

Four mathematical models of waves exist. They are as follows:

- **Stationary:** This wave doesn't change in time. It's represented by $y(x) = A \sin(k\,x + \delta)$.

- **Right-traveling:** This wave moves toward the right when you look at it from the side. It's represented by $y(x,t) = A \sin(k\,x - \omega t + \delta)$.

- **Left-traveling:** This wave moves toward the left when you look at it from the side. It's represented by $y(x,t) = A \sin(k\,x + \omega t + \delta)$.

- **Standing:** This wave has nodes that don't move in time. It's represented by $y(x,t) = A \cos(\omega t) \sin(k\,x + \delta)$.

In these formulas, y is the strength of the disturbance at the position x and time t. A is the amplitude of the wave, k is the wavenumber, ω is the angular frequency, and δ is the phase shift.

Adding linear superposition and interference

The *linear superposition principle* states that the net wave is the sum of all the waves present. Only the net amplitude changes with no other changes to the wave's characteristics.

If you have a speaker playing music and turn on a second speaker, then the music gets louder — the amplitude of the two sound waves combine to give a larger amplitude. Similarly, if you're reading *Biophysics For Dummies* by a single light, then you can increase the amplitude of the light wave by turning on a second light. In both cases, the amplitudes are combined to increase the net amplitude of the wave.

Additionally this principle states that no other characteristics will change. To understand, look at your friend's face. The light waves come in through the window, bounce off your friend's face, and then travel to your eyes. As the light waves travel toward your eyes, they pass through other light waves coming in through the window. The waves don't interact; if these waves did interact with each other as they pass through each other, then your friend's face would change color or become fuzzy, possibly disappearing every once in a while.

The *resultant (net) wave* is the combination of many waves. From the linear superposition principle, this means the resultant (net) wave is the sum of all the different waves' strengths, $y(x,t)$, at that point in space. Two possibilities can occur when the waves combine and interfere with each other:

- ✔ *Constructive interference* occurs when the waves combine to increase the amplitude of the *resultant (net) wave*.

- ✔ *Destructive interference* occurs when the waves combine to decrease the amplitude of the resultant (net) wave.

Consider the following example. Two stationary waves are interacting. The waves are described by the parameters:

- ✔ The amplitudes are $A_1 = A_2 = 1$.

- ✔ The wavelengths are $\lambda_1 = 1$ foot (or 1 meter) and $\lambda_2 = 2$ feet (or 2 meters).

- ✔ The wave vectors are $k_1 = 2\pi$ radians per foot (or 2π radians per meter) and $k_2 = \pi$ radians per foot (or π radians per meter).

- ✔ The phases are $\delta_1 = \delta_2 = \pi/2$ radians.

The resultant wave is

$$y(x) = y_1(x) + y_2(x) = \cos(2\pi x) + \cos(\pi x)$$

Figure 12-2 shows the resultant wave. The two waves have constructive interference at $x = 0$ feet (or 0 meters) and $x = 2$ feet (or 2 meters). The two waves have antinodes at $x = 1$ foot (or 1 meter), but they're destructively interfering with each other and the resultant wave has an amplitude of zero.

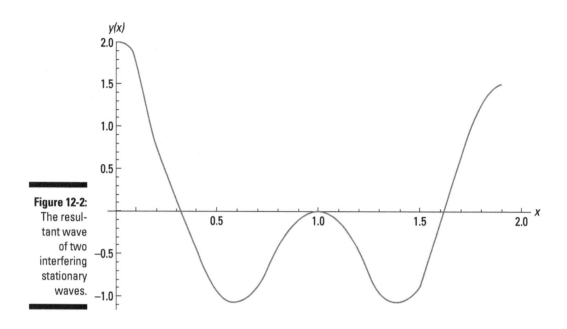

Figure 12-2:
The resultant wave of two interfering stationary waves.

Seeing the Effect of Boundaries on the Wave

Investigating the effects of the boundaries on waves is very important in biophysics because waves are almost always interacting with something, such as listening to the music or ultrasound imaging. These sections look at a wave traveling from one medium into another medium, at open and closed boundaries, and at the concept of resonance.

Traveling from a medium into a denser medium

This section looks at the type of boundary when the wave travels from one medium into a denser medium. You can think of light or sound traveling from air into glass. Before I go into the effects of these boundaries, you should note the following: *Denser medium* is in regards to properties of the medium that are relevant to the specific wave, and it doesn't necessarily mean the mass density. In the case of the light hitting a glass window, most of the light travels through the window, but I can take a wall that has the same mass density as my window and shine light on it; if so, none of the light will make it through.

When a wave strikes a boundary, part of the wave bounces back and part of the wave goes into the second material. You can think of this interaction with the boundary as three waves that are connected and dependent upon each other. The three waves and the connection of their properties to each other are

✔ *Incident wave* is the wave you started with that is traveling along and then strikes the boundary.

✔ *Reflected wave* is the portion of the wave that bounces back from the boundary. The speed, frequency, and wavelength of the reflected wave are the same as that for the incident wave. The only difference between the two waves is the direction of their motion, and the phase shift is almost π radians shifted from the incident wave.

The reason for the phase shift difference is so the amplitude of the reflected wave is inverted relative to the incident wave at the boundary and the waves destructively interfere with each other at the boundary. The second material is denser, so most of the wave is reflected and little of the wave is transmitted. If the incident and reflected wave are phase shifted by exactly π radians, then there is a node at the boundary and no disturbance will enter the second material and the entire wave is reflected.

✔ *Transmitted wave* is the portion of the wave that moves into the denser material. The frequency of the transmitted wave is the same as the frequency of incident wave. The phase shift is almost the same as the incident wave phase.

Traveling from a medium into a less dense medium

This section looks at the type of boundary when the wave travels from one medium into a less dense medium. You can think of light traveling from water into air. When a wave (the incident wave) strikes the boundary, part of the wave bounces back (reflected wave) and part of the wave goes into the second material (transmitted wave). Most of the properties are the same as the previous section, the big difference is this:

The reflected wave is the portion of the wave that bounces back from the boundary. The speed, frequency, and wavelength of the reflected wave are the same as the speed, frequency, and wavelength of the incident wave. Unlike the previous section, the phase shift is almost the same as the incident wave phase, so the amplitude of the reflected wave is reduced and it's not inverted relative to the incident wave. This allows more of the wave to be transmitted into the second medium.

Going to extremes: Open and closed boundaries

This section takes the boundaries to the extreme. The more dense material becomes infinitely dense (the wave can't disturb the material) and the less dense material becomes a vacuum. Essentially, these are the cases where the wave can't travel into the second material. Many examples and applications of these types of boundaries occur in biophysics. For example, in the case of a mirror, all of the light is reflected off it. In the case of a guitar string, the wave in the string travels to the end of the string then bounces off the end and goes back down the string.

A *closed boundary* forces the wave to have a node at that point in space. The phase shift of the reflected wave is exactly π radians shifted relative to the incident wave's phase shift, $\delta_r = \delta_i + \pi$, which is sometimes referred to as π-*reflection*. The reflected wave causes destructive interference with the incident wave at the boundary. An example is the vibrating string wave on a guitar string when the wave hits the tied down ends or when light strikes a mirror.

Meanwhile, an *open boundary* forces the wave to have an antinode at that point in space. The phase of the reflected wave isn't shifted relative to the incident wave ($\delta_r = \delta_i$). The reflected wave causes constructive interference with the incident wave. An example is a sound wave hitting the closed end of a clarinet.

Resonating with resonance

Resonant frequencies are special frequencies where a wave trapped between two (open and/or closed) boundaries bounces back and forth and interferes constructively with itself. The two boundaries restrict the value of the phase shift (δ) and the frequency (f/ω). The value of the wavelength ($\lambda = v/f$) and wavenumber ($k = \omega/v$) are also fixed. If the boundaries are separated by a distance of L, the resonant frequencies are as follows:

two closed boundaries or two open boundaries: $f_n = \dfrac{n\,v}{2L}$

one closed boundary and one open boundary: $f_{2n-1} = \dfrac{(2n-1)v}{4L}$

In both equations $n = 1, 2, 3 \ldots$, but in most cases n is small. f_1 ($n = 1$) is called the *fundamental frequency* or *fundamental harmonic*.

You can easily remember the frequency formulas if you keep in mind one simple rule. If the two boundaries are the same type, then you must have an integer number of half-wavelengths between the two boundaries, whereas if the two boundaries are different, then you must have an integer number of half-wavelengths plus one-quarter of a wavelength between the two boundaries. In addition, the resultant wave between the boundaries is a standing wave.

Use this information to solve the following problem.

You and your roommate are standing in a stairwell with concrete walls when your roommate starts singing. Assume the entire sound wave bounces back and forth between the walls. If the stairwell is 5.00 feet (1.52 meters) across, what is the fundamental resonant frequency? What does the mathematical sine wave look like if one wall is at $x = 0$ feet (0 meters)? The speed of sound is 1,120 feet per second (341 meters per second).

To figure out this problem, follow these steps:

1. **Draw a diagram.**

   Here your diagram should focus on the resonance of waves. Check out Chapter 13 for more details about the resonance of sound waves.

2. **Understand the physics and find the relevant formulas.**

   The first question wants you to find the fundamental frequency when you have two boundaries of the same type. (The walls act as two closed boundaries for the air molecules' harmonic displacement and have nodes at the walls.) You need the speed of the wave, the length between the two boundaries, and the integer n.

   Use the standing wave formula to find the angular frequency, the wavenumber, and the phase shift.

3. **Find the numbers for the fundamental frequency.**

   The speed of sound is $v = 1{,}120$ feet per second (341 meters per second).

   The distance between the boundaries is L = 5.00 feet (1.52 meters).

   The integer n = 1 for the fundamental resonant frequency.

4. **Solve for the fundamental resonant frequency.**

   Mathematically, fill in the numbers to the equation for resonance such as

   $$f_1 = (1)\left(\frac{1120 \text{ ft s}^{-1}}{10 \text{ ft}}\right) = 112 \text{ hertz}$$

   If your roommate has a loud voice and the walls are good at reflecting sound waves, then you're in trouble.

5. Find the numbers for the mathematical sine wave.

The problem tells you that one of the walls is located at $x = 0$ feet (0 meters), which must be a node, which means $\sin(\delta)$ equals 0, which is satisfied if the phase δ equals 0 radians.

The angular frequency is $\omega_1 = 2\pi f_1 = 2\pi (112\text{ Hz}) = 704$ radians per second.

The wavenumber is $k_1 = \omega_1/v = (704\text{ rads/s})/(1120\text{ ft/s}) = 0.628$ radians per foot (2.06 radians per meter).

6. Solve for the mathematical sine wave.

The mathematical formula for the sinusoidal standing wave describing the air molecules' displacement is

$$y_1(x,t) = A_1 \cos(\omega_1 t)\sin(k_1 x + \delta) = A_1 \cos\left(\left(704\frac{\text{rads}}{\text{s}}\right)t\right)\sin\left(\left(0.628\frac{\text{rads}}{\text{ft}}\right)x\right)$$

The wave is always zero at $x = 0$ and at $x = L = 5.00$ feet (1.52 meters). At $x = L/2 = 2.50$ feet (0.762 meters), this wave has an antinode. (You obtain the sound wave by switching the sine to a cosine. See Chapter 13 for more info.)

Chapter 13

Grasping How Animals and Instruments Produce Sound Waves

*A*nimals use sound for different functions, such as mating and echo-location. The range of sounds produced by humans is amazing. Most animals actually have a very limited range of sounds, but some like the crow can produce more than 50 distinct sounds, whereas the human species uses sounds for communication and entertainment.

Sound waves are pressure waves that propagate through material, which allows a person to transmit energy to another person. The energy interacts with the other person's ear and produces a signal, which the brain interprets. Hearing is one of the five senses, and how to produce sounds and music as opposed to noise is an important concept in biophysics.

This chapter focuses on longitudinal pressure waves, which most people refer to as *sound*. Here I look at the properties of sound waves and discuss the properties of making music and singing. I also discuss resonance in musical instruments and then focus on how the guitar and the human voice make sound.

Knowing the Nature of Sound and the Speed of Sound

You may have noticed that the sounds produced by people and animals sound different. This section discusses the properties of sound that make noises sound different and the physical properties of sound that remain the same between different sounds.

Vibrating the air and pressure waves

Sound waves within air are *longitudinal pressure waves,* which means that the air molecules are vibrating back and forth in the direction the wave is moving. This is contrary to water waves where the water moves up and down in the vertical direction as the waves move horizontally towards the shore.

These waves form alternating compression and rarefaction zones of the air molecules. A *compression zone* is a region where there are more air molecules than you would expect, and a *rarefaction zone* is a region with a shortage of air molecules. If you have a compression zone with too many air molecules, then the air molecules want to spread out creating a pressure (force per unit area).

The following mathematical models describe these pressure waves:

$$\text{right-traveling wave: } P^{(a)}(x,t) = P_{atm} + \Delta P \, \sin(kx - \omega t + \delta)$$

$$\text{left-traveling wave: } P^{(a)}(x,t) = P_{atm} + \Delta P \, \sin(kx + \omega t + \delta)$$

$$\text{standing wave: } P^{(a)}(x,t) = P_{atm} + \Delta P \, \cos(\omega t)\sin(kx + \delta)$$

In these models, $P^{(a)}$ is the net pressure with the superscript a meaning absolute pressure, and P_{atm} is the atmospheric pressure if no sound wave is traveling through the air. $P^{(g)} = P^{(a)} - P_{atm}$ is the gauge pressure. The standard is $P_{atm} = 1$ atmosphere = 14.69 pounds per square inch = 2,116 pounds per square foot = 101,300 pascals.

The sound wave propagates through space, but the air molecules vibrate with harmonic motion. *Harmonic motion* means motion that repeats over and over, such as a weight attached to a spring bouncing up and down. The air molecules move back and forth. Many properties of this harmonic motion are related to the pressure wave properties, because the harmonic motion of the air molecules are creating the pressure wave:

✔ The *wavenumber* (*k*) of the harmonic motion of the air molecules is the same as the wavenumber (*k*) of the sound (pressure) wave.

✔ The *angular frequency* (ω) of the harmonic motion of the air molecules is the same as the *angular frequency* (ω) of the sound (pressure) wave.

✔ The harmonic motion of the air molecules is a quarter wavelength ahead of the sound (pressure) wave. The *phase shift* is

$$\delta_{air} = \delta + \pi/2$$

To understand this shift, consider a single air molecule in the middle of a compression zone. An equal density of air molecules exists on both sides and the pressure (force) will balance. In the case of an air molecule in front of the compression zone, the density of air molecules behind the air molecule is greater than the density in front of it, so the air molecule will feel a net pressure (force) forward. An air molecule behind a compression zone will feel a net pressure (force) backwards because the density of air molecules in front is greater than the density of air molecules behind it. The situation for a rarefaction is reversed.

✔ The *average amplitude of the harmonic motion* of the air molecules is related to the amplitude of the sound (pressure) wave:

$$\Delta x = \frac{\Delta P}{2\pi f v \rho}$$

Δx is the average amplitude of the harmonic motion of air molecules, ΔP is the amplitude of the pressure (sound) wave, *f* is the frequency, *v* is the speed of the wave, and ρ is the mass density of the air.

For example, a 1,000 hertz sound wave will cause the eardrum to rupture if the pressure wave has an amplitude $\Delta P = 0.60$ pounds per square foot (29 pascals) or the air molecules have a harmonic oscillation amplitude of 3.36×10^{-7} feet (1.11×10^{-7} meters). The amplitude of the air molecules' harmonic motion is approximately one thousand times smaller than the diameter of a human hair.

Speeding ticket for sound

The speed of sound depends on the properties of the material and the temperature. These sections examine the relationship between the speed and the properties of the material and the relationship between the speed of sound in air and the temperature of the air.

Identifying the speed of sound in different materials

Knowing the speed of sound is important in different areas of biophysics. For example, for animals that use a frequency modulated echolocation, knowing the speed of sound allows them to estimate the distance to an object if they know the time lapse between the chirp and the echo. (Humans estimate distances by triangulating with their eyes.) In addition, the speed of sound through objects provides a lot of information about the properties of the material.

The *speed of sound* in a material is as follows:

$$v = \sqrt{\frac{K}{\rho}}$$

Here v is the speed and ρ is the density of the material. For liquids and gasses K is the adiabatic bulk modulus. The *adiabatic bulk modulus* is a measure of how easy it is to compress the material. In the case of solids, the situation can be more complicated. The sound wave can be a longitudinal wave or it can be a transverse wave like a water wave. If the motion is longitudinal and there is no transverse vibration of the solid, then K is replaced with *Young's modulus* and you can ignore the *shear modulus*. I discuss Young's modulus and the shear modulus in Chapter 6.

Focusing on the speed of sound at different temperatures

Degree Fahrenheit (degree Celsius) is usually used instead of kelvin when talking about everyday temperatures, but for the speed of sound, which depends on the temperature, you need to use kelvin units. In air, use this following formula:

$$v = \sqrt{\frac{\gamma R}{M} T_K} = \sqrt{4350\, T_K}\ \frac{\text{feet}}{\text{second}} \left(\sqrt{404\, T_K}\ \frac{\text{meters}}{\text{second}} \right)$$

$$T_K = \frac{5 T_F}{9} + 255 \text{ (degree Fahrenheit to kelvin)}$$

$$T_K = T_C + 273.15 \text{ (degree Celsius to kelvin)}$$

Here, γ is the ratio of the specific heat at constant pressure divided by the specific heat at constant volume, R is the ideal gas constant, M is the molecular mass, and T_K is the temperature in kelvin. I provide the conversions from degree Celsius and degree Fahrenheit to kelvin. T_C is the temperature in degree Celsius and T_F is the temperature in degree Fahrenheit. On a hot Texas day, $T_F = 130°$ Fahrenheit ($T_C = 54°$ Celsius) and the speed of sound is $v = 1{,}193$ feet per second (364 meters per second); whereas on a cold Siberian day, $T_F = -58°$ Fahrenheit ($T_C = -50°$ Celsius) and the speed of sound is $v = 985$ feet per second (300 meters per second).

Exploring the physical properties of sound

The power, intensity, and intensity level are physical properties of sound that allow you to quantify the sound instead of just having a qualitative description.

The *intensity* of a sound wave is a measure of the wave's potential to do work (for example, making the eardrum vibrate). The larger the intensity, the larger the pressure and the more work it can do on the eardrum. The intensity (*I*) is defined as

$$I = \frac{(\Delta P)^2}{2\rho v}$$

Here, ΔP is the *amplitude* of the pressure wave, ρ is the density of the medium, and v is the speed of the wave.

The intensity of sound waves can vary a lot, and it's surprising how large a range of intensities animals can hear. For example, humans can hear sound waves with an intensity less than 6.855×10^{-14} pounds per (second foot) (10^{-12} watts per square meter) up to sound waves with an intensity greater than 6.855×10^{-2} pounds per (second foot) (1 watt per square meter). Therefore, using a logarithmic scale is more convenient, which is called the *intensity level*. Intensity level (β) is defined as

$$\beta = (10 \text{ dB}) \log\left[\frac{I}{I_o}\right]$$

$$I = I_o \, 10^{\beta/(10 \text{ dB})}$$

dB is short for *decibels*, which is a dimensionless set of units similar to radians used in angle measurements. β is the *intensity level*, *I* is the *intensity,* and the reference intensity $I_o = 6.855 \times 10^{-14}$ pounds per (second foot) (10^{-12} watts per square meter) is a constant, which is equal to the human threshold of hearing at 1000 hertz.

Remember, the *power* is the work divided by the time. For sound waves, the power is also known as the *acoustic power*. The sound wave can do work moving the eardrum back and forth. Also, the power is related to the intensity; the *intensity* is a measure of the amount of power in the wave per unit cross-sectional area. The power is

$$P = I \, A$$

Here *P* is the power, *I* is the intensity, and *A* is the cross-sectional area.

In many problems, you can assume that the sound wave conserves energy, which means that the medium (air molecules) doesn't absorb any of the wave's energy. Therefore, the power is a constant that doesn't change, which gives a condition on how intensity (and intensity level) changes as a function of the area. The mathematical relationship for conservation of (sound) energy is

$$\frac{I_{final}}{I_{initial}} = \frac{A_{initial}}{A_{final}} = \frac{R_{initial}^2}{R_{final}^2}$$

$$\beta_{final} - \beta_{initial} = (10 \text{ dB})\log\left[\frac{A_{initial}}{A_{final}}\right] = (20 \text{ dB})\log\left[\frac{R_{initial}}{R_{final}}\right]$$

Here I is the intensity, β is the intensity level, A is the cross-sectional area, and R is the distance from the sound wave source.

Suppose your family is at a concert. A speaker is producing sound waves in a hemispherical shape that has an intensity level of 120 decibels at 1 foot (0.3048 meters). What are the intensity, power, and the amplitude of the sound wave at this distance? Suppose your family is standing 100 feet (30.48 meters) from the speaker. What are the intensity level, intensity, power, and the amplitude of the sound wave at this location?

To solve this problem, follow these steps:

1. **Understand what the problem wants you to solve.**

   The equations you need give relationships between the power (P), the intensity (I), the intensity level (β), the area (A), the amplitude of the pressure wave (ΔP), and the amplitude of the air molecules' harmonic motion (Δx). You also need the speed of sound and the density of air.

   You need to use *conservation of energy*, so that the total power at 100 feet is equal to the total power at 1 foot.

2. **Find the numbers you need to solve the problem.**

   The density of air is ρ = 0.00232 slugs per cubic foot (1.20 kilograms per cubic meter).

   The speed of sound is v = 1,130 feet per second (344 meters per second).

   A hemispherical area is A = $2\pi R^2$ (half a sphere).

   β_1 = 120 decibels at R_1 = 1 foot (0.3048 meters).

3. **Solve the problem at 1 foot.**

   First, you can use the intensity level to find the intensity. After you know the intensity and the area, you can find the power and the pressure wave amplitude.

$$I_1 = I_o 10^{\beta_1/(10 dB)} = \left(6.85 \times 10^{-14} \frac{lb}{ft\ s}\right) 10^{120 dB/10 dB} = 6.85 \times 10^{-2} \frac{pounds}{foot\ second}\left(1\ \frac{watt}{square\ meter}\right)$$

$$P_1 = I_1 A_1 = \left(6.85 \times 10^{-2} \frac{lb}{ft\ s}\right) 2\pi (1\ ft)^2 = 0.430 \frac{foot\text{-}pound}{second}(0.584\ watts)$$

$$\Delta P_1 = \sqrt{2\rho v I_1} = \sqrt{2\left(2.32 \times 10^{-3} \frac{slugs}{ft^3}\right)\left(1.13 \times 10^3 \frac{ft}{s}\right)\left(6.85 \times 10^{-2} \frac{lb}{ft\ s}\right)}$$

$$= 0.599 \frac{pounds}{square\ foot}(28.7\ pascals)$$

4. **Solve the problem at 100 feet.**

Using conservation of energy, the sound waves beating on your family's ears are described by the following properties:

$$P_{100} = P_1 = 0.430 \frac{foot\text{-}pound}{second}(0.584\ watts)$$

$$I_{100} = I_1 \left(\frac{R_1}{R_{100}}\right)^2 = \left(6.85 \times 10^{-2} \frac{lb}{ft\ s}\right)\left(\frac{1\ ft}{100\ ft}\right)^2 = 6.85 \times 10^{-6} \frac{pounds}{foot\ second}\left(10^{-4}\ \frac{watts}{square\ meter}\right)$$

$$\beta_{100} = \beta_1 + (20\ dB)\log\left[\frac{R_1}{R_{100}}\right] = 120\ dB + (20\ dB)\log\left[\frac{1\ ft}{100\ ft}\right] = 80\ decibels$$

$$\Delta P_{100} = \sqrt{2\rho v I_{100}} = \sqrt{2\left(2.32 \times 10^{-3} \frac{slugs}{ft^3}\right)\left(1.13 \times 10^3 \frac{ft}{s}\right)\left(6.85 \times 10^{-6} \frac{lb}{ft\ s}\right)}$$

$$= 0.00599 \frac{pounds}{square\ foot}(0.287\ pascals)$$

The power hasn't changed; the intensity is 10,000 times smaller, the intensity level is 80 decibels, which is still loud (normal conversation is between 65 and 70 decibels), and the amplitude of the pressure wave is 100 times smaller.

Resonating with Vibrations and Resonance

Consider a sound wave trapped between two boundaries. The sound will bounce back and forth, reflecting off boundaries and interfere with itself. Usually, the waves work against each other and reduce the amplitude of the resultant wave, which is called *destructive interference*. However, at certain frequencies, the sound will reflect off the boundaries and work together to enhance the amplitude, which is called *constructive interference*. These frequencies are called *resonant frequencies (resonance)*.

Changing the radius of the instrument can alter the timbre of a musical instrument. *Timbre* is the ability to tell the difference of a B-note being produced by a tuning fork, flute, and a clarinet. You can determine the timbre from two factors:

- ✔ **Spectrum:** The set of harmonics activated by the instrument.

- ✔ **Envelope:** The shape of the waveform created by the harmonics with differing amplitudes.

The air molecules' vibrations inside the musical instrument aren't a perfect harmonic oscillator, but a damped harmonic oscillator losing energy through viscosity with the pipe walls. When the instrument's radius is large, a very small percentage of the air is in contact with the walls so little energy is lost and only those frequencies close to the resonance condition will make it through the instrument.

As the radius shrinks, a larger percentage of the air molecules are in contact with walls, so more energy is lost to viscosity, which means that the amplitude of the sound making it through drops, but the range of frequencies, around each of the resonance frequencies, increases. In addition, the amplitude of fundamental frequency and the lower frequency overtones drop faster than the higher frequency overtones, giving an overall higher frequency sound when the radius is smaller.

The following sections introduce the concepts of resonance as applied to musical instruments. I focus on instruments with one open end and one closed end, such as single reed instruments like a clarinet. I then discuss instruments with both ends open, such as the pipe organ and flutes.

Resonating with a clarinet

The clarinet is part of a large group of the single-reed instruments with about a dozen different types and capable of playing frequency ranges of up to 4 octaves. (An *octave* is 12 semitones and is equal to double or half of the original frequency.) The most common clarinet is the B-flat soprano clarinet, which is the one people are usually talking about when they say "clarinet."

The *resonance condition* for a pipe with one open end and one closed end is

$$f_{2n-1} = \frac{v}{4(L + 0.8\,R)}(2n-1), \quad n = 1,2,3,\ldots$$

In this formula, v is the speed of sound, L is the length of the cylinder from the open end to the closed end, and R is the radius of the cylinder. This condition is based on fitting an integer number of half wavelengths plus an extra

quarter wavelength within the clarinet, which I have assumed to be cylindrical in shape. The first harmonic ($n = 1$) is called the *fundamental frequency* or
the *fundamental harmonic*. The second harmonic ($n = 2$) and higher are called
the *overtones*.

Many sources replace $2n–1$ with n and assume you know that n has to be odd.
Writing it as $2n–1$ and letting $n = 1, 2, 3…$ is easier so you don't have to remember to skip the even integers.

Be careful when looking at the location of the nodes and antinodes of musical
instruments because the nodes and antinodes of the air molecules' harmonic
vibrations are out of phase to the sound waves nodes and antinodes. Some
sources draw nodes for the air molecules' vibrations, whereas other sources
have antinodes for the sound wave at that the same location. Both views are
correct, so be careful which wave you're drawing.

The closed (reed) end of the clarinet is essentially a closed boundary for the
air molecules' vibrations. Because the air molecules can't vibrate through
a wall, it's a node, which means the reed end is an open boundary (an antinode) for the sound wave.

The open (bell) end of the clarinet is exposed to the outside air, so it's
approximately fixed at atmospheric pressure. It's a node (closed boundary)
for the sound wave, which means it's an open boundary (an antinode) for the
air molecules' vibrations.

By opening the holes in the clarinet, you're forcing nodes in the sound wave
or antinodes for the air molecules' vibrations at those locations. The location
of these open holes will damp out those harmonics that had sound wave antinodes at those locations.

What is the length of your clarinet if the fundamental frequency is $E_3 =$
164.81 hertz (Set R = 0)? What are the first four allowed harmonics? If you
open a hole a distance 0.979 feet (0.299 meters) from the reed, which harmonics are affected the most? If you set R = 0.3 inches (0.75 centimeters),
how much does the fundamental frequency change by?

This problem has many questions. The easiest way to solve this problem is
to solve each question individually in the order given. To solve this problem,
follow these steps:

1. Understand what the first question wants you to solve.

The first question is what is the length of your clarinet if the fundamental frequency is $E_3 = 164.81$ hertz (Set R = 0)?

All the information you need is here even though it isn't obvious. You need to look up the speed of sound. Use the following:

$R = 0$

$n = 1$ for the fundamental frequency.

$f_1 = 164.81$ hertz

$v = 1130$ feet per second (344 meters per second).

2. **Solve the first question to find the length of your clarinet (L).**

$$L = \frac{v}{4f_1} = \frac{1130 \text{ ft s}^{-1}}{4(164.81 \text{ s}^{-1})} = 1.71 \text{ feet } (0.522 \text{ meters})$$

3. **Solve the second question that asks to calculate the first four harmonics.**

The harmonics are an integer times the fundamental frequency, $f_{2n-1} = (2n-1) f_1$:

fundamental frequency: $f_1 = 164.81$ hertz (E_3)

first overtone: $f_3 = (3)164.81$ hertz $= 494.43$ hertz

second overtone: $f_5 = (5)164.81$ hertz $= 824.05$ hertz

third overtone: $f_7 = (7)164.81$ hertz $= 1153.67$ hertz

4. **Solve the next question.**

If you open a hole 0.979 feet (0.299 meters) from the reed, the harmonics affected the most are those that would be very close to having a sound wave antinode at 0.979 feet (0.299 meters) from the reed when the hole is closed. Opening the hole would dampen these frequencies.

Find the harmonics with half-integer wavelengths because the sound waves affected have an antinode at the reed and an antinode where the hole is being opened:

$$f_{hole, n_{hole}} = \frac{n_{hole} v}{2L_{hole}} = \frac{n_{hole}(1130 \text{ ft / s})}{2(0.979 \text{ ft})} = n_{hole} 577 \text{ hertz}$$

Compare these frequencies with the four preceding frequencies. Opening the hold won't significantly affect the first three frequencies, but the third overtone will be eliminated. If you compare this expression for the frequency with those of the clarinet, you can obtain an expression for the integers of the overtones that are affected:

$n = 0.5 + 3.5 n_{hole} = 4, 11, 18, ...$

5. Determine the fundamental change.

If you set R = 0.3 inches (0.75 centimeters), how much does the fundamental frequency change by?

To solve, substitute the new value for R into the equation:

$$f_1 = \frac{1130 \text{ ft s}^{-1}}{4(1.71 \text{ ft} + 0.8(0.025 \text{ ft}))} = 163.3 \text{ hertz}$$

This is only a 100 (1–163.3/164.81) = 0.919 percent change in the fundamental frequency.

Vibrating air in a flute

The pipe organs (except stopped pipes) and flutes have both ends open. The air molecules' vibrations have antinodes at both ends and the sound wave has nodes at both ends. The *resonance condition* for a pipe with both ends open is

$$f_n = \frac{v}{2(L+0.6R)}n, \quad n = 1,2,3,\ldots$$

In this formula, v is the speed of sound, L is the length of the cylinder from one open end to the other open end, and R is the inner radius of the cylinder. The first harmonic ($n = 1$) is called the fundamental frequency or the fundamental harmonic. The second harmonic ($n = 2$) and higher are called the overtones.

The flute has many holes, which allows the player to adjust the flute's frequency. Organs can't adjust their frequency, so a pipe organ needs many different pipes. A small pipe organ has a couple dozen pipes whereas the large organs can have tens of thousands.

Both ends of a pipe are at approximately atmospheric pressure and correspond to nodes for the sound wave, which means that the air molecules' vibrations have maximum displacement (antinodes) at the ends. Therefore, depending on the source, you may see the waves drawn in an open-ended pipe with either antinodes or nodes at the ends depending on what they're representing.

If you want to practice using the formula for resonance, you can use the problem in the clarinet section. The only difference is both ends are now open and you need to use this formula for the resonance condition.

Combining Cords: The Human Voice and Musical Instruments

The human voice is a very complex musical instrument, used for communication and singing. Understanding how humans control their body to produce these sounds is a very interesting area of biophysics. Here I highlight the similarities and differences between the human voice and different instruments.

These sections look at the case when a vibrating solid, such as a string or percussion, produces the initial sound wave. Here I focus on strings, such as in harps, pianos, and guitars and discuss how the waves on the string are translated into sound waves, what body resonance is and why it can be important, and what cavity resonance is.

Tying down the strings and cords

An object tied down at the ends, such as a guitar string, will vibrate at only certain frequencies because the waves propagate to the ends, bounce off, and come back, interfering with the initial wave. The interference is destructive except at certain frequencies. These sections examine the vibration of a string and its properties and the resonance condition of a vibrating string. I also look at vocal folds and the human voice.

Speeding waves on a string

The string tied to both ends of a guitar, piano, or harp, is made to vibrate. However, it vibrates only when the right-traveling and left-traveling waves combine constructively to form a standing wave on the string. The length of the string and hence the wavelength of the wave is fixed.

The only way to change the *frequency* of the wave is to change the speed. Remember the relationship between the frequency, wavelength, and the speed is: $f = v/\lambda$.

The *speed* of a wave on a string is

$$v_{string} = \sqrt{\frac{|\vec{F}_{T,string}|}{\mu_{string}}} = \sqrt{\frac{|\vec{F}_{T,string}|L}{m_{string}}}$$

The magnitude of the force is the tension in the string and μ_{string} is the mass (m_{string}) per unit length (L) of the string.

The heavier the string means the lower the speed of the wave on it and hence the lower the frequency. You may have noticed the bass strings on a guitar are thick and heavy so they can produce the lower frequencies. The more tension you add to the string, the greater the speed of the wave and hence the greater the frequency. The tension in the string allows a person to tune a guitar to the desired frequencies without having to change the length or mass of the strings.

The string's length and the wave's speed on the string don't have a direct relation to the wavelength and speed of the sound wave produced. But they do determine the frequency of the wave on the string, and the frequency of the string's vibration is equal to the frequency of the sound wave produced by the vibrating string, which is especially true for the electric instruments that use pickups. *Pickups* are an application of Faraday's law (see Chapter 16). The pickups are a permanent magnet wrapped by copper wire in a coil. The magnet magnetizes the steel string and the vibrating magnetized steel string induces an alternating current in the coil. The current has the same frequencies as the harmonics of the string.

Measuring an oscillating guitar string's resonance

The string is tied at both ends, which means the waves must have nodes at those points. Therefore the string's length must equal an integer multiple of half a wavelength:

$$L = \frac{n\,\lambda_n}{2}$$

The resonance frequencies (harmonics) of the standing waves on the string are

$$f_{n,string} = \frac{n\,v_{string}}{2L}$$

The fundamental frequency corresponds to $n = 1$, and the instrument has been tuned to the correct frequency by selecting the correct tension in the string for the choice of string mass. Normally, the string will vibrate at this set of frequencies, but touching the string at certain locations can modify the harmonics. The touching of the string dampens the harmonics that would normally have an antinode close to that position, thereby changing the sound. On the other hand, if you press the string firmly then that location becomes the new boundary condition (creating a new shorter L) and the frequencies are shifted upwards. Alternatively, instead of using your finger, you can use a capo (clamp) to hold down the string.

A 2.5-foot (0.762-meter) long guitar string has a weight per unit length of 2×10^{-3} pounds per foot (mass per unit length of 2.98×10^{-3} kilograms per meter). You tune the string to a fundamental frequency of an A_2 note (110 hertz). What are the properties of the waves on the string?

1. **Understand the question that the problem wants you to solve.**

   You need to figure out the properties of the waves on the string. These waves satisfy resonance, so you want to know what the resonance condition is, and the properties of waves include wavelength, frequency, and speed. Also, the string needs to be tuned, so you want to find the tension in the string. The numbers you need are

   The length of the string is $L = 2.5$-foot (0.762-meter)

   The mass per unit length is $\mu_{string} = 6.00 \times 10^{-5}$ slugs per foot (2.87×10^{-3} kilograms per meter)

   The fundamental frequency ($n = 1$) is $f_1 = 110$ hertz

2. **Solve the resonant conditions and the tension in the string.**

   You know the frequency, so all you have left to find is the wavelength from the resonant condition and the speed of the wave.

   The wavelength for $n = 1$ is

   $$\lambda_1 = \frac{v_{string}}{f_1} = \frac{2L}{n} = \frac{2(2.5ft)}{1} = 5.0 \text{ feet } (1.52 \text{ meter})$$

   The speed of the waves on the string is

   $$v_{string} = f_1\lambda_1 = (110 \text{ Hz})(5 \text{ ft}) = 550 \ \frac{\text{feet}}{\text{second}} \ \left(168 \ \frac{\text{meter}}{\text{second}}\right)$$

   The tension in string is

   $$\left|\vec{F}_T\right| = \mu\left(v_{string}\right)^2 = 6 \times 10^{-5} \ \frac{\text{slugs}}{\text{ft}}\left(550\frac{\text{ft}}{\text{s}}\right)^2 = 18.2 \text{ pounds } (80.7 \text{ newtons})$$

   The resonant condition gives the harmonics of the string

   $$f_n = n(110 \text{ hertz}) \text{ and } \lambda_n = \frac{5.0 \text{ feet } (1.52 \text{ meters})}{n}$$

Vibrating vocal cords – speaking frequency range

The human voice starts with the *glottis* (the vocal folds and the space between the folds). Similar to a vibrating guitar string, the vocal folds vibrate, producing several harmonics:

$$f_n = n f_1$$

The average fundamental frequency for a male is 125 hertz, whereas it's 200 hertz for the average female. The number of harmonics excited in the human vocal folds is around 20 to 25 ($n = 1, 2, 3 \dots 25$). The harmonics in the range from the fundamental frequency (which can be as low as approximately 50 hertz) to approximately 3,000 hertz are excited.

When a person prepares to make a noise, he closes the vocal folds and makes them tense. Air pressure builds up in the larynx and causes the vocal folds to open, the opening allows the air to escape past the vocal folds. The air pressure drops because of the moving air (Bernoulli's equation, which shows that the faster the air moves the lower the air pressure drops; you can read about it in Chapter 9), and the vocal folds close. The air pressure builds up and the cycle repeats. This opening and closing cycle is repeated several times every second and forms the initial sound wave, which is part of the answer of how humans can make so many different sounds.

With a vibrating string the speed of the wave is proportional to the square root of the tension. The length of the string is fixed, so the fundamental frequency is proportional to the square root of the tension. Similarly, people can change their fundamental frequency by how tense they make the vocal folds.

Checking body resonance in an acoustic guitar

Body resonance is very important for acoustic instruments because the energy within the vibration is used to increase the amplitude of the sound wave. *Body resonance* is simply the frequencies that an object will vibrate at easily. A body won't vibrate for long at most frequencies, but at certain frequencies it will vibrate while losing energy at a very slow rate. Here I take a closer look at resonance with the acoustic guitar (which is applicable to all string acoustic instruments).

A vibrating string produces sound at an intensity that can't be heard so a *soundboard* is added. The vibrating string forces the soundboard to vibrate at the same frequencies via a *bridge* that connects the string to the soundboard. The large surface area of the soundboard pushes more air back and forth creating a high amplitude wave. A person hears the soundboard and not the vibrating string. In the case of the acoustic guitar, the soundboard is the front cover of the box.

The soundboard vibrates at only a few of the string's harmonics and the other frequencies are damped. A high quality soundboard has resonance at the frequencies of the strings of the instrument, whereas a poorly designed soundboard won't have the correct resonances. Remember, an object doesn't like to vibrate unless it's vibrating at resonance; the vibrations are damped very rapidly at frequencies that aren't at resonance. The poorly designed soundboard produces distorted sounds, damping out the frequencies the player wants.

In addition, vibrating at the string's frequency, the soundboard has characteristic resonant frequencies (harmonics) of its own, which depends on the type of material, shape, size, and other physical parameters. These characteristic resonant frequencies add timbre to the sound wave. For example, an A note at the same amplitude from a guitar, violin, and a piano all sound different.

The energy required to drive the soundboard comes from the vibrating string, so the string won't vibrate for very long and needs to be plucked at regular intervals to keep the instrument producing the sound wave.

Collapsing cavities

The body resonance is the most important contribution in string instruments, but cavity resonance does play a role in the acoustic guitar as well. *Cavity resonance* is when the sound can travel through the cavity with little loss of amplitude. Cavity resonance occurs at specific frequencies, such as when you blow across the top of an empty bottle.

The following sections examine some different aspects of cavity resonance, one with guitars and one within the human body.

Seeing inside an acoustic guitar

A *Helmholtz resonator* is a container filled with air, which has a neck that is open to the atmospheric air (think of an open wine or beer bottle). When the outside air blows across the opening, the air inside the chamber vibrates with a characteristic frequency. The *characteristic frequency* is

$$f_1 = \sqrt{\frac{A v^2}{4\pi^2 LV}}$$

In this formula, A is the area of the opening, v is the speed of sound, L is the length of the neck, and V is the volume of the cavity.

Try this experiment with a beer bottle at your next Saturday night biophysics gathering. When a lawyer has drunk a quarter of their beer, blow across the top. It will make a loud low-frequency hum. Repeat when the lawyer's beer bottle is half-full, quarter full, and empty. The bottle will hum at a new lower frequency each time.

When the soundboard of an acoustic guitar vibrates, it pushes the air in and out of the box. The box acts similar to a Helmholtz resonator, and it enhances certain low-end frequencies if it combines with the soundboard wave constructively. If the wave from the cavity interacts with the soundboard's wave destructively, it will decrease the overall amplitude of the corresponding frequency. This addition and subtraction from the amplitudes add to the guitar's timbre.

Looking for cavities — no dentist required

The problem with vibrating vocal folds is they don't explain how humans can communicate with such a vast collection of sounds. The cavities within the human body allow people to control which harmonics are damped and which harmonics pass through.

The human voice is more like the clarinet (refer to the section, "Resonating with a clarinet" earlier in this chapter for more information). The vocal folds vibrate like the reed and the sound propagates down a long tube and out a big opening (in this case, the mouth). The vocal folds vibrate at a set of harmonics with them all having approximately the same amplitude. By varying the tension in the vocal folds, the fundamental frequency can vary over a range of frequencies from about 50 hertz to 200 hertz for a male. The sound wave travels through the body, the two most important cavities being the mouth cavity and the nasal cavity. The cavities have resonant frequencies, which are called *formants*. The harmonics close to the formants will travel through the cavities unimpeded, whereas the harmonics far from the formants will be damped out and go unheard. Changing the shape of the cavities changes the formants, allowing humans to make different sounds or sing different notes.

The range of speech (300 hertz to 3000 hertz) usually has three formants in this frequency range. Consider a sword-swallower with his or her throat and mouth all in a straight line, which resembles a stopped organ pipe. As an example, suppose the sword-swallower makes a 6-inch (0.152-meter) long and 1-inch (2.54-centimeter) diameter pipe. If so, then the resonant frequencies using $v = 1{,}130$ feet per second (344 meters per second) are

$$f_{2n-1} = \frac{v(2n-1)}{4(L+0.8R)} = \frac{1130 \text{ ft s}^{-1}(2n-1)}{4(0.5 \text{ ft} + 0.8(0.0417 \text{ ft}))}, \quad n = 1,2,3,...$$

The harmonics in the correct range of frequencies are $f_1 = 530$ hertz, $f_3 = 1590$ hertz, and $f_5 = 2650$ hertz.

If the sword-swallower is a male with an average fundamental frequency of 125 hertz, then the frequencies ($f_n = n\, f_1$) that will pass through most easily will be $f_4 = 500$ hertz, $f_{13} = 1{,}625$ hertz, and $f_{21} = 2{,}625$ hertz. All three are within 35 hertz and easily will pass out of the body.

If the sword-swallower is a female with an average fundamental frequency of 200 hertz, then the frequencies ($f_n = n\, f_1$) that will pass through most easily will be $f_3 = 600$ hertz, $f_8 = 1{,}600$ hertz, and $f_{13} = 2{,}600$ hertz. The harmonic f_8 is very close, but the other two are more than 50 hertz off and will be partially damped.

Chapter 14

Detecting Sound Waves with the Ear

....

In This Chapter

▶ Hanging things from the ear and more

▶ Powering the eardrum

▶ Understanding how important hearing is

....

*H*ow humans and other animals hear is very complex. Understanding the entire process from the sound waves entering the auditory canal to the processing of the electrical signals is important to know. In addition, the ear is extremely sensitive to sound waves. Humans can hear over an enormous range of intensities and over a very large range of frequencies. Hearing is an important area of biophysics with many applications in research and in society. For example, with an increased understanding of the ear, scientists can develop improvements in repairing people with damaged hearing.

This chapter examines the ear and one of its primary functions — hearing — discusses the power within a sound wave and its transfer to the ear, and explains different applications, such as complex waves and beats.

Understanding Hearing and the Ear

One of the primary functions of the ear is to convert energy from one form (sound waves) to another form (electrical impulses, which are transmitted to the brain). The human ear doesn't just convert the energy; it also tells the brain the frequency of the sound wave and the direction the sound is coming from and allows the brain to distinguish specific sounds within a noisy environment. (Your brain is good at listening to your friend talking while standing beside a very busy street.) The following sections look at the three parts of the ear.

Outer ear

The outer ear is the first of three parts of the ear. The outer ear consists of three parts:

- The *pinna* is the part of the ear that sticks off the head. It has become common practice in some cultures to stick pieces of metal or wood into it or puncture it with holes.

- The *auditory canal (ear canal)* looks like a stopped organ pipe. It's a round circular tube with one end open and one end closed.

- The *eardrum* (more technically referred to as the *tympanic membrane*) is the closed end of the auditory canal. The eardrum typically is the boundary between the outer ear and the middle ear.

The primary functions of the outer ear are to channel sound toward the eardrum and to amplify sounds within a certain frequency range. The greatest amplification occurs at the resonance frequencies of the outer ear canal. (The *resonance frequencies* are the special frequencies where a wave and its reflections off a pair of boundaries interfere constructively, thereby enhancing the amplitude of the wave.) The boundaries for the outer ear are the eardrum and the pinna.

To calculate the resonant frequencies, assume the eardrum is solid so it's a node for air displacement (antinode for sound waves) and the open end of the auditory canal is an antinode for air displacement (node for sound waves). I need some numbers, so I called Carrie over and measured her auditory canal and found:

- It had a length L = 1 inch (0.08 feet = 0.0244 meters).
- It had a radius R = 8.50×10^{-3} feet (0.00259 meters).

Remember, the speed of sound is 1,130 feet per second (344 meters per second). You can then look in Chapter 13 and find the formula for resonant frequencies with one open boundary and one closed boundary. The resonance frequencies for Carrie are as follows:

$$f_{2n-1} = \frac{v(2n-1)}{4(L+0.8R)} = \frac{1130\,\text{ft/s}(2n-1)}{4\left(0.08\text{ft} + 0.8\left(8.5 \times 10^{-3}\text{ft}\right)\right)}, \quad n = 1,2,3,...$$

The resonant frequencies for Carrie's ear in the audible range are f_1 = 3,250 hertz, f_3 = 9,760, and f_5 = 16,300 hertz. These frequencies cause the sound wave to resonate within the auditory canal. The typical adult human ear is most sensitive to sound waves in the range of 3,000 to 4,000 hertz, which corresponds to the fundamental resonant frequency of the outer ear. Note that the 88-key piano works in the frequency range 27 hertz to 4,200 hertz, so the upper four white keys are in the range of a human's most sensitive hearing.

Middle ear

The second area of the ear is the middle ear. The outer ear is filled with air at atmospheric pressure. The inner ear is filled with fluid. The middle ear and the eardrum separate the outer ear from the inner ear, allowing the human ear to hear and preventing the sound waves from bouncing off. The purpose of the middle ear is to transfer sound waves from air to the perilymph fluid within the inner ear.

The parts of the middle ear are as follows:

- ✔ The *eardrum (tympanic membrane)* is the barrier between the outer ear and the middle ear.

- ✔ The *oval window* is the barrier between the middle ear and the inner ear.

- ✔ The *tympanic cavity* is a cavity behind the tympanic membrane that is ventilated through the nose via the *eustachian tube*.

- ✔ The *malleus* is one of the three *auditory ossicles* within the middle ear. It connects to the eardrum.

- ✔ The *stapes* is one of the three *auditory ossicles* within the middle ear. It connects to the oval window.

- ✔ The *incus* is one of the three *auditory ossicles* within the middle ear. It connects the malleus to the stapes.

The tympanic cavity is necessary so the membrane will vibrate with the frequency of the incoming sound wave. The three auditory ossicles transfer the vibrations of the eardrum to the oval window. The *mechanical advantage* (the magnitude of the load force divided by the magnitude of the applied force [refer to Chapter 6], which in this case is the force on the oval window divided by the force of the eardrum on the malleus) of the auditory ossicles is approximately 2, and the surface area of the eardrum is 30 times larger than the surface area of the oval window, which means the ratio of the pressure at the oval window to the pressure at the eardrum is

$$\frac{P_{o.w.}^{(a)}}{P_{t.m.}^{(a)}} = \frac{F_{o.w.}}{F_{t.m.}} \frac{A_{t.m.}}{A_{o.w.}} = 2 \times 30 = 60$$

A factor of 60 amplifies the sound wave in this setup. The body fortunately is designed to protect the ear against loud sounds using the muscles of the middle ear.

The eardrum is very weak. A pressure difference between the auditory canal and the tympanic cavity of 4.1×10^{-3} pounds per square inch (0.60 pounds per square foot = 29 pascals) is sufficient to rupture the eardrum, whereas atmospheric pressure is 14.69 pounds per square inch (2,116 pounds per square foot = 1.013×10^5 pascals).

Inner ear

The third part of the ear is the inner ear, which plays an important part in human biophysics (biomechanics). The parts of the inner ear are as follows:

- The three *semicircular canals* detect the motion of the head and help with balance.
- The *vestibule* is the middle portion of the inner ear, which is connected to the semicircular canals and the cochlea. It helps with balance, motion, direction (gravity), and hearing.
- The *cochlea* houses the nerves required for hearing that send signals to the brain.

The middle ear is in contact with the vestibule with the stapes touching the oval window. The cochlea has two canals; as the *oval window* is pushed in and out, it creates a sound wave within the perilymph fluid. The fluid moves down one canal of the cochlea to the end and then down the other canal to the vestibule, where the *round window* is pushed in and out in sync with the oval window. The fluid is incompressible, and the pair of windows allows for the volume to remain constant.

The motion of the fluid in the cochlea causes the cochlear partition to vibrate at different locations depending on the frequency of the sound wave. The partition's motion causes the *stereocilia* (nerve hair cells) to bend and send a signal (nerve impulse) to the brain. The brain then processes the signal.

Realizing How Sensitive the Human Ear Is — the Power of Sound Waves

The human ear is most sensitive to sound in the 3,000 to 4,000 hertz frequency range. Outside this frequency range, the sound wave needs to be more intense for the average adult human to hear it. So, not only does the frequency of the sound have to be in the range of 20 to 20,000 hertz, but it also must have a minimum intensity to be heard. In these sections I introduce power and intensity, and the conversion to intensity level. I also discuss the reasons sound waves have a limited range.

Taking a closer look at ear power

A *sound wave* is a *pressure wave,* and the amplitude of the wave is related to the power of the wave. In addition, the amount of power available to move the eardrum decreases as the source moves away because a sound wave spreads out the power over a bigger area as it moves away from the source.

These sections focus on the intensity, intensity level, power, and pressure amplitude of sound waves. These sections allow you to quantify your analysis of sound waves interacting with the ear. If a sound wave doesn't have enough power, it won't be able to move the eardrum. If it has too much power, it will cause the eardrum to rupture.

Tuning into a sound wave

The *intensity* (I) of a sound wave is related to the *power* (P) in the sound wave and the amplitude of the sound wave:

$$I = \frac{P}{A} = \frac{\left(\Delta P^{(g)}\right)^2}{2\rho v}$$

Here I is the intensity of the sound with units of pounds per (foot second) (or watts per square meter), P is the power with units foot pounds per second (or watts), A is the cross-sectional area of the wave with units of square feet (or square meters), $\Delta P^{(g)}$ is the amplitude of the pressure wave with units of pounds per square foot (or newtons per square meter), ρ is the density of air with units of slugs per cubic foot (or kilograms per cubic meter), and v is the speed of sound with units of feet per second (or meters per second).

At a frequency of 1,000 hertz, an average adult human can hear sound waves with an intensity of 0.000000000000069 pound per (foot second) (0.000000000001 watts per square meter) or greater. This intensity is called the *threshold of hearing.* As the intensity increases, the sound wave will eventually become painful, which is called the *threshold of pain.* The threshold of pain is achieved at an intensity of 0.069 pound per (foot second) (1 watt per square meter) when the frequency is 1,000 hertz.

Sound waves at this intensity can cause permanent damage to the person's hearing. (Also, a person can have permanent damage done to his or her hearing at lower intensities if exposed to elevated noise levels over a long period of time.) The change in the intensities from the threshold of hearing to the threshold of pain is massive, so instead the *intensity level* (β) is used and it's related to the *intensity* (I):

$$\beta = (10dB) \log\left[\frac{I}{I_o}\right]$$

$$I = I_o 10^{\beta/(10dB)}$$

β is the intensity level with units of *decibels (dB)* and I is the intensity. In this formula, $I_o = 6.855 \times 10^{-14}$ pound per (foot second) (10^{-12} watts per square meter) is the reference intensity, which is a constant. It's the threshold of hearing for the average adult human at a frequency of 1,000 hertz.

An average human hears sound waves in the range of intensity levels, at a frequency of 1,000 hertz, starting at 0 decibels and increasing up to 120 decibels. People can't hear most sounds at 0 decibels. The threshold of hearing is a function of frequency. The threshold of hearing intensity level for an average adult increases as the frequency moves away from the 3,000 to 4,000 hertz range. A few examples of the threshold of hearing are $\beta_{threshold}$ (50Hz) = 55 dB, $\beta_{threshold}$ (100Hz) = 40 dB, $\beta_{threshold}$ (500Hz) = 5 dB, $\beta_{threshold}$ (700Hz to 1500Hz) = 0 dB and $\beta_{threshold}$ (3,000Hz to 4,000Hz) = –10 dB.

Grasping the eardrum and limit range

Even if you assume conservation of energy, you can't hear someone from a long distance away. This section looks at the reason why sound waves have a limit range.

To begin, you need to know how sound waves travel through the air. A machine such as a plane generates noise that spreads out in a sphere; the noise goes in all directions. If you follow one of the sound wave's compression zones (a *compression zone* is a region of space where more air molecules are packed into it than there are supposed to be present), it covers an area called the *cross-sectional area,* which is spherical in this case. In the case of an acoustic speaker, the sound waves usually travel only in one direction, so the cross-sectional area is a hemisphere. The cross-sectional area for sound waves produced by humans and animals is usually in a specific direction, which is less than hemispherical. You can test it by going to an open field and having someone speak normally while you stand behind her staring at her back. She's hard to hear when facing away from you because most of the sound wave is going in the opposite direction.

The formula for calculating the cross-sectional area of a sound cone is

$$A = R^2\Omega = R^2 2\pi\left(1 - \cos(\theta)\right)$$

A is the spherical surface area of the cone (think of the cone as part of a sphere) with units of square feet (square meters), R is the distance from the cone's apex to its base (the radius of the sphere) with units of feet (meters), Ω is called the solid angle with units of *steradians* (this is a dimensionless unit similar to radians), and θ is the polar angle with units of radians (or degrees). Consider the earth, and you can think of the cone starting at the center of the earth and going to the surface with the North Pole the center of the cone. R is the radius of the earth, and θ is the angle between the North Pole and the edge of the cone, so the total angle across the cone is $2\,\theta$.

For example, a machine will usually spread sound out in all directions, which has a polar angle $\theta = \pi$ radians. Substitution of this angle into the area formula gives A = $4\,\pi\,R^2$, which is the surface area of a sphere. In the case of the sound speaker, the sound will usually spread out with a polar angle, $\theta = \pi/2$. Substitution of this angle into the area formula gives A = $2\,\pi\,R^2$, which is the surface area of a hemisphere. An animal, such as a bat, can produce a focused sound wave that spreads out in a tight cone (θ is about 10 degrees).

If you assume no energy in the sound wave is lost to the air or the surroundings, then the sound wave keeps propagating, but the energy is spread over a larger surface as it moves outward. The intensity will eventually drop below the threshold of hearing, and a human won't be able to hear the sound. The *maximum range* is (assuming conservation of energy) as follows:

$$R_{max} = R_{start}\sqrt{\frac{I_{start}}{I_{threshold}}} = R_{start}\,10^{(\beta_{start} - \beta_{threshold})/(20\,dB)}; \text{ or}$$

$$I_{start} = I_{threshold}\left(\frac{R_{max}}{R_{start}}\right)^2; \text{ or } \beta_{start} = \beta_{threshold} + (20\,dB)\log\left[\frac{R_{max}}{R_{start}}\right]$$

The formulas look complicated, but that is because the expression has four equations. The sound is produced and spreads out in a cone. When it has traveled a distance R_{start}, the intensity (I_{start}) and intensity level (β_{start}) are measured. The sound wave continues traveling outward in a cone until you can't hear it anymore. The distance from the source where this occurs is R_{max}, and the threshold of hearing has an intensity $I_{threshold}$ and intensity level $\beta_{threshold}$. The first line in the formula allows you to calculate the maximum distance if you know the starting distance and the intensities (middle term) or intensity levels (right-side term). The second line allows you to calculate the starting intensity or starting intensity level if you know the starting distance, the maximum distance, and the threshold of hearing intensity or intensity level.

This formula is valid for any two distances and the corresponding intensities or intensities level. The formula does assume conservation of (acoustic) energy.

To solve this problem, use the previous formula and follow these steps:

Connie, a collie dog, lives on the balcony across from your place, and the sound waves from her mouth make a polar angle of 30 degrees when she barks. At 3 a.m. you measure her bark to have a frequency of 200 hertz and an intensity level of 70 decibels at 1.0 feet (0.3048 meters) from her mouth. How much power is in her bark? How far away do you have to be before you can't hear her bark?

To solve this problem, use the previous formula and follow these steps:

1. **Draw a diagram.**

   To save on space, I don't draw a diagram, but you should because doing so does help. In your diagram, include the dog, the cone 1 foot (0.3048 meters) away with the intensity level, and a second cross-sectional area for the cone farther from the dog. The second area is at R_{max}, $I_{threshold}$, and $\beta_{threshold}$.

2. **Find the formulas.**

   The first question in the problem asks you to find power. The previous section, "Tuning into a sound wave," tells you the relationship power equals intensity times the cross-sectional area of the cone. Therefore, you need to find the intensity and area first.

   You know the intensity level at 1.0 foot, so you can find the intensity.

   You know the polar angle at 1.0 foot, so you can calculate the area.

   The formula for maximum range requires you to know the starting distance, the starting intensity level, and the threshold of hearing.

3. **Find the numbers for the intensity at 1.0 foot.**

   The problem tells you that the intensity level is $\beta_{start} = 70$ decibels at $R_{start} = 1.0$ foot (0.3048 meters).

4. **Solve for the intensity at 1.0 foot.**

   $I_{start} = I_o \beta_{start/10} = (6.855 \times 10^{-14}$ pound per (foot second)$) 10^{70/10} = 6.855 \times 10^{-7}$ pound per (foot second) (10^{-5} watts per square meter).

5. **Find the numbers for the area at 1.0 foot.**

   The problem tells you that $R_{start} = 1.0$ foot (0.3048 meters), and $\theta = 30$ degrees.

6. **Solve for the area of the cone at 1.0 foot.**

   $$A = R^2 2\pi \left(1 - \cos(\theta)\right) = \left(1 \text{ ft}\right)^2 2\pi \left(1 - \cos\left(30^0\right)\right) =$$

   0.842 square feet $(0.0782$ square meters$)$

7. **Solve for the power in the bark.**

   Substitute the area and intensity into the formula for the power:

   $$P = I_{start} A_{start} = \left(6.855 \times 10^{-7} \frac{\text{lb}}{\text{ft s}}\right)\left[0.842 \text{ ft}^2\right] =$$

   $5.77 \times 10^{-7} \dfrac{\text{foot pound}}{\text{second}} \left(7.82 \times 10^{-7} \text{ watts}\right)$

8. **Find the threshold of hearing at 200 hertz.**

   The problem tells you the intensity level is β_{start} = 70 decibels and that R_{start} = 1.0 foot (0.3048 meters). The problem also says that the bark is at a frequency of 200 hertz, but it doesn't state what the threshold of hearing intensity level is at that frequency. Refer to the previous section to see that the $\beta_{threshold}$ = 25 decibels at 200 hertz.

9. **Solve for the maximum range.**

   $$R_{max} = R_{start} 10^{(\beta_{start} - \beta_{threshold})/20} = \left(1.0 \text{ft}\right)10^{(70-25)/20} = 178 \text{ feet } \left(54.2 \text{ meters}\right)$$

Grasping How Amazing Hearing Is

The ear is amazing at picking up different sounds. You can sit in a very noisy environment, and your brain will pick out the sounds you're interested in. This section looks at how hard a task that is, which your brain performs without any difficulty. These sections discuss how you can mathematically combine waves and introduce the beat frequency and how tuning by ear works.

Interacting complex waves

Waves are usually the combination of multiple waves. The *linear superposition principle* tells you the resultant wave is a linear sum of all the waves present. Figure 14-1 shows an example of a *complex wave*, which is the combination of three sound waves. This figure also includes an *envelope curve*, which shows how the three waves combine constructively sometimes and destructively at other times. Figure 14-1 shows the resultant of combining the following three single frequency sinusoidal sound waves, with the same amplitude, together as they enter a microphone: 20.1 hertz, 22.0 hertz, and 29.9 hertz.

$$\frac{P^{(a)}(t) - P_{atm}}{\Delta P} = \sin\left[t\left(126\frac{\text{rads}}{\text{s}}\right)\right] + \sin\left[t\left(138\frac{\text{rads}}{\text{s}}\right)\right] + \sin\left[t\left(188\frac{\text{rads}}{\text{s}}\right)\right]$$

You can do this for any kind of wave. In Chapter 12, I discuss the different mathematical functions that represent the different waves.

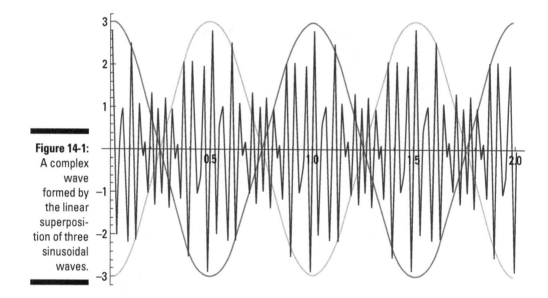

Figure 14-1:
A complex wave formed by the linear superposition of three sinusoidal waves.

The reverse is also true where complex waves can be studied as a linear superposition of simpler waves. You can use a *Fourier series* when you know the length (or duration) of periodicity or if the region of space is finite in size. The *harmonic waves* associated with the resonant frequencies are an example of a Fourier series. The technique can be expanded to *Fourier integrals*, which allows you to analyze general complex waves. Unfortunately, the methods of Fourier series and Fourier integrals are beyond the scope of this book.

Beating beats and tuning a guitar

The human ear is very sensitive and can detect changes in the intensity level as small as 0.5 decibels. This section is about how two waves sometimes interfere constructively and sometimes destructively. Suppose two standing sound waves have the same amplitude, then they can be combined together to form a single wave as the following demonstrates

$$\frac{P^{(a)}(t) - P_{atm}}{\Delta P^{(g)}} = \sin[\delta_1 - \omega_1 t] + \sin[\delta_2 - \omega_2 t] =$$

$$2\cos\left[\frac{\delta_1 - \delta_2 - (\omega_1 - \omega_2)t}{2}\right]\sin\left[\frac{\delta_1 + \delta_2 - (\omega_1 + \omega_2)t}{2}\right]$$

$P^{(a)}(t)$ is the absolute pressure, P_{atm} is the atmospheric pressure, $\Delta P^{(g)}$ is the amplitude of the sound (pressure) wave, the second set of terms is the combination of two sinusoidal waves, and the third set of terms (on the right-hand side) is the two sinusoidal waves rewritten with the help of a trigonometric identity. In the sine function, the average phase shift and the average angular frequency $(\omega_1 + \omega_2)/2$ are being used. For example, if you have a red light and a yellow light and combine them, the average frequency corresponds to orange light. The same thing happens with sound waves when they combine together.

The sine function is the sinusoidal pressure wave, while the cosine function combines with the ΔP to give an effective amplitude that varies slowly in time if the frequencies are close to the same value. The cosine causes the waves to combine constructively for a while, then destructively, then constructively, and so on. The cosine is the envelope of the wave, and it has antinodes (constructive interference) and nodes (destructive interference) with a slow variation in time (small frequency) as shown in Figure 14-1. Remember, a *node* is a place where the wave causes no disturbance (equilibrium), and an *antinode* is the location of maximum disturbance from equilibrium. This slow variation in time is called the *beat frequency*, which must produce at least a 0.5 decibel variation in the amplitude for the human ear to hear it.

The following formula shows the *beat frequency*:

$$f_b = |f_1 - f_2| = \frac{\omega_b}{2\pi} = \frac{|\omega_1 - \omega_2|}{2\pi} > 0$$

It's twice the frequency of the envelope. People who are good at tuning musical instruments can hear the beat frequency between two instruments playing the same note and adjust one of the notes until the beat is gone.

Consider this example: Doug is tuning a friend's guitar using his guitar as a reference. He vibrates his A_3 string (220 hertz) and then the A_3 string on his friend's guitar. He notices the volume gets quiet (nodes) five times every second, so the envelope frequency is $\frac{5}{2}$ hertz. Figure 14-2a shows how the two waves combine constructively and then combine destructively to form a beat frequency. The two individual sound waves, the resultant wave, and the envelope are shown for time t = 0.05 seconds to 0.10 seconds. Figure 14-2b shows the resultant wave along with the envelope for time t = 0.0 seconds to 0.5 seconds, which clearly shows the beat.

Doug concludes that the beat frequency is 5 hertz (twice the envelope) and the guitar string is tuned to either 225 hertz or 215 hertz. In this case, the guitar is tuned to 225 hertz. If Doug increases the tension in the string, then the beat frequency increases. If he decreases the tension in the string, then the beat frequency gets smaller and will vanish when the strings are in tune.

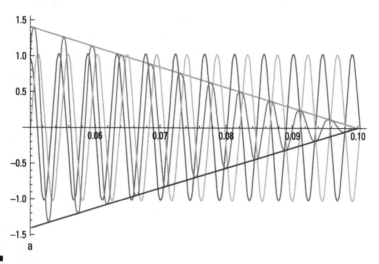

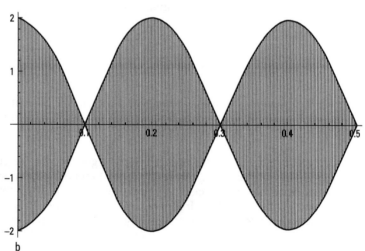

Figure 14-2: A 220 hertz sound wave combined with an unknown sound wave for time t = 0.05 seconds to 0.10 seconds in (a) and for time t = 0.0 seconds to 0.5 seconds in (b).

Chapter 15

Listening to Sound — Doppler Effect, Echolocation, and Imaging

· ·

In This Chapter

▶ Dancing with Doppler

▶ Finding the range of echolocation

▶ Sounding the inside with ultrasound

· ·

Sound is one of the main methods animals use to interact with their surroundings and allows the animals to understand what is going on around them, especially nocturnal animals. Animals use sound for communication, navigation, and tracking. These applications of sound make the understanding of sound and its applications an important field within biophysics.

This chapter discusses some very interesting applications of sound waves, including the Doppler Effect, the echolocation technique, and ultrasound imaging.

Forecasting with the Doppler Effect

The *Doppler Effect* is the name given to the phenomenon of the frequency of a wave changing when the source of the wave or the observer is in motion. Imagine you're at the beach. If you're just standing in the water, the crests of waves will hit you with some frequency, but if you're walking into the waves, they'll hit you at a faster rate. Many fields of science including biophysics use the Doppler Effect; even some animals have evolved to take advantage of it.

The following sections explain why the Doppler Effect occurs by first looking at the listener moving, then the source moving, and finally when both the source and the listener are moving. These sections also introduce the Doppler Effect for electromagnetic radiation (light) because the behavior is slightly different for light.

Moving on the receiver's end

If the source of the wave is stationary and the receiver is moving ($v_{receiver}$), then the frequency at the receiver ($f_{receiver}$) will change relative to the frequency originally produced (f_{source}):

$$f_{receiver} = f_{source} \left[1 \mp \frac{v_{receiver}}{v_{wave}} \right]$$

The minus sign is if the receiver is moving away from the source, and the plus sign is for when the receiver is moving toward the source. This modification to the frequency occurs because the distance between the *crests* (the wavelength) is unchanged, but because the receiver is moving, it will take longer (if it's moving away from the source) or less time (if it's moving toward the source) for each crest to reach the receiver.

Moving on the source's end

If the source of the wave is moving (v_{source}) and the receiver is stationary, then the frequency received ($f_{receiver}$) will change relative to the frequency originally produced (f_{source}) as such:

$$f_{receiver} = \frac{f_{source}}{\left[1 \pm \dfrac{v_{source}}{v_{wave}} \right]}$$

The plus sign is if the source is moving away from the receiver, and the minus sign is for when the source is moving toward the receiver. This modification to the frequency occurs because the distance between the crests is shrunk in the direction the source is moving, whereas the distance between the crests is stretched in the direction opposite to the direction the source is moving.

Moving sources and receiver

In other cases, both the source of the wave is moving and the receiver is moving. I assume one-dimensional motion, where they're either moving toward or away from each other. The frequency received ($f_{receiver}$) changes relative to the frequency originally produced (f_{source}) as such:

$$f_{receiver} = f_{source} \frac{\left[v_{wave} \mp v_{receiver} \right]}{\left[v_{wave} \pm v_{source} \right]}$$

The $-v_{receiver}$ is for when the receiver is moving away from the source, the $+v_{receiver}$ is for when the receiver is moving toward the source, the $+v_{source}$ is for when the source is moving away from the receiver, and the $-v_{source}$ is for when the source is moving toward the receiver.

The easiest way to remember the signs in the equation is to remember that an object moving toward another causes the frequency to increase, whereas an object moving away causes the frequency to decrease.

Erin is walking toward you at 3 miles per hour (4.40 feet per second = 1.34 meters per second) while playing her out-of-tune violin. You run away in terror. She tries to play an A_4 note, but the frequency is 450 hertz. How fast do you have to run so the note sounds in tune (440 hertz)?

To solve this problem, follow these simple steps:

1. **Find the correct formula and determine what you need to find.**

   You want to find your speed, which you can find by using the formula for the Doppler Effect when both the receiver and source are moving. You need to find several numbers for different variables in this formula.

2. **Find the numbers for the two frequencies, speed of sound, and Erin's speed.**

   The source of the sound is the violin, and it's moving toward you at Erin's speed. You're the listener (receiver), and you're moving away from the source. You're given

   $f_{source} = f_{violin} = 450$ hertz

   $f_{receiver} = f_{listener} = 440$ hertz

   $v_{source} = 4.40$ feet per second (1.34 meters per second). Use the negative sign in front of the source's speed.

   You have to look up the speed of sound, and you find it is $v_{wave} = v_{sound} = 1{,}130$ feet per second (344 meters per second).

   You're running away from Erin, so you need to use the negative sign in front of the receiver's speed in the Doppler Effect formula.

3. **Solve the Doppler Effect formula for the receiver's speed.**

   Rearrange the Doppler Effect formula (both signs are negative) and solve for $v_{receiver}$:

   $$v_{receiver} = v_{wave} - \frac{f_{receiver}}{f_{source}}\left[v_{wave} - v_{source}\right] = 1130\,\text{ft/s} - \frac{440\text{Hz}}{450\text{Hz}}\left[1130\,\text{ft/s} - 4.40\,\text{ft/s}\right]$$

Doing the calculation, you see you have to be running at 29.4 feet per second (8.97 meters per second) for the violin to sound in tune. Changing units, this corresponds to running at 20.1 miles per hour (32.3 kilometers per hour).

Considering the special case — light

All waves require a medium to move through, and the Doppler Effect is the same for all waves with the exception of light. Light doesn't need a medium and has other strange properties, which makes the Doppler Effect different relative to other waves. You can't tell if the receiver or the source is moving. This motion is relative, and the Doppler Effect in this case is

$$f_{receiver} = f_{source} \sqrt{\frac{v_{light} - v_{relative}}{v_{light} + v_{relative}}}$$

$v_{relative}$ is positive if the objects are moving apart, which lowers the frequency (red shifted), and $v_{relative}$ is negative if the objects are moving together, which increases the frequency (blue shifted). Light goes from red (low frequency) to blue (high frequency). (The light from all the distant galaxies is shifted toward the red [red shifted], which is how scientists know the universe is expanding.)

Finding Your Way in the Dark — Echolocation

All mammals use their ears to locate the source of sound waves, but bats, porpoises, and a few other animals create a noise and use the echo to locate objects, prey, and predators. This concept is called *echolocation*. These sections take a closer look at echolocation and discuss how some animals use a constant frequency sound wave combined with the Doppler Effect for echolocation. I also discuss how some animals use a frequency modulated sound wave for echolocation and why echolocation has a limited range even when you assume conservation of (acoustic) energy.

Echolocating with constant frequency sound waves and the Doppler Effect

Some animals, when avoiding objects and doing a general search for food, emit a long chirp at a constant frequency. This frequency-shifted echo, because of the Doppler Effect, is as follows:

$$f_{heard} = f_{chirp} \frac{\left[v_{sound} + v_{bat} \right]\left[v_{sound} - v_{insect} \right]}{\left[v_{sound} - v_{bat} \right]\left[v_{sound} + v_{insect} \right]}$$

In the equation, I assume the bat is flying toward its supper and the supper is trying to get away. If the bat is flying away from the insect, then v_{bat} is negative, and if the insect is flying toward the bat, then v_{insect} is negative. There are two sets of speeds in the formula because the sound wave leaves the bat's mouth and travels to the insect. The sound wave hitting the insect has the frequency shifted because of the Doppler Effect. This new frequency sound wave then bounces off the insect and travels back to the bat. The reflected wave traveling to the bat has its frequency shifted because of the Doppler Effect as well. In the first part, the bat's mouth is the source, and the insect is the receiver. In the second part, the bat's ear is the receiver, and the insect is the source. Note that I call the source of the sound and the receiver of the echo a bat, but it's true for any animal that uses echolocation.

The C. parnelli bat uses this form of echolocation. It has a threshold of hearing between 30 and 40 decibels from 25 kilohertz to 75 kilohertz with one exception. At a frequency of 61.8 kilohertz, the bat has a threshold of hearing of 0 decibels. What is the bat's typical flying speed assuming the bat emits chirps at a frequency of 60.0 kilohertz?

To solve this problem, follow these steps:

1. **Understand the problem by finding the appropriate equation.**

   You want to find the bat's speed, so you need to use the echolocation equation. You need to find both frequencies, the speed of sound, and the speed of the object the sound bounces off of.

2. **Finding the numbers for both frequencies, the speed of the object the sound bounces off of, and the speed of sound.**

   - You're told the frequency the bat emits sound at: $f_{chirp} = 60000$ hertz

   - You're told the frequency of the echo the bat hears: $f_{heard} = 61800$ hertz

   - Most objects don't move, so you can assume a speed of zero for the insect: $v_{insect} = 0$ feet per second (0 meters per second)

   - You can look up the speed of sound: $v_{sound} = 1{,}130$ feet per second (344 meters per second)

3. **Substitute the numbers into the formula after rearranging it for v_{bat}.**

   $$v_{bat} = v_{sound} \frac{\left[f_{heard} - f_{chirp} \right]}{\left[f_{heard} + f_{chirp} \right]} = (1130\,\text{ft/s}) \frac{\left[61.8\ \text{kHz} - 60.0\ \text{kHz} \right]}{\left[61.8\ \text{kHz} + 60.0\ \text{kHz} \right]} =$$

   $$16.7 \frac{\text{feet}}{\text{second}} \left(5.09 \frac{\text{meters}}{\text{second}} \right)$$

The C. parnelli bat has an average flying speed of 16.7 feet per second (5.09 meters per second), which is 11.4 miles per hour (18.3 kilometers per hour). At this speed the echo off stationary objects will be at a frequency the bat's hearing is most sensitive to.

You can only see small objects to a certain size before they become too small. In the case of waves, any object smaller than half a wavelength is invisible to the wave, which is called the *diffraction-limit*. This means for the C. parnelli bat the insect must be bigger than a diameter equaling wavelength/2 = v/(2f) = (1130 ft/s)/(60,000 Hz) = 0.0188 feet = 0.113 inches (2.87 millimeters) in order for the bat to detect the insect.

Triangulating with frequency modulated sound — echolocation

Humans rely on their eyes to find objects and judge the distance. In fact, the brain is very good at using the two eyes to *triangulate* on an object and estimate the distance, which helps a lot when reaching for your glass. How does your brain triangulate on an object: Your two eyes combine with the object to form a triangle. Using the angles and the distance between your eyes, your brain can calculate the distance to the object.

Animals can use sound to do the same thing. To begin, refer to the equation in the previous section about the diffraction-limit. If I calculate the diffraction-limit for the human head assuming a wavelength equal to the size of a human head (crest at each ear), I obtain the following:

$$f = \frac{v}{\lambda} \sim \frac{v}{\text{diameter}} \approx \frac{1130 \text{ ft s}^{-1}}{0.67 \text{ ft}} = 1700 \text{ hertz}$$

For frequencies below 1,700 hertz (longer wavelength), the sound wave doesn't notice the head and travels around it. The brain picks up the lag in the crests reaching one ear compared to the other ear, allowing the brain to determine the direction of the source of the sound. When the frequency is greater than 1,700 hertz (shorter wavelength), the head acts like a brick wall reflecting sound waves and preventing the sound waves from reaching the ear on the backside. The difference in the volume of the sound allows the brain to determine the direction of the source of the sound wave. 1,700 hertz is an arbitrary frequency because every head is slightly different and sounds don't come directly from the side. The brain uses both techniques for sound between 1,000 hertz and 4,000 hertz.

This technique is applicable to mammals in general, but the frequency range changes for each animal. Therefore, animals can determine the direction of the source of the sound. Animals that use frequency modulated echolocation go beyond this and use the echo to estimate the direction and distance. Their ears tell them the direction the echo is coming back from, and the brain estimates the distance by knowing the time delay between the chirp from the mouth and the echo reaching the ear:

$$R = \frac{v\,\Delta t}{2}$$

The factor of 2 is present because the chirp has to leave the animal's mouth, travel to the object (prey), bounce off the object (prey), and travel back to the animal's ears.

Understanding the Limited Range of Echolocation

When an animal is using sound waves to locate dinner, the echo needs enough power to drive the eardrum, which means the echolocation technique has limited range. Here I discuss a bat, although it's true for any animal using echolocation. To understand this limited range, a few reasonable assumptions are necessary:

- ✔ The bat's mouth is shaped to produce a small sound wave cone with a cross-sectional area:

  $A_{wave} = \Omega\,R^2$, where $\Omega = 2\pi\big[1 - \cos(\theta)\big]$

  θ is the polar angle. R is the distance from the hungry bat to the bug (the bat's dinner). Ω is typically about 0.1 for a bat. A_{wave} is the area over which the compression of a sound wave is spread out.

- ✔ The cross-sectional area of the bug is A_{bug}.

- ✔ All the power in the sound wave striking the bug goes into the echo, and none of the energy is lost. *Conservation of sound energy* occurs for the entire event from the time the wave leaves the bat's mouth until the echo reaches the bat's ears. The power in the sound wave leaving the bat's mouth is $P_{initial}$.

- ✔ The echo leaves the bug in a uniform hemispherical wave.

The intensity of the echo at the bat's ears is

$$I_{heard} = \frac{P_{initial} A_{bug}}{2\pi \, \Omega \, R^4}$$

which must be greater than the threshold of hearing. Notice that the intensity drops off as the inverse of the separation to the fourth power. This is a very rapid drop-off, making echolocation a very short-range technique, especially considering how small the typical bug is.

BRAINTEASER

Suppose a bat produces a 10^{-3} foot-pound-per-second (1.36 milliwatts) chirp. The solid angle of the sound wave cone is $\Omega = 0.1$. What is the maximum distance that can be between the bat and the bug if the bat is to find its dinner? The bug has a cross-sectional area of 10^{-4} square feet (0.0144 square inches = 9.29 square millimeters). The bat's threshold of hearing is 6.86×10^{-14} pound per (foot second) (10^{-12} watts per square meter).

To figure out this problem, stick to these steps:

1. **Examine the problem to see what you need.**

   The problem wants you to find the maximum distance between the bat and dinner, which means you need to use the preceding formula. Looking at the formula, you need: I (given), P (given), Ω (given), and A_{bug} (given). Therefore, you know everything.

2. **Find the numbers.**

   $I_{heard} = 6.86 \times 10^{-14}$ pound per (foot second) (10^{-12} watts per square meter)

   $P_{initial} = 10^{-3}$ foot pound per second (1.36×10^{-3} watts)

   $A_{dinner} = 10^{-4}$ square feet (0.0144 square inches = 9.29×10^{-6} square meters)

   $\Omega = 0.1$

3. **Substitute the numbers into the formula and solve.**

   $$R = \left[\frac{P_{initial} A_{dinner}}{2\pi \, \Omega \, I_{heard}} \right]^{1/4} = \left[\frac{\left(10^{-3} \text{ ft lb/s}^{-1}\right)\left(10^{-4} \text{ft}^2\right)}{2\pi \; 0.1 \left(6.86 \times 10^{-14} \; \text{lb}/\left(\text{ft s}\right)^{-1}\right)} \right]^{1/4}$$

   $= 39.0$ feet $(11.9$ meters$)$

Seeing the Unseen: Ultrasound Imaging

Sound (pressure) waves are typically split into three regions based on the human hearing:

- ✔ **Infrasound (or infrasonic):** For frequencies less than 20 hertz

- ✔ **Sound (or acoustic or sonic):** For frequencies between 20 hertz and 20,000 hertz

- ✔ **Ultrasound (or ultrasonic):** For frequencies greater than 20,000 hertz

The *diffraction-limit* shows that objects half a wavelength and larger can be imaged using waves, which means ultrasound waves can be used to image very small objects; usually frequencies between 10^6 hertz and 10^{10} hertz are used. Ultrasound waves have been used for more than half a century in industry, the sciences, and the medical field to image objects. The technique is referred to as *acoustic microscopy* in industry and the sciences, whereas it is referred to as *sonography* (or medical sonography) in the medical profession. One of the most important features is low-intensity ultrasound waves have no apparent harmful effects.

When a wave strikes a boundary, some of the wave is reflected and some of the wave is transmitted, so objects even with little difference in their densities will reflect some of the incident sound wave. (Think of your roommate singing in the shower. Some of the sound wave travels through the wall.) The *intensity of the reflected wave* at *normal incidence* is

$$I_{reflected} = I_{incident} \left[\frac{\rho_1 v_1 - \rho_2 v_2}{\rho_1 v_1 + \rho_2 v_2} \right]^2$$

Normal incidence means the incident wave and reflected wave travel perpendicular to the surface. In this equation, $I_{reflected}$ is the intensity of the reflected sound wave, $I_{incident}$ is the intensity of the incident sound wave, ρ_1 is the weight (or mass) density of medium 1, ρ_2 is the weight (or mass) density of medium 2, v_1 is the speed of sound in medium 1, v_2 is the speed of sound in medium 2, medium 1 is the material the incident sound wave and the reflected sound wave are traveling through, and medium 2 is the material the transmitted sound wave is traveling through.

You're at the beach with your MP3 player. The music leaves the speaker, travels through the air, and strikes Frank's smooth, bare stomach. What ratio of the sound waves is reflected from Frank's stomach?

To solve this problem, follow these steps:

1. **Examine the problem to see what you need.**

   The problem wants you to find how much of a sound wave bounces off Frank's stomach. You know which formula to use, which is the easy part. The hard part is that problem doesn't give any numbers, so you need to find the numbers.

2. **Find the numbers for the density of air, the speed of sound in air, the density of human skin, and the speed of sound in human skin.**

   You have to look up these numbers in your favorite reference source. I help by giving you the numbers:

   - The sound is traveling through air, so ρ_1 is the weight density of air is 0.0752 pounds per cubic foot (the mass density of air is 1.2041 kilograms per cubic meter).

   - The speed of sound in air is $v_1 = 1,126$ feet per second (343.2 meters per second).

   - The sound strikes skin, which has a density slightly greater than water, so ρ_2 is the weight density is 65.40 pounds per cubic foot (the mass density is 1,047 kilograms per cubic meter).

   - The speed of sound in skin is $v_2 = 5,151$ feet per second (1,570 meters per second).

3. **Substitute these numbers into the equation and solve for the ratio of the intensities.**

   You know everything except the two intensities, so you can solve using the earlier formula:

   $$\frac{I_{reflected}}{I_{incident}} = \left[\frac{\left(0.0752 \ \text{lb/ft}^3\right)\left(1126 \ \text{ft/s}\right) - \left(65.4 \ \text{lb/ft}^3\right)\left(5151 \ \text{ft/s}\right)}{\left(0.0752 \ \text{lb/ft}^3\right)\left(1126 \ \text{ft/s}\right) + \left(65.4 \ \text{lb/ft}^3\right)\left(5151 \ \text{ft/s}\right)} \right]^2 = 0.9990$$

   The formula shows that almost the entire wave is reflected when the sound wave travels between air and a liquid (or a solid); hence in medical clinics the paddle is kept in contact with the skin so the densities and speeds are closer to each other with less reflection. This is also the reason for the middle ear in your body.

One danger of ultrasonic sound waves is that the wavelength is smaller than a cell and will cause stresses on the cell. If the intensity is too great, it will cause the cell to rupture.

Part V

Interacting Subatomic Particles' Influence on Biological Organisms

An Electrical Circuit

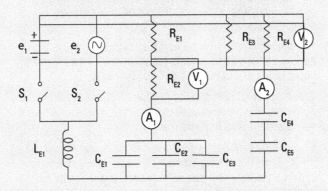

In this part . . .

✔ Discover the electromagnetic force, why it's the dominant force in biophysics, and how the force binds matter together to form organisms.

✔ Learn how to build electrical circuits, as well as uncover how to store energy in electrical and magnetic fields. You can also see the advantages and disadvantages of alternating and direct circuits.

✔ Find out why electromagnetic radiation can be split into non-ionizing radiation and ionizing radiation and all the interesting applications and benefits of both types.

✔ Grasp the ideals of radioactivity, how elements decay, and the applications in biophysics such as carbon dating (and how it works and what its limitations are).

✔ Explore why radiation is dangerous to biological systems, both the short-term and long-term effects, and discover how to estimate the danger of the radiation and if you should be concerned.

✔ Discover how biophysics is applied in the medical profession through medical physics by discovering what nuclear medicine is, how diagnostic images are made, and what radiotherapy is.

Chapter 16

Charging Matter: The Laws of Physics for Electricity, Magnetism, and Electromagnetism

....

In This Chapter

▶ Interacting objects via electric and magnetic fields

▶ Supplying power and energy to the system

▶ Building electrical circuits

....

*T*he *electromagnetic force* is the primary force in biophysics, and it's the force that makes biological systems work and binds them together. Only four fundamental forces exist in nature: the strong force, the weak force, the electromagnetic force, and the gravitational force. Most things are based on the electromagnetic force, from signals in nerves, thoughts in the brain, to a muscle contracting, to the binding of atoms to form molecules. This force forms and controls biological organisms.

This chapter allows you to understand electricity, magnetism, and electromagnetism, and how they interact with matter. These sections describe the laws of electromagnetism and how charge, electric fields, and magnetic fields interact; power, energy, and the storage of energy; and circuits and neural networks.

You may have notice that the same symbols keep coming up but with different subscripts. The symbols are shorthand, and the same symbols keep being used, so you must be careful when reading different sources. Usually, the meaning of the symbol is clear in the discussion.

Forcing Matter in Biological Systems to Interact

Forces are the quantitative description of the interaction between particles. Biological systems are formed by molecules, which are formed by atoms that are made of electrons, protons, and neutrons. These particles create electric fields, magnetic fields, and electromagnetic radiation. In addition, electric and magnetic fields create forces on other particles.

The following sections introduce the laws that describe how charged particles, electric fields, and magnetic fields interact with each other, including the Lorentz force, Coulomb's law, Gauss' law, the Biot-Savart law, the Maxwell-Ampere law, and Faraday's law.

The four laws, Gauss' law for electric fields, Gauss' law for magnetic fields, the Maxwell-Ampere law, and Faraday's law, are known as *Maxwell's equations,* and combined with the Lorentz force are a complete description of electricity, magnetism, and electromagnetism (electromagnetic radiation) and the foundation of optics and electrical circuits.

Describing matter by their properties: The Lorentz force

You can classify materials by how they behave in electric, magnetic, and electromagnetic fields. The *Lorentz force* states that electric fields and magnetic fields will produce forces on charged particles such as electrons and protons.

The magnitude of the force is equal to the strength of the electric field times the magnitude of the particle's charge. The electric field forces positively charged particles (protons) in the same direction as that of the electric field, and forces negatively charged particles (electrons) in the opposite direction of the electric field's direction.

The magnetic field produces a force on moving charges. The magnitude of the force is equal to the magnitude of the charge times the speed times the magnitude of the magnetic field. The direction of the force is perpendicular to the magnetic field and perpendicular to the direction of the charged particle's velocity.

The charge of a neutron is $Q_n = 0$ coulombs, the charge of a proton is $Q_p = +e = 1.602177 \times 10^{-19}$ coulombs, and the charge of an electron is $Q_e = -e = -1.602177 \times 10^{-19}$ coulombs. The mathematical representation of the Lorentz force acting on a charged particle from an electric field and magnetic field is

$$\vec{F} = Q\left[\vec{E} + \vec{v}_Q \times \vec{B}\right]$$

The particle has charge Q and velocity $\vec{v}_Q$, and it feels a force $\vec{F}$ from an electric field $\vec{E}$ and magnetic field $\vec{B}$. The $\vec{v}_Q \times \vec{B}$ is the cross product. It has a magnitude of $|\vec{v}_Q||\vec{B}|\sin(\theta)$.

In this formula, θ is the angle between the direction of the velocity and the direction of the magnetic field. The direction of the cross product is determined by the right-hand rule. (The *right-hand* points in the direction of the velocity, the fingers (or palm) point in the direction of the magnetic field, and the thumb points in the direction of the cross-product.)

The standard unit is the *ampere* in SI units, which is actually the unit for electric current. The unit for charge is the *coulomb*, which is equal to one *ampere* times one *second*. Other common units in electromagnetism are 1 *volt* = 1 joule per coulomb = 0.7376 foot pound per coulomb. The electric field has the units of 1 volt per meter (= 1 newton per coulomb) or 1 volt per foot (= 0.7376 pounds per coulomb). The magnetic field has the unit tesla where 1 tesla = 10^4 Gauss = 1 kilogram per (coulomb second) = 0.06852 slugs per (coulomb second).

All matter is made up of charged particles (electrons and protons), and materials can be classified depending on how they behave under the influence of electric and magnetic fields. Some types of materials include the following:

- **Electrical conductors:** Some of the charged particles within the material are bonded weakly to the atoms through the Lorentz force. When an external electric field is placed on these types of materials, the Lorentz force is sufficient to cause electrical charge to move even with a very small electric field. Two examples are

  - **Superconductors:** Superconductors have no electrical resistivity, so the charged particles (electrons) begin to move with any kind of electric field.

  - **Metals:** Good metals, such as copper and silver, have very small electrical resistivity, $\rho_{E,Cu} = 5.51 \times 10^{-8}$ ohm feet (1.68×10^{-8} ohm meters) and $\rho_{E,Ag} = 5.22 \times 10^{-8}$ ohm feet (1.59×10^{-8} ohm meters).

    Electrical resistivity is a measure of how much a material doesn't like having charged particles flowing through the material, and the material is trying to stop them from moving. The electrical resistivity arises because most particles in a material don't move; only a very small fraction of the charge particles are moving. All the stationary particles are creating electric and magnetic fields inside the material and through the Lorentz force are trying to stop the moving charges.

✔ **Electrical semiconductors:** This group of materials is sometimes split into semi-metals and semi-insulators, depending on whether the material is more like a conductor or an insulator. These materials require more force than the conductors to move the charge carriers. The resistivity of semiconductors varies quite a bit: Saltwater is $\rho_E = 0.14$ ohm feet (0.044 ohm meters), wet skin is $\rho_E = 3,000$ ohm feet (10^3 ohm meters), and silicon is $\rho_E = 8,200$ ohm feet (2,500 ohm meters). Semiconductors are the foundation of electronics. A few examples of where these materials are used include transistors, integrated computer chips (IC chip), and diodes (LED TV).

✔ **Electrical insulator:** Materials that are electrical insulators don't like to conduct electricity. No material is a perfect insulator, but some materials such as air and glass are examples of very good electrical insulators. The resistivities of a few insulators are as follows: Pure water is $\rho_E = 8.2 \times 10^5$ ohm feet (2.5×10^5 ohm meters), dry skin is $\rho_E = 3.3 \times 10^5$ ohm feet (10^5 ohm meters), and glass is $\rho_E = 8.2 \times 10^{12}$ ohm feet (10^{12} ohm meters). If you apply a strong enough force, anything will conduct.

Note that dry skin is 10^{13} times more resistant than a good metal. Old cars were made of metal and not plastic, which is why they're one of the safest places to be during a lightning storm. Even if a lightning bolt hits the car, it will travel through the metal and not through your body. This is referred to as a _Faraday cage_.

✔ **Dielectrics:** They're materials where an electric field will polarize the molecules within an insulator. The polarization of the material reduces the strength of the electric field.

A _polarized_ material produces an electric field because it has the molecules within the material rotated and aligned with the external electric field. Some molecules have an uneven distribution of charges (like water where the oxygen atom has a negative charge and the two hydrogen atoms have positive charge). These molecules shift to align their charges with the electric field.

The _dielectric strength_ is the strongest electric field the dielectric can withstand before it becomes a conductor. The dielectric strength for air is 9×10^5 volts per foot (= 76 volts per mil = 3×10^6 volts per meter). This can occur when static charge has built up on an object or person. For example, if someone tries to give you a kiss on the cheek, instead of receiving lips touching your cheek, you get a lightning bolt piercing your cheek.

✔ **Ferroelectric materials:** They're materials that acquire a spontaneous polarization below a certain temperature dubbed the _Curie temperature_, which will produce an electric field.

✔ **Piezoelectric materials:** They're materials that change their size in an electric field or become polarized from an external stress. Bones, some proteins, and DNA are piezoelectric materials.

✔ **Ferrimagnetic materials:** They're permanent magnets. Lodestone is the classic example of a ferrimagnet (ferrites) and has been known for several thousand years.

✔ **Ferromagnetic materials:** They're materials that are *magnetized* (the electrons, proton, and neutrons within the material are aligned to produce a magnetic field). Ferromagnets are permanent magnets, but usually different parts of the material are magnetized in different directions. An external magnetic field will align all the domains in the same direction. In the 1980s, neodymium rare-earth magnets were developed and can now produce magnetic fields up to 1.5 teslas.

✔ **Paramagnetic materials:** They're materials that acquire a magnetization parallel to an external magnetic field and enhance the magnetic field. Metal coins are an example of paramagnetic materials.

✔ **Diamagnetic materials:** These materials acquire a magnetization that opposes the external magnetic field. The magnetic effects of most materials are small. Biological systems are also diamagnetic. In fact, it was experimentally shown that frogs will float in air if placed in a strong enough magnetic field. Superconductors are perfect diamagnetic materials (no magnetic field enters a superconductor) because the material produces a magnetic field that exactly cancels out the external magnetic field inside the material.

Sticking balloons on the wall: Coulomb's law and static charge

Coulomb's law tells you how two *static* (stationary) charged objects interact with each other. The law states

✔ There are only two kinds of charge: positive and negative.

✔ Two like charges will repel each other and two opposite charges will attract each other.

✔ The direction of the force lies along the line between the two charges.

✔ The magnitude of the force between two static (stationary) charges is

$$\left|\vec{F}_c\right| = \frac{k_E Q_1 Q_2}{R_{1,2}^2}$$

In this formula, k_E (which equals 2.18×10^{10} pounds square foot per square coulomb = 9.0×10^9 newtons square meters per square coulomb) is the *electric constant*, Q_1 is the charge of object 1, Q_2 is the charge of object 2, and $R_{1,2}$ is the shortest distance between objects 1 and 2.

When you rub a balloon on your hair, some electrons are transferred between your hair and the balloon, leaving hair positively charged and the balloon negatively charged. The rubber of the balloon is an insulator so the charge doesn't move or dissipate from the balloon. As you move the balloon toward the wall, the like charges in the wall are repelled and the opposite charges are attracted, leaving a net amount of opposite charge in the wall close to the balloon. The charges attract each other and hold the balloon in place against the force of gravity.

Producing electric fields

The Lorentz force states the electric field will produce a force on a charge particle. You can use this statement to define the electric field. If I take a test charge (inside my testing equipment) and measure the force acting on my charge, then the electric field is equal to the force divided by the charge of my test charge. Every charged particle produces an electric field and the electric field my test charge measures is the vector sum of all the individual electric fields.

Finding the electric field from Coulomb's law

You can arrange the Lorentz force to give the electric field in terms of the force, and Coulomb's law gives the force between two stationary charges, so I can combine them to define the electric field of a static (stationary) charge Q:

$$\left|\vec{E}\right| = \frac{k_E Q}{R^2}$$

In this formula, k_E (= 2.18×10^{10} pounds square foot per square coulomb = 9.0×10^9 newtons square meters per square coulomb) is the electric constant, Q is the charge of the object, and R is the distance from the object to the location where you're measuring the electric field.

The electric field points away from an object with charge Q if the charge Q is positive (proton), and the electric field points toward an object with charge Q if the charge Q is negative (electron). The total electric field at any given point in space is the vector sum of all the contributing electric fields.

Figure 16-1a shows an electric field for a single positive charge. If the charge was negative, the electric field lines would point toward the charge. Figure 16-1b shows the net electric field for two charges (equal magnitudes but opposite signs). The Lorentz force states that the force is equal to the charge times the electric field, so if you place a charge on a net electric field line (either the lines in Figure 16-1a or Figure 16-1b), then the charge will feel a force in the direction of the line and will want to follow that line. Remember, a positive charge will move in the direction of the electric field line, and a negative charge will move in the direction opposite to the electric field line.

Figure 16-1:
The electric field of a single charge (a). The net electric field of a positive charge and a negative charge (b).

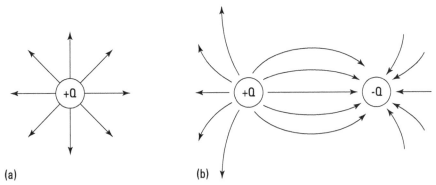

(a)   (b)

Creating electric fields using Gauss' law

In biophysics, the biological systems have many charges, so the formula for calculating the electric field for each individual charge may not be the best approach. In that case, it is usually better to use *Gauss' law*, which gives the relationship between the *electric flux* (a measure of the amount of electric field passing through an area [A]) and the *total charge enclosed* (which is the sum of all the charges [electrons and protons] within the volume surrounded by the area A):

$$\Phi_E^{(o)} = 4\pi k_E Q_{enc}$$

In this formula, k_E, which is equal to 2.18×10^{10} pounds square foot per square coulomb (9.0×10^9 newtons square meters per square coulomb) is known as the electric constant. (The *electric constant* is a number that is related to the speed of light.) Q_{enc} is the total charge enclosed by the closed surface. Φ_E is the electric flux, where the superscript (o) means the surface is closed. *Closed* means you can't get from one side of the surface to the other side without passing through the surface, such as an inflated balloon or a box with a closed lid. An open surface would be like the page in a book.

The *permittivity of free space* (ε_o), also known as the *vacuum permittivity*, is related to the electric constant by the following relationship: $k_E = 1/(4\pi\varepsilon_o) =$ 8,987,551,787.368176 newtons square meters per square coulomb.

Gauss' law states that the charges within the volume produce the electric flux. Also, electric flux is related to the electric field, and from the Lorentz force the electric field is related to the force acting on other charged particles. The two relationships, Gauss' law and Lorentz force, tell you how charged particles create forces on other charged particles.

You know the relationship between the force and the electric field and you know the relationship between the electric flux and the charges creating it; all you need now to complete the connection is the relationship between the electric flux and the electric field. If the electric field is a constant over the surface enclosing the charges, then the electric flux is

$$\Phi_E = \vec{E} \bullet \vec{A} = \left|\vec{E}\right|\left|\vec{A}\right|\cos(\theta)$$

In this formula, $\left|\vec{A}\right|$ is the magnitude of area of the surface, $\left|\vec{E}\right|$ is the magnitude of electric field and Φ_E is the electric flux. The vector area has a direction normal to the surface. The angle θ is the angle between the direction of the electric field and the direction of the area.

A nice feature of Gauss' law is that the surface can have any shape so you can always pick a shape that works best for the problem. You can use the symmetry of the problem to make your life easier.

In Figure 16-1a, you can draw a box or a sphere around the charge. In either case the flux is equal to $4\pi k_E Q$. However, in Figure 16-1a, you can see the electric field is *spherically symmetrical,* which means if you choose the surface to be a sphere of radius R, then the electric field has a constant magnitude on the sphere. The area of a sphere A equals $4\pi R^2$ and Gauss' law gives $E4\pi R^2 = \Phi_E = 4\pi k_E Q$, which gives the magnitude of the electric field for a point charge: $E = k_E Q\, R^{-2}$.

Figure 16-1b has one negative charge and one positive charge. If you enclose the two charges in any closed surface, then the total charge enclosed is zero and the total electric flux must be zero according to Gauss' law. This means the amount of electric field flowing out through the surface is equal to the amount of electric field flowing in through the surface, which is true for any surface containing both charges.

Understanding the electric potential

Conservative forces have a corresponding potential energy. (Chapter 4 discusses potential energy.) Similarly, the *electric field* (the means by which a charged particle creates a force on another charged particle) has a corresponding *electric potential* (the potential of an electric field to do work on a charge particle). The *electric potential difference* is defined as the change in the electric potential energy divided the charge (q) of a test particle that is moved from the initial location to the final location. Remember, the change in the potential energy is equal to the negative of the work done on the particle, the definition of the work done is it equals the displacement times the force parallel to the displacement, and from the Lorentz force the force is equal to the charge of the particle times the electric field. All these relationships can be combined into a single mathematical expression:

$$\Delta V = \frac{\Delta E_p}{q} = \frac{-W}{q} = \frac{-\vec{F}\bullet\Delta\vec{s}}{q} = -\vec{E}\bullet\Delta\vec{s} = -\left|\vec{E}\right|\left|\Delta\vec{s}\right|\cos(\theta)$$

In this formula, ΔV is the electric potential difference and ΔE_p is the change in the potential energy, W the work done on the particle, F is the force, Δs is the displacement, and E is the electric field. The unit for the electric potential difference is *volts*. You can use this expression in Gauss's law to find the electric flux and the enclosed charge from the electric potential.

The electric potential difference measures the change in the potential energy in space. Related to this is the work that is needed to move the charges. The *electromotive force (emf)* is defined as the external work per unit charge (W/Q). The emf is the power source in a circuit, such as batteries (electrochemical cells), solar cells, piezoelectrical materials under stress, or electrical generators.

The relationship between the electric potential and the electric field is the same as the relation between the potential energy and the conservative force. The average strength of the electric field is equal to the electric potential difference divided by the displacement. A couple of consequences of this are as follows:

✔ The electric field points in the direction of maximum decrease in the electric potential.

✔ A surface of constant electric potential (called an *equipotential*) has an electric field pointing perpendicular to the surface.

✔ A metal with no current flowing in it is an equipotential with no electric field inside. Any excess charge in a conductor will be located on the surface. If there were excess charge inside a conductor, it would create an electric field that will force charge to move.

✔ Fish produce an electric field to help them navigate. Objects cause changes to the electric field that the fish can detect.

✔ Sharks are very sensitive to the electric fields produced by animals, which explains how they track fish and other sea life, even if they're hidden under rocks or sand.

Producing magnetic fields and the Biot-Savart law

Charges create electric fields, and the motion of charges creates magnetic fields, which the symbol B denotes. The *average current* is defined as the amount of charge flowing through a cross-sectional area divided by the time elapsed over which you measured the amount of charge:

$$\bar{I} = \frac{\Delta Q}{\Delta t}$$

The current flows in the direction of positive charge. In good metals like copper, the charge carriers are electrons, which flow in the direction opposite to I because they have negative charge.

Electric field lines start at positive charges and end at negative charges. Magnetic field lines have no beginning or end but form circles around currents. The direction of the magnetic field lines are determined by the right-hand rule. Figure 16-2 shows a wire with current flowing upwards. If you place the thumb of your right hand in the direction of the current, then your fingers curl around the wire in the same direction as the magnetic field.

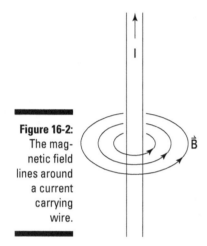

Figure 16-2: The magnetic field lines around a current carrying wire.

In the case of magnets with no current flowing through them, the electrons in the atoms are orbiting the nuclei and create a small magnetic field around each atom. To help visualize this, take the wire in Figure 16-2 and connect the top to the bottom. The magnetic field can combine with the magnetic fields of the other atoms to give the material a net magnetic field. The magnetic field comes out of the material at what is called the *North (Magnetic) Pole*, circles around, and goes back into the material at the *South (Magnetic) Pole*.

The earth also creates a magnetic field. The South (Magnetic) Pole was located in Canada from at least the 1800s until the early part of the 21st century. Recently, the South (Magnetic) Pole has been moving and is now located in the Arctic Ocean, moving toward Siberia. Many animals such as the salmon and sea turtles use the earth's magnetic field. The effects of the earth's magnetic poles shifting could have consequences on the animals that use the earth's magnetic field.

The *magnetic version of Gauss' law* gives this expression:

$$\Phi_B^{(o)} = 0$$

Φ_B is the magnetic flux, where the superscript (o) means the surface is closed. This law states that the number of magnetic field lines flowing out a surface equals the number of magnetic field lines flowing into the closed surface. This just reiterates what I said earlier that magnetic field lines don't start or end, but go in circles.

The *magnetic flux* is a measure of the amount of magnetic field passing through the surface. If the magnetic field is a constant over a surface, then the magnetic flux is defined as

$$\Phi_B = \vec{B} \bullet \vec{A} = \left| \vec{B} \right| \left| \vec{A} \right| \cos(\theta)$$

In this formula, Φ_B is the magnetic flux, $\vec{B}$ is the magnetic field, $\vec{A}$ is the area of the surface. The vector area has a direction normal to the surface. The angle θ is the angle between the direction of the magnetic field and the direction of the area.

The *Biot-Savart law* is a formula for calculating the magnetic field due to a straight piece of current carrying wire of length L. The right-hand rule gives the direction of the magnetic field. The magnitude of the magnetic field is as follows:

$$B = \frac{k_B I \left| \vec{L} \times \vec{R} \right|}{R^3} = \frac{k_B IL \sin(\theta)}{R^2}$$

In this formula, B is the magnitude of the magnetic field at a point in space a distance R from the wire, k_B is the magnetic constant which equals 10^{-7} tesla meter per ampere = 2.25×10^{-8} pounds per square ampere, I is the current in the wire of length L, and θ is the angle between the direction of R and the direction of L.

Three special applications of the *Biot-Savart law* are

✔ The magnetic field produced by a current I in an infinitely long straight wire is as this formula shows

$$B = \frac{2k_B I}{R}$$

R is the perpendicular distance from the wire.

✔ The magnetic field along the axis perpendicular to a loop of wire with current I and radius d is as follows:

$$B = \frac{2\pi k_B I d^2}{\left(R^2 + d^2\right)^{3/2}}$$

R is the perpendicular distance from the plane of the loop. $R = 0$ is the center of the loop.

✔ The magnetic field inside a long thin *solenoid* (many loops all connected and stacked one beside the other) is as follows:

$$B = 4\pi k_B I n$$

In this formula, B is the magnitude of the magnetic field, k_B is the magnetic constant, I is the current within the solenoid, n is the number of loops per unit length. This result is exact for an infinitely long solenoid and is a good approximation near the center of a long thin solenoid.

Changing electric fields create magnet fields: Maxwell-Ampere law

Current is convenient when studying currents in thin wires. When you have charge flowing through a material, using the current density is more convenient. *Current density* is the amount of current per cross-sectional area flowing through a material. The current can be expressed in terms of the current density times the cross-sectional area that the charge is flowing through as this expression shows:

$$I = \vec{J} \bullet \vec{A} = \left|\vec{J}\right|\left|\vec{A}\right| \cos(\theta)$$

The *Maxwell-Ampere law* states that the rate charge flows through an area (A) plus the rate of change in the electric field through the same area (A) is proportional to the magnetic field times the circumference of the area (Δd) as this expression shows:

$$\vec{B} \bullet \Delta \vec{d} = \left[4\pi k_B \vec{J} + \frac{1}{v_{light}^2} \frac{\Delta \vec{E}}{\Delta t} \right] \Phi \vec{A}$$

In this formula, v_{light} = 186,282 miles per second = 983,571,056 feet per second = 299,792,458 meters per second is the speed that light travels at. The Maxwell-Ampere law shows that moving charges (current density) and changing electric fields create a magnetic field.

Creating electric fields: Faraday's law

Faraday's law states that the *average electromagnetic force (emf)* (ε) (the external work done per unit charge) induced in a circuit is equal to the negative of the **average rate** of change in the *magnetic flux* (the measure of the magnetic field passing through an area (A)). In other words, a changing current will produce a changing (time dependent) magnetic field, or a changing magnetic field will produce a current in the material. You can write this expression as

$$\bar{\varepsilon} = -\frac{\Delta \Phi_B}{\Delta t}$$

The emf is proportional to the electric field along the circumference (Δd) of the area the magnetic field is going through. If the area is a constant and the magnetic field is the same throughout the area, then Faraday's law can be written as

$$\vec{E} \bullet \Delta \vec{d} = -\frac{\Delta \vec{B}}{\Delta t} \Phi \vec{A}$$

The negative sign in Faraday's law is a consequence of Lenz's law. *Lenz's law* states the induced current always opposes the change in the magnetic flux. For example, suppose the magnetic flux is increasing, then the induced current in the circuit is flowing such that the magnetic field created by the current opposes the magnetic field within the area. Faraday's law is the principle of electrical generators and led to Faraday making the first generator.

In biological systems, electricity, magnetism, and electromagnetism are fundamental to all forms of energy production and storage. This form of energy is what makes biological organisms work. A few examples of electrical energy, currents, and the storage of energy include the electrical impulses in the muscles, the nerves sending electrical pulses to the brain, and photosynthesis.

Resisting AC/DC — the resistance of the human body and other resistors

The *electrical resistivity* (ρ_E) is a measure of a material's ability to resist the flow of electrical charge through it. The larger the resistivity, the quicker the electrical energy is converted into heat energy. The *electrical conductivity* (a measure of how easy it is for charge to flow through the material) is equal to one divided by the resistivity ($\sigma_E = 1/\rho_E$). Metals have a very small resistivity, which means a very large conductivity, whereas insulators have a very large resistivity.

Resistance is related to the resistivity of the material combined with the shape and size of the object. The resistance of a material gives a measure of the amount of electrical energy lost by passing a current through the object and it's expressed this way:

$$R_E = \frac{\Delta V}{I}$$

In this formula, ΔV is the electrical potential difference across the material. I is the current passing through the material. The unit of the resistance R_E is ohm, which equals volt per ampere. **Note:** Measuring your body's resistance by passing a current through your body isn't a good idea! (Just think of what a taser does to a person.)

Materials with a constant resistance R_E are called *ohmic*. Metals are ohmic materials with very small resistance R_E. Ohmic materials with large resistance R_E are called *resistors*. Diodes are an example of *nonohmic* devices where the resistance isn't a constant. In the case of a cylindrical object of length L and cross-sectional area A, the resistance parallel to the axis of the cylinder is expressed as such:

$$R_E = \frac{\rho_E L}{A}$$

Plants and animals are constantly generating electrical currents throughout their biological systems. Depending on where an external source is producing a current within the biological system, it will have a different resistance and different effect.

In the human body, the resistivity of wet skin is 3,000 ohm feet (10^3 ohm meters), the resistivity of dry skin is 3.3×10^5 ohm feet (10^5 ohm meters), and the body has an average internal resistivity of 15 ohm feet (4.6 ohm meters). This difference is because things such as body fluids are good electrical conductors with a resistivity of 0.50 ohm feet (0.15 ohm meters). The human skin is a protective blanket to protect muscles and organs from external sources of electrical currents.

During a lightning storm, hiding inside a metal car is smart because if the car gets struck by lightning or an electrical power line is lying across the car, then the current flows through the metal (low resistance). Your body is the opposite; as soon as the current gets through your skin (high resistance), it wants to travel inside your body (low resistance). Open wounds and wet skin aren't good to have around electrical devices.

Your body sends electrical pulses to the muscles, which cause them to contract. Turning on and off the electrical pulses allows you to control your muscles. If an external current of less than 0.01 amperes enters the body, then you still have control over the muscle; however, if the current is greater than

0.02 amperes, then the muscle contracts and you have no control. That is why you can't let go when you're getting a shock. If the current is increased to around 0.05 amperes, then it causes serious damage inside the body. The heart is a massive muscle and there is a chance it will go into ventricular fibrillation for currents above 0.03 amperes. The most dangerous currents are around 0.1 amperes because at this current the heart has the greatest chance of going into ventricular fibrillation, which means it won't pump sufficient blood to the body.

If the heart does go into ventricular fibrillation, then a defibrillator may be needed. A defibrillator actually does the opposite of what most people think. The device cranks up the juice (more than 0.3 amperes), which makes the heart completely contract and stop beating (flat-line). The heart then starts beating in a regular rhythm again when the current stops flowing through it. The new defibrillators have a better than 95 percent success rate of fixing ventricular fibrillation on the first zap.

Storing energy with charge: Capacitors

Capacitors are devices that store energy within an electric field by having two oppositely charged conductors within the device. A measure of the amount of charge a capacitor can hold is the capacitance. The *capacitance* is the amount of charge on each conductor within the device divided by the potential difference between the plates. It's expressed as:

$$C_E = \frac{Q}{\Delta V}$$

Capacitors are two conductors not in contact, with one conductor having a charge Q and the other conductor having a charge –Q. ΔV is the electric potential difference between the two conductors. The unit of the capacitance is the *farad* or in most real capacitors microfarad = 10^{-6} farads or picofarad = 10^{-12} farads.

Meanwhile, *dielectrics* are materials that are insulators and become polarized in an electric field. A dielectric is usually placed between the conductors in a capacitor, which increases the capacitance of the capacitor. Dielectrics make the following two changes to the laws of electromagnetism:

$$k_E \to \frac{k_E}{K} \text{ and } v_{light}^2 \to \frac{v_{light}^2}{K}$$

K is called the *dielectric constant* or *relative permittivity*. It's a measure of the effect of the polarization of the dielectric material. These two changes in the equations mean the electric field is smaller in a dielectric material by a factor of 1/K, light travels slower through a dielectric material by a factor of $1/K^{\frac{1}{2}}$ and the capacitance increases by a factor of K (K > 1).

Here are a few useful values of the *dielectric constant K* that arise in biophysics: 1 in a vacuum, 1.00054 in air at 68° F (20° C), 80.1 in water at 68° F (20° C), 78 in water at 77° F (25° C) and 8 for an *unmyelinated axon membrane* at 98.6° F (37° C).

Myelin is a dielectric material that forms a protective insulating boundary between the axon and the surroundings. The presence of the myelin increases the electrical resistance of the membrane and decreases its capacitance.

Capacitors store energy that can be used in electrical circuits. The *potential energy of a capacitor* (the energy that can be used to do work) is:

$$E_p = \frac{Q\,\Delta V}{2} = \frac{C_E\,(\Delta V)^2}{2}$$

In this equation, Q is the total charge on one of the two conductors and ΔV is the electric potential difference.

Anything can be made into a capacitor as long as you can keep the charge separated. The simplest capacitor is two conducting plates that are parallel with a gap between them, called a parallel-plate capacitor. A *parallel-plate capacitor* has the following electrical properties between the two plates:

$$\text{electric field: } E = \frac{4\pi k_E Q}{KA}$$

$$\text{electric potential difference: } \Delta V = \frac{4\pi k_E Q\,\Delta x}{KA}$$

$$\text{capacitance: } C_E = \frac{KA}{4\pi k_E \Delta x}$$

$$\text{potential energy: } E_P = \frac{2\pi k_E Q^2\,\Delta x}{KA}$$

The plates have an area A and a distance Δx between the plates. The direction of the electric field (E) is normal to the conducting plates from the positive plate to the negative plate. ΔV is the electric potential difference across the plates, E_p is the electric potential energy, Q is the charge on a plate ($-Q$ on the other), and K is the relative permittivity of the dielectric between the plates.

For example, Harold has graciously volunteered to give us an unmyelinated axon membrane to experiment with during your Saturday biophysics party. This party will be held in the lab because the experiment needs some specialized equipment. You want to calculate the electric field, the capacitance, the charge, and the potential energy.

You meet your biophysics friends at the local biophysics lab. You take Harold's unmyelinated axon membrane and spread it flat, so it looks like a parallel-plate capacitor. You and your friends pull out the equipment and make some measurements.

The results of those measurements are:

✔ You measure the electric potential difference ΔV = 0.1 volts.

✔ You measure the relative permittivity (dielectric constant) K = 8.

✔ You measure the thickness of the membrane Δx = 4.0 × 10⁻⁷ inches (3.3 × 10⁻⁸ feet = 1.0 × 10⁻⁸ meters).

✔ You measure the area of the membrane A = 2.0 × 10⁻³ square inches (1.4 × 10⁻⁵ square feet = 1.3 × 10⁻⁶ square meters).

You can now substitute these measurements into the formulas for a parallel plate capacitor and determine the unmyelinated axon's physical properties:

electric field: $E = \dfrac{\Delta V}{\Delta x} = \dfrac{0.1\ \text{V}}{3.3 \times 10^{-8}\ \text{ft}} = 3.03 \times 10^{6}\ \dfrac{\text{volts}}{\text{foot}} \left(1.00 \times 10^{7}\ \dfrac{\text{volts}}{\text{meter}}\right)$

capacitance: $C_E = \dfrac{KA}{4\pi k_E \Delta x} = \dfrac{8\left(1.4 \times 10^{-5}\ \text{ft}^2\right)}{4\pi\left(2.18 \times 10^{10}\ \text{lb ft}^2/\text{C}^2\right)3.3 \times 10^{-8}\ \text{ft}} =$

$1.24 \times 10^{-8}\ \dfrac{\text{square coulomb}}{\text{foot pound}}\left(9.20 \times 10^{-9}\text{farads}\right)$

charge: $Q = C_E\ \Delta V = \left(9.20 \times 10^{-9}\ \text{F}\right)0.1\ \text{V} = 9.20 \times 10^{-10}\text{coulombs}$

potential energy: $E_P = \dfrac{Q^2}{2C_E} = \dfrac{\left(9.20 \times 10^{-10}\text{C}\right)^2}{2\left(1.24 \times 10^{-8}\text{C}^2/(\text{ft lb})\right)} =$

$3.41 \times 10^{-11}\text{foot pound}\left(4.60 \times 10^{-11}\text{joules}\right)$

If you thought the electric field is large, then you're correct. This is the same size of electric field needed for air to go from being an insulator to a conductor, and a spark could travel through the air. In addition, the stored energy may seem small, but you have to remember how small the membrane is.

Connecting Electric Circuits

A large area of biophysics is involved with studying electrical circuits within biological organisms, which are all around (and in you). Humans and animals are made up of electrical circuits throughout the body.

Biophysicists are interested in the electrical nerve pulses between the senses and the brain, or when the brain sends electrical signals to a muscle and causes it to contract. In fact, the brain is a large complex computer with electrical signals traveling throughout it, so biophysicists are interested in the electrical wiring (circuitry) of the brain (neural networks).

You can use the information from this chapter to form the properties of electrical circuits with applications to neural networks or other aspects of biophysics. In these sections, I discuss Ohm's law, energy, and power; illustrate how to draw and read electrical schematics; and peruse the conservation laws of complex electrical circuits with devices in parallel and series.

Conserving energy: Ohm's law and the power dissipation of devices

Any biophysical system always has some sort of energy or work (power) put into the system and energy or work (power) has to come out. You always need to ask, where did the energy go and what did it do. In electrical systems within the body or in a circuit, something is supplying emf (work/power) and then the body or the devices use the electrical potential energy (charge times electrical potential). *Ohm's law* states how much the electrical potential decreases across a device when a current flows through the device.

The electrical potential difference (ΔV) across a device is equal to the direct current (I) flowing through the device times the resistance (R_E) of the device. Mathematically, *Ohm's law* is $\Delta V = I R_E$. A device is said to be ohmic if the resistance (R_E) is a constant and nonohmic otherwise.

Remember, the power supplied by an *electrical power source* is equal to the emf times the current. Mathematically, $P_{input} = \varepsilon I$. Now, the *power dissipated* by an electrical device is: $P_{output} = R_E I^2$ (*ohmic*) or $P_{output} = \Delta V I$ (*nonohmic*). The power expressions combined with Ohm's law should be your starting point for any circuit problem.

Drawing road maps for electrons: Circuits and circuit diagrams

Current flows in neural networks and in wires (usually copper) to the devices connected in a circuit to a power supply. These wires are like roadways for the electricity, and if you have roadways, then having a map is nice. You may be required to draw your own map or read a map, so this section introduces the symbol conventions used for the devices and power supplies. Figure 16-3 shows an example of a *circuit schematic* with the common symbols.

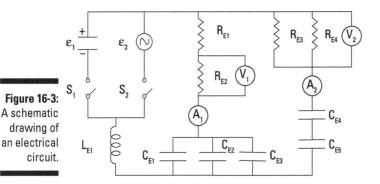

Figure 16-3:
A schematic
drawing of
an electrical
circuit.

The symbols in Figure 16-3 are as follows:

✔ **Straight lines:** The *straight lines* are the conducting wires connecting the devices. The resistance of the wires for copper ($\rho_E = 5.51 \times 10^{-8}$ ohm-foot = 1.68×10^{-8} ohm meters) is expressed as:

$$\frac{R_E}{L} = \frac{\rho_E}{A} = \begin{cases} 2.53 \times 10^{-3} \frac{\text{ohm}}{\text{foot}} \left(8.29 \times 10^{-3} \frac{\text{ohm}}{\text{meter}}\right) & \text{14-gauge} \\ 1.59 \times 10^{-3} \frac{\text{ohm}}{\text{foot}} \left(5.21 \times 10^{-3} \frac{\text{ohm}}{\text{meter}}\right) & \text{12-gauge} \end{cases}$$

An L = 100 feet, 14-gauge copper wire has a resistance of 0.253 ohms, which is very small. Your *skin* has a resistivity of $\rho_E = 3.3 \times 10^5$ ohm-feet (10^5 ohm-meters) when dry, which is 10^{13} times larger than copper! The body fluids within your body have a resistivity of approximately $\rho_E = 0.50$ ohm-feet (0.15 ohm-meters), which is 10^7 times larger than copper but 10^{-6} times smaller than your skin. In circuits where the devices are connected with copper wires, you can ignore the resistance of the copper because the resistance is so small compared to the devices. The wires are just pathways for the current.

✔ **Switches:** The *switches* are S_1 and S_2. Both switches are shown in the open position so no current is flowing through the circuit. Closed switches have the bar touching both poles.

✔ **Ammeters:** The *ammeters* are uppercase A's with a circle around them. These devices measure the current flowing through the circuit and must be placed in series with the device. Figure 16-3 shows two:

  • Ammeter A_1 is measuring the current passing through $R_{E,1}$ and $R_{E,2}$.

  • Ammeter A_2 is measuring the current passing through $C_{E,4}$ and $C_{E,5}$.

✔ **Voltmeters:** The *voltmeters* are upper case V's with a circle around them. These devices measure the electric potential difference across devices and must be placed in parallel with the device. Figure 16-3 has two:

- Voltmeter V_1 is measuring the electric potential difference across $R_{E,2}$.

- Voltmeter V_2 is measuring the electric potential difference across both $R_{E,3}$ and $R_{E,4}$.

✔ **DC emf power source:** The *DC emf power source* is ε_1 in Figure 16-3. The symbol consists of two parallel lines of unequal length:

- The long line represents the positive pole that the current comes out of.

- The short line represents the negative pole where the current flows back into the power source.

The positive and negative sign are usually added to help distinguish it from the capacitor. Remember the electrons flow opposite to the current, which represents positive charge flow.

✔ **AC emf power source:** The *AC emf power source* is ε_2 in Figure 16-3. It's a circle with a wave inside. The wave indicates that the current is oscillating back and forth: $\varepsilon_2 = \varepsilon_0 \cos(2\pi f t)$, where ε_0 is the emf amplitude, f is the frequency, and t is the time.

✔ **Resistor:** Figure 16-3 denotes the *resistor* by a jagged line. $R_{E,1}$, $R_{E,2}$, $R_{E,3}$, and $R_{E,4}$ are resistors. For a single resistor connected to an emf power source (nothing else in the circuit except a single loop of wire with a switch), the following is true:

DC emf: ε_1, $R_{E,1}$, $\Delta V = I\, R_E$ and $\varepsilon_1 - \Delta V = 0$.

AC emf: $\varepsilon_2 = \varepsilon_0 \cos(2\pi f t)$, $Z_{rms} = X_R = R_{E,1}$ and $I_2 = (\varepsilon_0/Z_{rms}) \cos(2\pi f t)$. Z_{rms} is the *root mean square (rms) impedance* of the circuit. The impedance is a measure of the device's resistance to the current. Many devices have impedance that depends on the resistance, capacitance, and inductance of the device. These quantities don't add together, but combine in a complicated manner. The root mean square impedance is the magnitude of the impedance. X_R is the *reactance* of the resistor. Note X_R is usually ignored because it is equivalent to R_E.

✔ **Capacitor:** Two straight parallel lines denote the *capacitor*. The parallel lines represent the simplest capacitor, two parallel conducting plates with an insulating gap between them. $C_{E,1}$, $C_{E,2}$, $C_{E,3}$, $C_{E,4}$, and $C_{E,5}$ in Figure 16-3 are capacitors. For a single capacitor connected to an emf power source, the following is true:

DC emf with the switch S_1 closed at time 0: In Figure 16-4, consider the circuit C_1 and R_1 and everything else removed: ε_1, $C_{E,1}$,

$R_{E,1}$, and $I(t) = (\varepsilon_1/R_{E,1}) \exp[t/(R_{E,1}C_{E,1})]$. [Remember, $\exp(x) = e^x = (2.7182818...)^x$.] The capacitor initially acts like a wire, but after a long time it behaves like a broken wire because as time goes on, the charge builds up in the capacitor, which repels further current from flowing into the capacitor.

AC emf: In Figure 16-3, consider the circuit $C_{E,1}$ and everything else removed: $\varepsilon_2 = \varepsilon_0 \cos(2 \pi f t)$, $Z_{rms} = X_C = 1/(2 \pi f C_{E,1})$, $I_2 = -(\varepsilon_0/Z_{rms}) \sin(2 \pi f t)$. X_C is the reactance of the capacitor (called the *capacitive reactance*), which is the effective resistance of the capacitor. The reactance is dependent on the frequency of the AC emf power source.

✔ **Inductor:** A looping wire denotes the *inductor*. The looping wire represents the solenoid. $L_{E,1}$ in Figure 16-3 is the inductor. For a single inductor connected to an emf power source, the following is true:

DC emf with the switch S$_1$ closed at time 0: In Figure 16-3, consider a single loop of wire with $L_{E,1}$ and $R_{E,1}$ and everything else removed: ε_1, $L_{E,1}$, $R_{E,1}$ and $I(t) = (\varepsilon_1/R_{E,1}) (1 - \exp[-t R_{E,1}/L_{E,1}])$. The inductor acts initially like a broken wire, but after a long time it behaves like a piece of wire.

AC emf: In Figure 16-3, consider the circuit $L_{E,1}$ and everything else removed: $\varepsilon_2 = \varepsilon_0 \cos(2 \pi f t)$, $Z_{rms} = X_L = 2 \pi f L_{E,1}$, $I_2 = (\varepsilon_0/Z_{rms}) \sin(2 \pi f t)$. X_L is the reactance of the inductor (called the *inductive reactance*), which is the effective resistance of the inductor. The reactance is dependent on the frequency of the AC emf power source.

Conserving energy and charge within a circuit: Kirchhoff's laws

In any biophysical system (or physical systems) you look at how the energy and work don't change, but only change form. This section is about power in equals power out combined with particles (mass and charge) in equals particles (mass and charge) out. This section is about how these things don't change in a neural network and electrical circuit. They're called *Kirchhoff's laws*. Here are Kirchhoff's laws and what they mean:

✔ **Kirchhoff's first law:** The current flowing into any point of a circuit is equal to the amount of current flowing out. Kirchhoff's first law is stating *conservation of charged particles*. The number of charged particles in an electrical circuit can't change. In Figure 16-3, the current flowing through inductor $L_{E,1}$ is equal to the sum of the currents flowing $R_{E,1}$, $R_{E,3}$, and $R_{E,4}$. Alternatively, the sum of the currents flowing into a junction must equal the sum of the currents flowing out of the junction. (A junction is where many wires come together — see Figure 16-3.)

✔ **Kirchhoff's second law:** The sum of the electrical potential differences around any closed path (loop) in a circuit must be zero. To add up the electrical potential difference around a closed circuit you take

- The emf of a power supply as positive if going in the direction of the current and negative if going in the direction opposite to the current.

- The electrical potential difference across a device is negative if going in the direction of the current and positive if going against the current.

Kirchhoff's second law is the conservation of the electrical potential difference (and the conservation of electrical potential energy). The electrical potential difference is related to the electrical potential energy, which is related to the conservative force. All conservative forces are path independent, so the change in the electrical potential difference around any closed circuit must be zero.

Alternatively, you can look at Kirchhoff's second law as the conservation of power (energy), which is a very important concept in the study electrical systems. Remember, the electrical potential energy (E_P) is equal to the charge (Q) times the electrical potential (V). You can't create or destroy energy, so the power put into a circuit must equal what comes out.

Chapter 17

Tapping into the Physics of Radiation

. . . .

In This Chapter

▶ Getting the lowdown on nuclear physics and radioactivity

▶ Tackling common myths about radioactivity

▶ Looking at radioactivity in biological systems

. . . .

S ome materials emit radiation and others are efficient at absorbing certain types of radiation. These properties are of interest to people involved in medical physics, health physics, and some areas of biophysics. Radiation can have harmful effects on the body, but there are also many benefits of radiation.

This chapter examines nuclear physics, radioactivity, and the different types of radiation. The chapter explores how radiation is produced, the different types of electromagnetic radiation and the pros and cons of this radiation, and radioactive material inside people and organic material.

Understanding What Nuclear Physics and Radioactivity Are

Radioactivity is a natural process with many benefits in society, but most people unfortunately think of cancer and nuclear weapons whenever radiation and nuclear physics are mentioned.

Nuclear physics deals with the study of the nucleus of atoms. I am not interested in the specific details within the *nucleons* (the particles within the nucleus) or the *nucleus* (the central core of the atom where most of the mass and all the positive charge reside) in general except for the consequence of what is going on has on biological organisms or molecules. The nucleons are called *protons* and *neutrons*. For our purposes, the protons are the positively charged particles within the nucleus with a mass approximately 1,800 times larger than an electron's mass. The neutrons have no charge and have a mass approximately the same as a proton's.

In most of biophysics, the electromagnetic force is the primary force, but nuclear physics does play an important role in some areas of biophysics. Nuclear physics considers three primary forces within the nucleus:

- ✔ **Strong force:** This force keeps the neutrons and protons together.

- ✔ **Electromagnetic force:** This force causes the protons to repel each other. (See Chapter 16 for a more detailed discussion.)

- ✔ **Weak force:** This force is responsible for *beta decay* and *beta radiation*. I discuss beta decay and beta radiation in more detail in the following section.

These sections explain nuclear physics and radioactivity and dispel some of the myths and fears surrounding these terms.

Explaining radioactivity

Radioactivity is everywhere and is relevant to everyone. This section discusses what radioactivity is and what the consequences are for matter. I also introduce the most common forms of radioactive decay that you will find important in biophysics.

If you take any organic or inorganic material and break it apart (break apart the molecules too), you will discover *atoms* composed of a nucleus with protons and neutrons surrounded by electrons. An atom with a specific number of protons is called an *element*, because the number of protons in the nucleus determines the physical properties of the atom. The *atomic number Z* is the number of protons in the element's nucleus. For example, the element hydrogen has one proton, so its atomic number is one ($Z = 1$), the element carbon has six protons ($Z = 6$), and the element oxygen has eight protons ($Z = 8$).

The *atomic mass A* of an element is equal to the total number of protons plus neutrons in the nucleus. Elements with a small atomic number have approximately the same number of neutrons as protons, but as the atomic number grows, then the number of neutrons grows faster than the number of protons so that uranium has approximately 1.6 neutrons for every proton in the nucleus. The mass of the proton is 1,836 times greater than the mass of an electron, and the mass of the neutron is 1,839 times greater than the mass of an electron, which means you can ignore the mass of the electrons when calculating the mass of an atom.

The *mass of an atom* is $M = A\,u$, where $u = 1.138 \times 10^{-28}$ slugs (1.661×10^{-27} kilograms) is the atomic mass unit and A is the atomic mass.

The number of neutrons in the nucleus isn't as important for the physical properties of the element (except for the mass of the element and its stability). An element can have a different number of neutrons, which are called *isotopes*. The number of neutrons equals the atomic mass (A) minus the atomic number (Z). The three most common elements in biophysics are

- ✔ Hydrogen ($Z = 1$): Hydrogen-1, (also known as Hydrogen or Protium) has the symbol H or ^{1}H ($A = 1$), Hydrogen-2 (also known as Heavy Hydrogen or Deuterium) has the symbol D or ^{2}H ($A = 2$), and Hydrogen-3 (also known as *Tritium*) has the symbol T or ^{3}H ($A = 3$).

- ✔ Carbon ($Z = 6$): The symbol is ^{A}C, where Carbon-12 or ^{12}C ($A = 12$) is the most abundant isotope. Carbon-13 and Carbon-14 are also common in nature. The carbon isotopes vary from $A = 9$ to $A = 17$.

- ✔ Oxygen ($Z = 8$): The symbol is ^{A}O, where Oxygen-16 or ^{16}O ($A = 16$) is the most abundant isotope. Oxygen-17 and Oxygen-18 are also relatively stable. The oxygen isotopes vary from $A = 13$ to $A = 21$.

Elements like to have a certain combination of protons and neutrons. If there are too many or too few neutrons within the nucleus (or too many protons), then the element is unstable. Radioactivity is the process by which unstable isotopes of elements change into different elements in an attempt to become stable. Several quantities must be preserved during the process of radioactivity: conservation of energy, conservation of momentum, conservation of charge, and conservation of nucleons. As a result, radioactivity can be restricted to a few different radioactive processes:

- ✔ **Positive beta decay:** If the isotope has not enough neutrons (or too many protons), then it will change a proton into a neutron by *positive beta decay*. The radioactive process is

$$\,_1^1 p \rightarrow \,_0^1 n + \,_1^0 \beta + v$$

$$\,_Z^A E_{old} \rightarrow \,_{Z-1}^A E_{new} + \,_1^0 \beta + v$$

In the first line, p is the symbol for the proton, n is the symbol for the neutron, the β is the positive beta, and the v is the symbol for a neutrino. The β^+ is a positron that has been emitted from the nucleus. A positron is an antielectron, which is exactly like an electron except it has the opposite charge. The *neutrino* is a fundamental particle of the universe, which means it isn't made up of smaller particles. It has no charge and its mass is almost zero. The superscripts and subscript on the beta particle are usually dropped and the beta particle is written as β^+. The second line shows how an element is changed into a new element through beta-positive decay. An example of this type of decay is Nitrogen-12 decaying into Carbon-12.

✔ **Negative beta decay:** If the isotope has too many neutrons (or not enough protons), then it will change a neutron into a proton by *negative beta decay*. The radioactive process is

$$_0^1 n \rightarrow {}_1^1 p + {}_{-1}^0 \beta + \bar{v}$$

$$_Z^A E_{old} \rightarrow {}_{Z+1}^A E_{new} + {}_{-1}^0 \beta + \bar{v}$$

In the first line, the n is the symbol for the neutron, the p is the symbol for the proton, the β^- is an electron that has been emitted from the nucleus, and the v-bar is the symbol for an antineutrino. The *antineutrino* is a fundamental particle, and it's the anti-particle of the neutrino. Their properties are very similar except they'll annihilate each other when they come into contact. The superscripts and subscript on the beta particle are usually dropped and the beta particle is written as β^-. The second line shows how an element is changed into a new element through beta-negative decay. An example of this type of decay would be a Carbon-14 decaying into Nitrogen-14.

Note the first line shows a neutron changing into a proton. Neutrons like being in a nucleus; they don't like being free. A neutron outside the nucleus (a *free neutron*) will decay into a proton with a half-life of 882 seconds (14.7 minutes).

✔ **Alpha decay:** If the element has more protons than bismuth (Z = 83), then the nucleus is too big and it is unstable. In this case a chunk of the nucleus usually breaks off, which is called *alpha decay*. An α particle is a Helium-4 nucleus consisting of two protons and two neutrons. The radioactive process is

$$_Z^A E_{old} \rightarrow {}_{Z-2}^{A-4} E_{new} + {}_2^4 \alpha$$

The superscripts and the subscripts are usually dropped off the alpha particle, because the numbers are always 4 and 2.

✔ **Gamma decay:** If the nucleus of the element is in an *excited* state (high energy state), then it will lose the excess energy through *gamma decay*. A γ particle is a high energy photon, and the radioactive process is

$$_Z^A E_{old}^* \rightarrow {}_Z^A E_{old} + \gamma$$

The * superscript indicates the nucleus is in an excited state, and no superscript * indicates the nucleus is in its ground state.

An example of this type of decay would be the decay of Caesium-137. Ninety-five percent of the β^- decays produce Barium-137 in an excited state (^{137}Cs becomes ^{137}Ba* + β^- + v-bar). The Barium-137 changes into the lowest energy state (ground state energy) by emitting a gamma particle with an energy of 661.7 kiloelectron volts (7.819×10^{-14} foot pound = 1.06×10^{-13} joules). ^{137}Ba* becomes ^{137}Ba + γ. The energy of this single gamma photon doesn't seem like much, but the energy is about 250,000 times larger than the energy of a single blue-light photon.

Decaying of elements – the physical half-life

An unstable isotope is called a *radionuclide*. If you have a material with radionuclides within, it's important to be able to predict how many of the radionuclides are remaining at any given moment. This section also shows the relationship between the number of radionuclides and how fast the unstable material is decaying (activity).

The number of radionuclides present drops by half in a time called the *physical half-life*. In the mathematical relationships, working with the *decay constant,* which is equal to the natural logarithm of 2 divided by the half-life, is more convenient. The *activity* is the rate that the number of atoms is decaying, which has units of *becquerel (Bq)* (1 becquerel equals 1 disintegration per second = 2.70×10^{-11} *curies* (Ci)). Mathematically, the number of radionuclides and the activity as a function of time are as follows. I also provide the mathematical relationship between the decay constant and the half-life.

$$\text{number of radionuclides: } N_{final} = N_{initial}e^{-\lambda \Delta t} \text{ or } \ln\left[\frac{N_{final}}{N_{initial}}\right] = -\lambda \Delta t$$

$$\text{activity: } A_{final} = \lambda N_{final} = \lambda N_{initial}e^{-\lambda \Delta t}$$

$$\text{decay constant: } \lambda = \frac{\ln[2]}{T_{1/2}}$$

Several sources on nuclear radioactivity introduce a second time scale called the mean time. Don't confuse mean time with the half-life. The *mean time* is the inverse of the decay constant; the mean time is equal to the half-life times 1.44. The mean time is the average time it takes a radionuclide to decay, whereas the half-life is the time it takes half of the radionuclides to decay. The mathematical relationship between the mean time, the decay constant, and the half-life is

$$T_{mean} = \frac{1}{\lambda} = \frac{T_{1/2}}{\ln(2)}$$

For example, Caesium-137 has a half-life, $T_{1/2}$, of 30 years and the decay constant, λ, is 0.023 per year, which means that all the Caesium-137 had radioactively decayed before life formed on the planet. Therefore, humanity made all the Caesium-137 currently on the planet. The Chernobyl nuclear reactor in April 1986 was one of the worst nuclear accidents where the initial activity of the Caesium-137 released into the air was estimated to be 8.5×10^{16} becquerels, which corresponds to $N_{initial} = A_{initial}/\lambda = 8.5 \times 10^{16}$ becquerels $\times (3.156 \times 10^7$ seconds per year)/0.023 per year = 1.17×10^{26} Caesium-137 atoms ejected into the atmosphere.

Identifying the three types of isotopes

Radionuclides (or *radioactive isotopes*) emit radiation, which comes in four types: *alpha* (α) particles, positive and negative *beta* (β) particles, and *gamma* (γ) rays (a type of electromagnetic radiation). They're the most common forms of radiation from radionuclides that you'll encounter in biophysics. (Other types of radiation that you will see in biophysics are non-ionizing electromagnetic radiation, ionizing electromagnetic radiation [ultraviolet and X-rays], and neutrons.)

The following sections examine alpha particles, beta particles, and gamma rays. Many materials around you have the potential to be radioactive, and the type of radiation coming off the material will be one or more of these types of radiation.

Examining alpha particles

When a specific isotope decays by emitting an alpha particle, the alpha particle will have the same amount of energy (*monoenergetic*). This means that all the alpha particles will travel approximately the same distance into a material plus or minus a few percent before stopping. Alpha (α) particles have a mass of *4 u* (4.55×10^{-28} slugs = 6.64×10^{-27} kilograms) and a charge of *2 e* (3.20×10^{-19} coulombs). In other words, alpha particles have a large mass and a large charge, plowing through material like a bulldozer.

Note that the *atomic mass unit u* is 1.138×10^{-28} slugs (1.661×10^{-27} kilograms) and each nucleon in a nucleus is assigned a mass of 1 u. Also, all *electrical charge* comes in discrete units of e (1.60×10^{-19} coulombs) with the proton having a charge of +1 e, the neutron has a charge of 0 e, and the electron has a charge of –1 e. (I ignore the fractional charge of the quarks.)

To get a feel for this bull in a china shop, consider the following example. High energy alpha particles with an energy of 4.0×10^7 electron volts (4.73×10^{-12} foot pound = 6.41×10^{-12} joules) will travel about 3.25 feet (0.99 meters) in air and about 3.5×10^{-3} feet (1.1×10^{-3} meters) in skin. Most alpha particles have a much lower energy than this.

Understanding beta particles

In elements with less than 84 protons (Polonium), the most common type of decay is beta decay in radionuclide with too many or too few neutrons. In this section, I briefly mention the ability of beta radiation to penetrate the human body.

Beta particles share their energy with neutrinos (or antineutrinos), so they don't have a fixed energy when coming out of a specific radionuclide, contrary to alpha particles that have the same energy when emitted from the same radionuclide. An example of the penetration depth for beta particles with an energy of 4.0×10^7 electron volts (4.73×10^{-12} foot pound = 6.41×10^{-12} joules): they'll travel about 0.66 feet (0.20 meters) in skin. These beta particles would travel about 0.5 inches (1.27 centimeters) into lead.

You probably noticed that alpha particles are stopped by air and can't penetrate skin. If you stand a yard (meter) from an alpha source, then you'll be safe. On the other hand, air has very little effect on beta particles, and they will travel a long distance. Even in skin, which has a lot higher density, the beta particles are traveling 10 to 1,000 times farther than the alpha particles.

The two types of beta particles are as follows:

- ✔ The *beta-negative* (β^-) particles are electrons, but they're called beta particles because they're created through the weak force inside the nucleus. They have a mass of 6.24×10^{-32} slugs (9.11×10^{-31} kilograms) and have a charge $-e = -1.60 \times 10^{-19}$ coulombs.

- ✔ The *beta-positive* (β^+) particles are *positrons,* which are anti-electrons. They have a mass of 6.24×10^{-32} slugs (9.11×10^{-31} kilograms) and have a charge $e = 1.60 \times 10^{-19}$ coulombs.

Comprehending gamma rays

The *gamma* (γ) rays are electromagnetic radiation. They're similar to X-rays, but X-rays are produced through atomic processes whereas gamma rays are produced within the nucleus. Usually gamma rays have more energy than X-rays. Just like X-rays, the majority of the gamma rays will pass right through your body. The photons that do interact with the body can be split into three categories:

- ✔ The photons with an energy below 10^5 electron volts (1.18×10^{-14} foot pound = 1.60×10^{-14} joules) will interact with the body through the photoelectric effect. The *photoelectric effect* is where the atom absorbs the photon and the excess energy causes an electron to be ejected.

- ✔ The photons with an energy between 10^5 electron volts (1.18×10^{-14} foot pound = 1.60×10^{-14} joules) and 10^7 electron volts (1.18×10^{-12} foot pound = 1.60×10^{-12} joules) will interact with the body through Compton scattering. *Compton scattering* is when the photon scatters with an electron. The photon loses some energy, which is transferred to the electron, which is ejected from the atom. Note that gamma rays in this energy range usually interact with matter through the Compton scattering, but some interact by the photoelectric effect and some interact through pair production.

✔ The photons with an energy greater than 10^7 electron volts (1.18×10^{-12} foot pound = 1.60×10^{-12} joules) will interact with the body primarily through pair production. With *pair production,* the nucleus absorbs the photon, creating an electron-positron pair, and ejects the pair.

The intensity of the gamma radiation drops off exponentially as a function of the thickness of the material that the radiation is traveling through. The mathematical formula for *intensity* of the radiation as a function of the penetration into the material is

$$I_{final} = I_{initial}\, e^{-\alpha\, \Delta x} \text{ or } \ln\left[\frac{I_{final}}{I_{initial}}\right] = -\alpha\, \Delta x$$

In this formula, $I_{initial}$ is the intensity of the radiation at the surface of the material, Δx is the displacement into the material, and α is the *absorption coefficient* of the material. Similar to half-life, a *half-intensity length* can be defined as $\ln(2)/\alpha$. The half-intensity length is sometimes called the *half value layer* or the *halving thickness* because the intensity of the radiation is half the initial value at this depth.

Be careful when looking up values for the absorption coefficient and make sure you know what you're reading. Here are three things to keep in mind:

✔ Some sources will provide numbers for the absorption coefficient whereas others provide numbers for the half-intensity length. Sometimes they aren't clear on what numbers they are providing, so be careful.

✔ The absorption coefficient is a measure of the amount of radiation absorbed by the material, whereas the *attenuation coefficient* (and the *skin depth*, δ) is a measure of the total drop in the intensity caused by absorption and scattering of the radiation. They aren't the same thing.

✔ The sources use one of two common sets of units for the absorption coefficient

• **Length:** The absorption coefficient has units of per length, which is per foot (per meter) or something related, such as per inch. My formula is written for this set of units.

• **Mass per unit length:** Some sources make the distinction and call this the *mass absorption coefficient.* The conversion to the length units is: mass absorption coefficient equals the absorption coefficient (α) divided by the mass density (ρ) of the material.

Blocking out electromagnetic radiation

A common shield used to protect people from electromagnetic radiation is lead. A property of electromagnetic radiation is the fact that the energy is proportional to the frequency, so the higher the energy of the photon, the shorter its wavelength will be. This means that X-rays and gamma rays have a wavelength shorter than the distance between the atoms. These photons see matter as mostly empty space. Lead is an effective shield of X-rays and gamma rays because it has a high atomic mass and high density. The chance of the radiation hitting an atom and being absorbed is much greater than with other materials. The following numbers show the difference between air, skin, and lead:

- Lead's absorption coefficient $\alpha = 20$ per foot (67 per meter). The mass density of lead is 22.1 slugs per cubic foot (11,400 kilograms per cubic meter) and the weight density of lead is 712 pounds per cubic foot (112,000 newtons per cubic meter).

- Air's absorption coefficient $\alpha = 0.0014$ per foot (0.0046 per meter). The mass density of air is 0.0023 slugs per cubic foot (1.20 kilograms per cubic meter) and the weight density of air 0.075 pounds per cubic foot (11.8 newtons per cubic meter).

- Human skin's absorption coefficient $\alpha = 1.2$ per foot (3.9 per meter).

To help you get a feel for these numbers, suppose you have some gamma rays and they travel 3 feet (0.914 meters) through a material, the gamma rays' intensity will be $8.8 \times 10^{-25}\%$ ~ 0% (in lead), 99.6% (in air), 2.7% (in skin) its original intensity. Essentially, gamma rays and x-rays will not penetrate a lead wall that thick.

Debunking Misconceptions about Electromagnetic Radiation

Many misconceptions about electromagnetic radiation exist. Electromagnetic radiation continuously bathes humans. This radiation propagates through space in clumps of energy called photons.

Electromagnetic radiation can be split into two categories:

- **Non-ionizing radiation:** This radiation has low energy photons.

- **Ionizing radiation:** This radiation has high energy photons. Ionizing radiation has photons with sufficient energy that when a molecule absorbs a photon, an electron is ejected from it leaving the molecule ionized. For most molecules the minimum energy required for ionization is 5 electronvolts, which is ultraviolet radiation.

The energy of a photon is related to the other physical parameters of the electromagnetic radiation with this formula:

$$E_{e/m} = h f = \frac{h \, v_{light}}{\lambda}$$

Here the photon energy is $E_{e/m}$, the frequency is f, the wavelength is λ, the *speed of light* is $v_{light} = 9.8357 \times 10^8$ feet per second (299792458 meters per second), and *Planck's constant* is $h = 4.1357 \times 10^{-15}$ electronvolt second (4.8875×10^{-34} foot pound second = 6.6262×10^{-34} joule second). Note that 1 electronvolt (eV) = 1.181×10^{-19} foot pounds (ft lb) = 1.602×10^{-19} joules (J).

These sections dispel some myths related to electromagnetic radiation and highlight some benefits associated with it.

Understanding non-ionizing radiation

With non-ionizing radiation, the photons don't have enough energy to ionize a molecule by causing an electron to escape. The photons either scatter off the molecules or they are absorbed and excite the molecule, but they can't ionize the molecule.

The main types of non-ionizing radiation include the following:

- **Radio waves:** The lowest energy photons from 0 electronvolts to 5×10^{-6} electronvolts are called *radio waves* because AM, FM, and TV signals are transmitted at these frequencies through the air.

- **Microwaves:** The next range of energies from 5×10^{-6} electronvolts to 10^{-3} electronvolts is called *microwaves.*

- **Terahertz radiation:** The next range of energies from 10^{-3} electronvolts to 10^{-2} electronvolts is called *terahertz radiation.*

- **Infrared radiation:** Photons with energies between 10^{-2} electronvolts and 1 electronvolt are called *infrared radiation.*

- **Light:** Photons with energies between 1 electronvolt and 3.1 electronvolts are called *light.* The specific colors are: 1.6 electronvolts (red), 2.0 electronvolts (orange), 2.1 electronvolts (yellow), 2.3 electronvolts (green), 2.7 electronvolts (blue), 3.1 electronvolts (violet).

- **Near ultraviolet radiation:** Photons with energies between 3.1 electronvolts and 3.9 electronvolts are called *UVA radiation.* Photons with energies between 3.9 electronvolts and 4.4 electronvolts are called *UVB radiation.* UVA and UVB combined are known as *near ultraviolet radiation.*

I include near ultraviolet because many animals such as honeybees see in the near ultraviolet. I give discrete values for the mean energies of the light colors, but in reality, the energies vary over a continuous range of energies around these values.

The importance of electromagnetic radiation can't be understated. In fact, it's one of the four fundamental forces in the universe (along with strong force, weak force, and gravitational force) and the most important one in biological systems. The list of uses of non-ionizing radiation in biological systems and in society could go on for many pages, so I only mention a few:

- ✔ **Light:** One of the five senses, *sight*, depends on the interaction of light with your eyes. Producing red, yellow, and green lights at traffic intersections and lighting your home are a few examples.

- ✔ **Laser surgery:** It has become a very important tool in the medical field, especially in correcting a person's vision.

- ✔ **Magnetic Resonance Imaging (MRI):** It has become a very important tool in medical diagnostics with its ability to image soft tissue. The MRI is based on nuclear magnetic resonance (NMR) and uses magnetic fields and radio frequency (RF) electromagnetic radiation (10^{-6} electronvolts).

- ✔ **Temperature control:** One of the methods the human body uses to control its temperature is to emit infrared radiation to dissipate heat energy. Infrared radiation is also used for heating and cooking.

- ✔ **Remote controls:** They're very important during those major sporting events or when you don't want to watch commercials.

- ✔ **Radar:** It tracks weather storms and air traffic.

- ✔ **Microwave ovens:** They're a convenient and quick way to cook foods. Microwave photons with an energy of 1.03×10^{-5} electronvolts are used in microwave ovens because water molecules have an *absorption band* at this energy, which, means that the molecule can easily convert the electromagnetic radiation energy into heat energy.

- ✔ **Cellphones and WiFi:** They use microwave radiation. In fact, there are more cellphones on earth than humans.

- ✔ **Radio and television:** Society relies on them for daily communication.

Comprehending ionizing radiation

Ionizing radiation is radiation that leaves a path of charged molecules and atoms in its path. The molecules in the path of the radiation lose electrons and are all electrically charged, which makes them interact strongly with their surroundings. As a result, they can cause severe damage to living cells.

Ionizing radiation includes not only ultraviolet, X-rays, and gamma rays (electromagnetic radiation with photon energies greater than light), but also alpha particles, beta particles, and neutrons. Here I break down these types:

- **Electromagnetic radiation:** *Electromagnetic radiation* with high energy photons can cause ionization of the molecules within animals. The photons have sufficient energy that electrons in the molecules can be ejected. This type of radiation includes ultraviolet radiation, X-rays and gamma rays. Atomic processes generate ultraviolet radiation and X-rays, whereas nuclear processes within the nucleus produce gamma rays.

  Near ultraviolet radiation (UVA and UVB) don't ionize molecules, but they do break chemical bonds. Animals are composed mostly of water so the radiation can produce H and OH (*free radicals*), which are highly reactive with the other molecules within the cell. The World Health Organization (WHO) classifies this type of radiation as carcinogenic.

  Middle ultraviolet radiation (UVC), far ultraviolet radiation, and low energy X-rays interact with atoms through the photoelectric effect where an atom absorbs the photon and then the atom ejects an electron with the excess energy.

  For high-energy X-rays, photons, and low- to mid-energy gamma photons, the primary process is Compton scattering. These photons keep some of their energy and transfer some energy to the atoms, which eject their electrons.

  The high energy gamma rays enter the atom's nucleus and create an electron-positron pair that escapes the nucleus. This is called electron-positron pair production.

- **Alpha particles:** These positive ions entering the body have lots of energy and scatter off atoms, transferring some energy and momentum to the atomic electrons. The energy transferred to the electron can be sufficient to eject the electron from the atom. These collisions continue until the ions have lost their excess energy and then interact with the local molecules.

- **Beta particles:** Similarly, the *beta particles* (electrons and positrons) scatter off the atoms, causing the atom to become excited or ionized if an electron escapes the atom. The beta particle's mass is small, so the rate of energy loss during each collision is small and the particles travel much farther into the animal. In the case of the positron, when the energy becomes low enough, it will collide with an electron and the two will be annihilated, creating gamma radiation.

- **Neutrons:** Neutrons have no charge, but they do have a *magnetic dipole* field and mass, which means that neutrons will travel deep into the animal, and they don't ionize the molecules directly but only bounce (scatter) off the molecules. Eventually, the neutrons do lose a sufficient amount of energy and then they will be absorbed within a nucleus. The change in the element's isotope usually makes it unstable, and leaves the nucleus in an excited state as well, so a gamma ray is emitted shortly after absorption. If the new isotope is unstable, then the isotope will undergo beta-decay creating a new element. In addition, free neutrons have a half-life of 11 minutes, so each free neutron can decay into a proton, a negative-beta particle, and an antineutrino.

Ionizing radiation is harmful to living cells, but if treated with care, it has many benefits to humans. A few examples include the following:

- ✔ **Safety devices:** Smoke detectors use ionizing radiation. They use an alpha-emitter because alpha particles can't penetrate the detector's casing. The alpha particles easily interact and ionize the air within the detector. The ionized air is attracted to charged plates, which completes the circuit. Smoke particles interact with the alpha particles, reducing the amount of ionized air and lowering the current in the circuit and setting off the alarm.

- ✔ **Food preservation:** Food is treated with ionizing radiation (*food irradiation*). The ionizing radiation damages the DNA within cells, killing microorganisms, bacteria, and viruses, and slowing down enzyme processes, which delays ripening. Typical forms of radiation used with food preservation are X-rays, gamma rays, and electrons (beta particles). The beta particles don't penetrate very far into the food and are effective for surface sterilization.

- ✔ **Insect control:** This method is related to food preservation. The irradiation of the food kills insects, such as fruit flies and other organisms on the food, without cooking the food.

- ✔ **Instrument sterilization:** This technique is ideal for sterilizing instruments, such as those used in operating rooms.

- ✔ **Medical diagnoses:** Making diagnoses with ionizing radiation has two different approaches:

    - *Diagnostic radiology* involves using an external source of ionizing radiation (usually X-rays) and studying the radiation intensity that comes out the other end.

    - *Nuclear medicine* involves producing radionuclides for medical purposes, which are intended for placing inside the body. *Radionuclides* are radioactive isotopes usually with a short half-life. As a diagnostic tool, the detectors surrounding the body detect the emission of the ionizing radiation.

✔ **Medical treatment:** This method is similar to medical diagnoses, but the dosages are higher.

- *External beam radiotherapy* uses several beams of ionizing radiation; all the beams meet at a specific point within the body, producing a very high intensity source at that point (cancer tumor), killing the cells.

- In *brachytherapy*, pharmaceuticals bind to radionuclides that then transport the radionuclides to the target cells. A large intensity of radiation is then produced at the target cells.

Seeing How Radioactivity Interacts with Biological Systems

The interaction of radioactive materials with biological organisms is an important area of biophysics, medical physics, and health physics. Many applications and uses exist in many different fields of science. As a biophysicist, you want to understand this interaction. These sections look at how radioactivity interacts within biological systems. These sections focus on the Carbon-14 death date technique, discuss the elimination of radioactive material within a living animal, and estimate how radioactive the human body is.

Finding a date in archaeology — call Carbon-14

Carbon-14 has a half-life of 5,730 years and under goes β^- decay. Only 10 carbon atoms out of every 10^{13} is a Carbon-14 isotope when a biological system dies and 5,730 years after its death only 5 carbon atoms out of every 10^{13} are Carbon-14 atoms. The ratio drops to zero the older the object gets because no new Carbon-14 is being put into the material to replenish the Carbon-14. Knowing the half-life of Carbon-14 and the ratio of Carbon-14 in an organic material allows you to calculate how long it has been since the organism died.

Mathematically, the *half-life* is $T_{1/2} = 5{,}730$ years and the inverse of the *decay constant* is $1/\lambda = T_{1/2}/\ln(2) = 8{,}270$ years. The formula for calculating how long it has been since an organic material has died is

$$\Delta t = \frac{1}{\lambda}\ \ln\left[\frac{N_{initial}}{N_{final}}\right]$$

In September 2011, the body of King Richard III was discovered. (Many authors including William Shakespeare have written about him.) In 2012, Carbon-14 dating was done on the body and his death date was determined to be between 1455 and 1540. King Richard III died during battle in 1485, what was the ratio of Carbon-14 to Carbon-12 in his bones in 2012?

Use these steps to help you solve this problem.

1. **Determine the formula and understand the problem.**

   The problem gives the time and wants you to find the final number, N_{final}. You need to rearrange the formula and isolate for N_{final}, which means you need to find numbers for the decay constant, the elapsed time, and the initial number of Carbon-14 to Carbon-12 ratio.

2. **Search for the numbers you need.**

   I stated before the problem that the decay constant for Carbon-14 is $\lambda =$ ln(2)/(5730 years) = 1.21×10^{-4} per year.

   The time of death and the time of the Carbon-14 test are given in the problem, so the elapsed time is $\Delta t = 2012 - 1485 = 527$ years.

   The long-term average ratio of Carbon-14 to Carbon-12, or the initial ratio is $N_{initial}$ = 1 ppt (parts-per-trillion) = 1000 ppq (parts-per-quadrillion) = 1 Carbon-14 to 10^{12} Carbon-12.

3. **Solve the problem.**

   Rearrange the formula and substitute the numbers into the formula and solve.

   $$N_{final} = N_{initial}e^{-\lambda \Delta t} = \left(10^3 \, \text{ppq}\right)e^{-1.21 \times 10^{-4} \, y^{-1}(527 \, y)} = 938 \, \text{ppq}$$

The ratio of Carbon-14 to Carbon-12 has dropped from 1000 ppq to 938 ppq, or dropped by 6.2 percent in the 527 years. The number of Carbon-14 compared to Carbon-12 is very small, and this is a small change, which is why the uncertainty is so large in Carbon-14 dating.

Keep the following issues in mind as you solve this problem:

- ✔ Carbon-14 dating is only good for dating organic material that has died in the last 60,000 to 70,000 years. Note that 12 half-lives is equal to 68,760 years and $2^{-12} = 2.4 \times 10^{-4}$ is the fraction of the original Carbon-14 left. When starting with only 1 part-per-trillion, that is a very small number.

- ✔ Carbon-14 undergoes negative beta-decay to Nitrogen-14, and the activity is less than the background radiation for the small sample sizes, which makes detection very difficult.

✔ After 4.5 billion years, no natural Carbon-14 is left within the planet, only the stable isotopes Carbon-12 (98.9 percent) and Carbon-13 (1.1 percent). All the natural Carbon-14 present today comes from cosmic rays interacting with the atmosphere. The atmosphere is 78 percent nitrogen, and a nitrogen atom can capture a neutron changing into Carbon-14 plus a proton. The Carbon-14 reacts with oxygen gas to form carbon dioxide. Plants then absorb the carbon dioxide.

✔ *Calibration* is needed for accurate dating because the amount of cosmic radiation entering the atmosphere is fluctuating and therefore the amount of Carbon-14 entering organisms is not a constant. To *calibrate* means that the measurements are adjusted to a standard. Tree rings are an excellent standard for calibrating dates, because when the tree was born and died are well known and provide information about the amount of Carbon-14 in the atmosphere during different ages, which extends back in time several thousands of years.

✔ Humans have been disrupting the Carbon-14 ratio. In the 1800s, the burning of coal lowered the Carbon-14 ratio by a few percentages and then atmospheric nuclear tests in the 20th century increased the ratio. By the mid 1960s, the Carbon-14 ratio was double at 2 parts per trillion. Today, the ratio has dropped and is only a few percentage points above the historical average.

Eliminating radioactive material within the body — biological half-life

The human body will remove half of the carbon within fat in 35 days, which is called the *biological half-life* (T_b). The human body doesn't care if the carbon is Carbon-11, Carbon-12, Carbon-13, or Carbon-14. However, Carbon-11 and Carbon-14 are radioactive and decay with a half-life called the *physical half-life* (T_p). The physical half-life is a property of the isotope and doesn't care about the biological system, whereas the biological half-life depends on the type of element (not the isotope) and the type of biological system.

The combination of the biological half-life and the physical half-life gives the *decay constant* (λ) and the *effective half-life* (T_{eff}):

$$\lambda = \frac{\ln(2)}{T_{eff.}} = \frac{\ln(2)}{T_p} + \frac{\ln(2)}{T_b}$$

As an example of how you would apply this, consider the following biophysical application. Iron-59 is used to test blood disorders. You place Iron-59 with an activity of 10^4 becquerels (2.70×10^{-7} curies) in a vial and you give a patient the same amount of Iron-59. After exactly 30 days, you measure the activity in the vial as 6,267 becquerels (1.69×10^{-7} curies), and in the patient you measure the activity as 4,551 becquerels (1.23×10^{-7} curies). You want to determine the physical and biological half-life of Iron-59.

Earlier in the "Decaying of elements – the physical half-life" section, I provide the relationship between activity, decay constant, and time. This formula provides the relationship between the decay constant and the effective half-life. In addition, it provides the relation between the physical half-life, the biological half-life, and the effective half-life.

The activity in the vial is determined solely by the physical half-life, which means you can find the physical decay constant and the physical half-life from this data.

In the patient, the decay is the combination of the physical half-life and the biological half-life, so you can find the decay constant and the effective half-life. After you know the effective half-life and the physical half-life from the vial, you can find the biological half-life.

The solutions are

1. $\dfrac{\ln(2)}{T_p} = \lambda_p = \dfrac{-1}{\Delta t}\ln\left(\dfrac{A_{final}}{A_{initial}}\right) = \dfrac{-1}{30d}\ln\left(\dfrac{6267Bq}{10000Bq}\right) \Rightarrow T_p = 44.5 \text{ days}$

2. $\dfrac{\ln(2)}{T_b} + \dfrac{\ln(2)}{T_p} = \dfrac{\ln(2)}{T_{eff}} = \lambda_{eff} = \dfrac{-1}{\Delta t}\ln\left(\dfrac{A_{final}}{A_{initial}}\right) \Rightarrow$

$\dfrac{\ln(2)}{T_b} = -\dfrac{\ln(2)}{44.50} - \dfrac{1}{30d}\ln\left(\dfrac{4551Bq}{10000Bq}\right) \Rightarrow T_b = 65.0 \text{ days}$

The effective half-life of Iron-59 within a human is 26.4 days, which is almost half the physical half-life. This means that after 26.4 days, half the Iron-59 is gone from the body with approximately one quarter being excreted from the body and another quarter radioactively decaying through negative beta-decay into Cobalt-59, which is a stable isotope of Cobalt. In addition, notice that the effective half-life is less than the physical half-life and the biological half-life. In fact, the effective half-life is always smaller than the smaller of the physical half-life and the biological half-life.

Determining how radioactive humans are

Everything is radioactive to some extent, which is why there is a constant background radiation. To illustrate, I calculate how radioactive the human body is by examining Mike.

Mike isn't a Baby Boomer and didn't live during the atmospheric atomic tests of the 1950s and 1960s and so didn't ingest radioactive Strontium-90 when he drank his milk. In addition, Mike wasn't in the area affected by the Chernobyl accident in April 1986 or near the Fukushima nuclear accident in March 2011. Therefore, I assume Mike doesn't have excess amounts of radionuclides from human sources and that all the radionuclides in his body are from natural sources. Mike weighs 200 pounds (890 newtons). His mass is 6.21 slugs (90.7 kilograms).

Approximately 10^{28} atoms are present in the human body with the majority being water. Mike's mass in terms of elements is as follows:

- ✓ Oxygen makes up 65 percent of the body's weight. The weight of the oxygen in Mike's body is 130 pounds. (The mass is 59.0 kilograms.) The relative abundance of each isotope is 99.76 percent Oxygen-16, 0.04 percent Oxygen-17, and 0.20 percent Oxygen-18. All three isotopes are stable.

- ✓ Carbon makes up 18 percent of the body's weight. The weight of the carbon in Mike's body is 36.0 pounds. (The mass is 16.3 kilograms.) The relative abundance of each isotope is 98.89 percent Carbon-12 and 1.11 percent Carbon-13. These two isotopes are stable. In addition, there is 1 part per trillion (weight is 3.60×10^{-11} pounds and the mass is 1.63×10^{-11} kilograms) of Carbon-14, which undergoes negative beta decay with a half-life of 5,730 years.

- ✓ Hydrogen makes up 10 percent of the body's weight. The weight of the hydrogen in Mike's body is 20.0 pounds. (The mass is 9.07 kilograms.) The relative abundance of each isotope is 99.985 percent Hydrogen-1 and 0.015 percent Hydrogen-2. These two isotopes are stable.

- ✓ Nitrogen makes up 3 percent of the body's weight. The weight of the nitrogen in Mike's body is 6.00 pounds. (The mass is 2.72 kilograms.) The relative abundance of each isotope is 99.63 percent Nitrogen-14 and 0.37 percent Nitrogen-15. These two isotopes are stable.

- ✓ Calcium makes up 1.5 percent of the body's weight. The weight of the calcium in Mike's body is 3.00 pounds. (The mass is 1.36 kilograms.) The relative abundance of each isotope is 96.941 percent Calcium-40, 0.647 percent Calcium-42, 0.135 percent Calcium-43, 2.086 percent Calcium-44, 0.004 percent Calcium-46, and 0.187 percent Calcium-48. All the calcium isotopes are stable with the exception of Calcium-48, which has a half-life of 4.4×10^{19} years.

Calcium-48 has a very large half-life because it decays through a double beta-decay, which is rare. The double beta-decay process consists of two neutrons being converted into two protons and two β⁻ particles without antineutrinos.

✔ Phosphorus makes up 1.2 percent of the body's weight. The weight of the phosphorus in Mike's body is 2.40 pounds. (The mass is 1.09 kilograms.) One hundred percent of the abundance is the stable isotope Phosphorus-31.

✔ Potassium makes up 0.2 percent of the body's weight. The weight of the phosphorus in Mike's body is 0.400 pounds. (The mass is 0.181 kilograms.) The relative abundance of each isotope is 93.258 percent Potassium-39, 0.012 percent Potassium-40, and 6.730 percent Potassium-41. The only unstable isotope is Potassium-40 (^{40}K) with a half-life of 1.28×10^9 years. Eighty-nine percent of the decays have Potassium-40 changing into Calcium-40, a β⁻ particle, and an antineutrino. The other 11 percent of the decays have Potassium-40 changing into Argon-40, a β⁺ particle and a neutrino.

✔ Sulfur, chlorine, and sodium each make up 0.2 percent of the body's weight. The weight of each of these elements in Mike's body is 0.400 pounds (0.181 kilograms mass). All the isotopes for these three elements are stable. The relative abundance of sulfur's isotopes is 95.02 percent Sulfur-32, 0.75 percent Sulfur-33, 4.21 percent Sulfur-34, and 0.02 percent Sulfur-36. The relative abundance of chlorine's isotopes is 75.77 percent is Chlorine-35 and 24.23 percent is Chlorine-37. Sodium-23 is the only stable sodium isotope.

✔ The last 0.5 percent of Mike's mass is made up of trace elements.

The total internal activity within Mike's body can be calculated for Carbon-14, Calcium-48, and Potasium-40 by using the formula in the "Decaying of elements – the physical half-life" section earlier in this chapter that states that the activity is equal to the decay constant times the number of particles. This example gives the mass and the abundance fraction. You can find the average atomic mass in the periodic table to find the number (N) and then the activity:

$$A = \lambda N = \frac{\ln(2)}{T_{1/2}} \frac{M_{total}}{\overline{M}_{atom}} (\text{abundance fraction})$$

Carbon-14: $A_{C-14} = \dfrac{\ln(2)(36.0 \text{ lb})10^{-12}}{1.81 \times 10^{11} \text{s}\left(12.0107 \times 3.66 \times 10^{-27} \text{lb}\right)} = 3136 \text{ becquerels}$

Calcium-48: $A_{Ca-48} = \dfrac{\ln(2)(3 \text{ lb})0.00187}{1.39 \times 10^{27} \text{s}\left(40.078 \times 3.66 \times 10^{-27} \text{lb}\right)} = 1.91 \times 10^{-5} \text{ becquerels}$

Potassium-40: $A_{K-40} = \dfrac{\ln(2)(0.4 \text{ lb})0.00012}{4.04 \times 10^{16} \text{s}\left(39.0983 \times 3.66 \times 10^{-27} \text{lb}\right)} = 5755 \text{ becquerels}$

Mike's radiation output here is a minimum value, and it's probably larger because of other sources of radioactivity that have entered into the food chain. In addition, the amount of Carbon-14 in the food chain is still higher than the historic average. Note that this radiation is referred to as *internal radiation*, which accounts for approximately 10 percent of the *natural background radiation* the human body receives.

Chapter 18

Fighting the Big C — But Not All Radiation Is Bad

....

In This Chapter

▶ Taking a closer look at radiation

▶ Seeing the effects of radiation on the body

....

Radiation is a word that incites fear in people, but many people don't understand why and how it can be dangerous to living organisms. As a budding biophysicist, you need to be able to explain to people when they should be concerned about the radiation and when the radiation level is safe.

This chapter focuses on quantifying radiation and exploring the harmful effects of radiation on biological systems. Here I examine organisms being exposed to and absorbing radiation, as well as the effect upon the organism and quantifying the health effects of the radiation.

Investigating Radiation within Biological Systems

The primary types of radiation that interact with biological systems are electromagnetic radiation, alpha particles, and beta particles. The *alpha* and *beta* particles are produced during the radioactive decay of unstable nuclides (radionuclides). Here I introduce the activity of these radionuclides.

The radionuclides change to new nuclides in an attempt to become stable. The majority of them decay by emitting alpha and beta particles. The radiation activity from a radionuclide is a measure of the rate that alpha and beta particles are being produced. The activity is also related to the number of radionuclides present. The number of radionuclides and the activity can be described mathematically by:

$$\text{number of radionuclides: } N_{final} = N_{initial}e^{-\lambda \ t} \ (\text{units: moles or particles})$$

$$\text{activity: } A = \lambda N \ (\text{units: becquerel or curie})$$

$$\text{decay constant: } \lambda = \frac{\ln(2)}{T_{eff}} = \frac{\ln(2)}{T_p} + \frac{\ln(2)}{T_b} \ (\text{units: per second})$$

The formula in the first line contains four symbols: N_{final} is the final number of radionuclides, $N_{initial}$ is the initial number of radionuclides, λ is the decay constant, and Δt is the elapsed time between $N_{initial}$ and N_{final}. The formula in the second line contains: A is the activity, λ is the decay constant, and N is the number of radionuclides. In this formula, N has units of particles and not moles. Remember to convert from moles to particles you have to multiply by Avogadro's number (1 mole = 6.022142×10^{23} particles). The units of the activity in this formula are becquerels (Bq), but sometimes using curies (Ci) is more convenient (1 *curie* = 3.7×10^{10} *becquerels* = 3.7×10^{10} disintegrations per second). The formula in the third line shows you how to calculate the decay constant and the effective half-life of a radionuclide. The symbols in the formula are: λ is the decay constant, T_{eff} is the effective half-life, T_p is the physical half-life, and T_b is the biological half-life.

To clarify some terms for you, here are some easy definitions to remember:

✔ *Physical half-life, T_p,* is the time it takes half of the radionuclides to decay into new nuclides.

✔ *Biological half-life, T_b,* is the time it takes a biological organism to remove half of the nuclides from its body through natural processes. The body doesn't care if the nuclide is radioactive or not. For example, the human body treats all types of carbon the same.

✔ *Effective half-life, T_{eff},* is the combination of the physical half-life and the biological half-life. It's the time it takes a biological organism to remove half of the radionuclides within its body through radioactive decay and natural processes.

✔ *Decay constant* is equal to the natural logarithm of 2 divided by the effective half-life, so it's a measure of how fast the radionuclide will decay. The larger the number, the faster the material decays.

✔ *N* is the number of radionuclides. $N_{initial}$ is the number of radionuclides you start with, and N_{final} is the final number of radionuclides at the end. The units are either moles or particles and Avogadro's number, $N_A = 6.022142 \times 10^{23}$ particles/mole, is the conversion between the two sets of units.

- *Elapsed time*, Δt, is the time it takes from the start to the finish.

- *Activity*, A, is the speed that the nuclides are disappearing. The units are *becquerel* = disintegration per second, or *curie*.

The following sections discuss how to measure the interaction of radiation with a biological organism and how to measure the harmful effects of the radiation on the biological organism. Refer to Chapter 17 for more information about the basic parts of radiation.

Interacting radiation with matter

The interaction of radiation with biological systems is *stochastic* in nature, which means the interaction is random. You can't predict what it will do, but you can describe all the radiation interacting with the biological system using probabilities. For example, an X-ray photon may pass through an organism without doing anything or an atom may absorb it, and the organism repairs the damage or the damage can't be repaired. Each event has a certain probability of occurring. When the biological organism is exposed to a large dosage of radiation, the effects of the radiation become more deterministic.

Radiation has energy and when matter absorbs it, the matter gains that energy. A couple of common methods of quantifying the energy absorbed are as follows:

- **Exposure:** *Exposure* (X) is a measure of the amount of X-ray or gamma ray radiation entering an object. It measures the amount of ionization that will occur in the material; it's restricted to photon energies between 40 electron volts (4.7×10^{-18} foot pound = 6.4×10^{-18} joules) and 3×10^{6} electron volts (3.5×10^{-13} foot pound = 4.8×10^{-13} joules). These photons interact with matter through the photoelectric effect and Compton scattering. (*Photoelectric effect* means an atom absorbs the photon and then the atom emits an electron, whereas the *Compton scattering* means the atom absorbs the photon and the atom emits both an electron and a photon with a smaller frequency; refer to Chapter 17 for more information.)

  The S.I. unit of measurement for exposure is the *coulomb per kilogram*. The older unit of measure is 1 *roentgen* (R) = 3.76×10^{-3} coulomb per slug = 2.58×10^{-4} coulomb per kilogram. The exposure is a property of the X-rays and gamma rays and has nothing to do with the material the radiation is striking. The exposure states that 1 roentgen of X-rays produces -2.58×10^{-4} coulombs of negatively charged ions and 2.58×10^{-4} coulombs of positively charged ions within 2.2 pounds (1 kilogram) of material.

✔ **Absorbed dose:** *Absorbed dose* (*D*) is the amount of energy from the radiation that is absorbed by the material per unit mass. The unit for the absorbed dose is 1 *gray* (*Gy*) = 1 joule per kilogram = 10.76 foot pound per slug. Another unit for the absorbed dose is the *rad*, where 1 rad = 0.01 grays. The absorbed dose is used for all types of radiation unlike exposure, which is used only for X-rays and gamma rays.

To help you get a feel for these units: a 1 roentgen exposure of X-rays on soft tissue produces an absorbed dose of approximately 0.01 gray = 1 rad, and 100 grays (10,000 rads) of radiation completely destroys living tissue.

Some books introduce the kinetic energy of radiation absorbed per unit mass (Kerma), which is very similar to the absorbed dose, and it's used only for uncharged radiations, such as photons and neutrons. Calculating it is easier than calculating the absorbed dose, but I don't calculate the Kerma.

Hurting cells with radiation — mechanisms of cell damage

Cells are mostly water. When water is ionized, it becomes H and OH free radicals that are highly reactive. The other ionized molecules can also become highly reactive within the cell. Some of the free radicals interact with chromosomes, damaging them; and some of the radiation directly damages chromosomes. Recent research has also shown that an absorbed dose of 1 gray will cause 35 breaks of the deoxyribonucleic acid (DNA) in each cell. A *break* means both strands of the double helix have been completely separated. These double-strand breaks are the most dangerous and the probability of the repair being incorrect is greater. The cells that are badly damaged will die, but the cells that were repaired incorrectly could possibly be functional and pass on mutations to future generations, which can lead to cancer.

Meteors are a perfect example of how radiation works. Think of the Tunguska, Siberia, Russia meteor in 1908 or the Chelyabinsk, Russia meteor in 2013. The 2013 meteor exploded over the city of Chelyabinsk and injured 1,500 people. It had a weight of 7,000 tons (a mass of 6,350 tonnes), a diameter of 50 feet (15 meters), and an energy release of 500 kilotons of TNT (1.543×10^{15} foot pounds = 2.092×10^{15} joules). If this energy was spread uniformly over the entire surface of the Earth, then the energy per unit area would be 0.280 pounds per foot (4.09 joules per square meter). You wouldn't even notice this flux of energy. Instead, the energy was mostly localized in a circular region with a radius of approximately 50 miles (80 kilometers), so the energy per unit area was actually 7,050 pounds per foot (1.03×10^5 joules per square meter) over the city. The real amount of energy per unit area in Chelyabinsk, Russia was 25,180 greater than if the energy had been spread over the entire surface of the earth!

Radiation is like the meteor with all the energy focused in a small region. The non-ionizing ultraviolet radiation does have sufficient energy to cause some ionization and break chemical bonds in molecules. Therefore, low energy ultraviolet radiation has the same carcinogenic mechanism as ionizing radiation.

Use the preceding information in this section and the previous section to solve this problem.

The previous subsection states that 100 grays of radiation will kill living tissue. If all the energy is converted to heat energy, what is the temperature rise in the tissue?

To solve this problem, follow these steps:

1. **Determine which formula is needed to solve the problem.**

   The problem tells you that the radiation energy is converted into heat energy and causes a rise in the temperature. The relationship between the temperature and heat energy is $\Delta Q = m \, c \, \Delta T$. (Refer to Chapter 10 for additional information.)

2. **Determine the numbers needed, which are given in the problem.**

   The problem states that the energy per unit mass is

   $$\frac{\Delta Q}{m} = 100 \text{ grays} = 1{,}076 \; \frac{\text{foot pound}}{\text{slug}}$$

3. **Find the remaining numbers you need to solve the problem.**

   You need the specific heat. You can look for the number in a reference source that has the number. (I can help here.) Assume the specific heat of soft tissue is the same as water. I found the value for the specific heat to be: $c_{tissue} = c_{water} = 25{,}000$ foot pound per (slug degree Fahrenheit) = 4,180 joules per (kilogram kelvin).

4. **Substitute the numbers into the equation and solve.**

   $$\Delta T = \frac{1}{c}\left(\frac{\Delta Q}{m}\right) = \frac{1 \text{ slug } {}^{0}\text{F}}{25000 \text{ ft lb}}\left(1076\frac{\text{ft lb}}{\text{slug}}\right) = 0.0430 \text{ degree Fahrenheit}$$

   $$\left(0.0239 \text{ degree Celsius}\right)$$

The energy spread over the tissue causes an insignificant rise in the temperature, so what kills the tissue? The answer is: Radiation behaves like particles. The massive particles are alpha radiation, beta radiation, neutron radiation, and cosmic radiation. Similarly, the electromagnetic radiation (X-rays and gamma rays) behaves as particles as well, called *photons*. These particles strike the cell at specific locations and transfer all the energy to specific atoms. The energy is localized and large enough that the radiation can do damage by ionizing the atom that absorbs the radiation, similar to the meteor example previously mentioned.

Exposing the Body to Radiation

Radiation affects humans, although the process to quantify the radiation's effects is a difficult process because different types of radiation act differently on the body even if the energies are the same. Also, different tissues react differently to the radiation. The location where the interaction between the radiation and the tissue takes place is important as well. For example, 10 grays of radiation through my finger would be bad for my finger, but my body as a whole should be fine, whereas 10 grays of radiation through all my organs could be very bad for my overall health.

In these sections, I introduce some of the quantities needed to describe the effects of radiation, quantify some of the negative effects of radiation, discuss the fact that there is radiation everywhere, and examine a preventable cancer.

Estimating the effects from radiation

Comparing the effects of different types of radiation on different parts of the body is important to help you determine the health risk in different situations, given the absorbed dose of each type of radiation on each part of the body.

The *equivalent dose* (H_T) is defined as

$$H_T = \sum_R w_R D_{R,T}$$

Here the subscript T represents the tissue or organ being irradiated, the R represents the type of radiation, the large sigma means you have to sum over all types of radiation irradiating the tissue, w_R is called the *radiation-weighting factor* (It measures the damage to the tissue by the radiation. For example, 1 gray of X-rays will cause a different amount of damage than 1 gray of alpha particles. The radiation-weighting factor is independent of the type of tissue or organ.), and $D_{R,T}$ is the absorbed dose of radiation R in tissue T and gives the amount of energy transferred to the tissue or organ.

The equivalent dose has units of sievert (Sv) or rem (old unit), where 1 *sievert* = 100 *rem*. The radiation weighting factor has units of sievert per gray (or rem per rad). Typical radiation-weighting factors are 1 sievert per gray for X-rays, gamma rays, and beta particles; 5 sieverts per gray for protons; 20 sieverts per gray for alpha particles and heavy nuclei; and 5 to 20 sieverts per gray for neutrons. The neutrons' radiation-weighting factor varies depending on the kinetic energy of the neutron with the peak of 20 sievert per gray for 10^6 electron volt neutrons.

Meanwhile, the *effective dose* (ε) is defined as

$$\varepsilon = \sum_{T} w_T H_T$$

Here the subscript T represents the sum over all tissues and organs being irradiated and w_T is the tissue-weighting factor. The effective dose is an equivalent whole-body equivalent dose, so you can compare the biological damage done by different radiation events.

The tissue-weighting factor is independent of the type of radiation. The sum of all the tissue-weighting factors is 1, so if the entire body is exposed to the equivalent dose, then the effective dose is equal to the equivalent dose. The individual factors are being updated on a regular basis as data becomes available. For example in the case of breasts, w_{breast} = 0.05 in 1990 and was changed to w_{breast} = 0.12 in 2007. The International Commission on Radiological Protection (ICRP) (www.icrp.org) has up-to-date values for the tissue-weighting factor.

Suppose Nancy is being exposed to 1 gray of gamma radiation over the entire body and 1 gray of alpha particles from an external source. What is the effective dose?

To figure out this problem, follow these steps:

1. **Figure out what equations you need.**

   The problem tells you the type of radiation present and the absorbed doses, so you need to calculate the equivalent doses first. After you know the equivalent doses, then you can calculate the effective dose.

2. **Find the numbers for the equivalent dose calculation.**

   The gamma radiation is in all the organs and tissues, so $D_{gamma,body}$ = 1 gray.

   The radiation-weighting factor for X-rays, gamma rays, and beta particles is w_{gamma} = 1 sievert per gray.

The alpha particles can only reach the skin. Remember, alpha particles can't penetrate the skin, so $D_{alpha,skin}$ = 1 gray and $D_{alpha,rest}$ = 0 gray.

The radiation-weighting factor for alpha particles is w_{alpha} = 20 sievert per gray.

3. **Calculate the equivalent dose.**

   Find the equivalent dose in the skin and the remainder of the body separately. The reason is that the gamma radiation is throughout the body whereas the alpha is only in the skin.

   The equivalent doses are

   $$H_{skin} = w_\alpha D_{\alpha,skin} + w_\gamma D_{\gamma,skin} = (20\,Sv/Gy)(1Gy) + (1Sv/Gy)(1Gy) = 21\ \text{sievert}$$
   $$H_{rest} = w_\alpha D_{\alpha,rest} + w_\gamma D_{\gamma,rest} = (20\,Sv/Gy)(0Gy) + (1Sv/Gy)(1Gy) = 1\ \text{sievert}$$

4. **Find the numbers for the effective dose calculation.**

   The equivalent doses are known, so now you can calculate the effective dose once you know the tissue weighting factors:

   The tissue weighting factor for skin is w_{skin} = 0.01.

   The tissue weighting factor for the rest of the body is w_{rest} = 1 − w_{skin} = 0.99.

5. **Calculate the effective dose.**

   The effective dose is

   $$\varepsilon = w_{skin}H_{skin} + w_{rest}H_{rest} = (0.01)(21Sv) + (1-0.01)(1Sv) = 1.20\ \text{sievert}$$

   The alpha particles are nasty, but the gamma rays cause more damage because the alpha particles can't get through the skin to the organs inside the body.

Measuring the unhealthy effects of radiation

Having a firm understanding of what the numbers mean in the formulas I present in the previous section is important. At high dosages the radiation is deterministic. For example, 100 grays of radiation kills living tissue. At moderate dosages you need to consider probabilities. At low dosages you need to consider statistical averages over a population. For example, the regular use of a tanning salon before the age of 30 increases the chances of getting melanoma by 75 percent. Like smoking, it depends on the person. These results are based on a collection of studies of populations. It's a general average based on people who never use tanning beds versus one-timers to weekly users.

Keeping track of older terminology

The terminology has changed over the years and some of the older terms are still in use so I should bring some of these terms to your attention:

✔ *Relative Biological Effectiveness (RBE)* is defined as

$$RBE = \frac{D_{X,T}}{D_{R,T}}$$

Here $D_{X,T}$ is the absorbed dose of 250 kiloelectron-volt X-rays in the tissue or organ T (the X-rays are the reference radiation) and $D_{R,T}$ is the absorbed dose of radiation R in the same tissue or organ T that will produce the same effect as the X-rays.

The relative biological effectiveness isn't really old terminology because it's still used. You can usually experimentally determine this quantity; it takes into account all factors affecting the biological system.

✔ *Dose equivalent* (*H*) (notice the order of the words is changed from equivalent dose) is defined as the *quality factor* (*QF*) times the *absorbed dose* (*D*). The dose equivalent was introduced for the development of a set of standards for radiation risk.

✔ The *quality factor* (*QF*) is the predecessor for the *radiation-weighting factor* (w_R). The absorbed dose was taken at a specific point in space and multiplied by the quality factor at that point to give the dose equivalent at that point in space. The quality factor is then defined for a specific radiation in terms of the *stopping power* or *linear energy transfer* (*LET*).

✔ The *effective dose equivalent* is the predecessor of the effective dose. It's the sum similar to the effective dose after averaging the dose equivalent over each tissue or organ.

In addition, the formulas in the previous section didn't take time into account. For example, is an absorbed dose of 1 gray of gamma radiation over one hour the same as an absorbed dose of 1 gray of gamma radiation over ten hours? The time over which the exposure to the radiation occurs plays an important role in the affects of radiation on the body. At low rates of exposure, the body can repair most of the damage, whereas at high rates of exposure the body can't repair the damage fast enough.

Walking into a high dose of radiation

A high dose of ionizing radiation will have immediate effects on the body (within one day) and is known as *radiation poisoning, radiation sickness, radiation toxicity,* or *acute radiation syndrome*. This situation occurs when the person received a whole body absorbed dose of radiation between 0.25 grays and 500 grays at a rate greater than 0.1 grays per hour. A single blast of 100 grays kills living tissue, and a whole body absorbed dose of 500 grays is immediate death.

The radiation poisoning has four stages:

✔ **Prodromal stage:** This stage starts within minutes to a few hours after exposure and will last less than 48 hours. The prodromal stage typically includes nausea and vomiting, fatigue, fever, and headaches.

✔ **Latent stage:** The time duration of the latent stage depends upon the dosage and can be several weeks long for a whole body dosage less than one gray, whereas at high dosages it doesn't occur.

✔ **Manifest illness stage:** This stage can be split into three syndromes:

- The *hematopoietic symptoms* occur from exposure of the blood forming cells. The symptoms are a drop in the white blood cell count, the red cell count, and platelets.

- The *gastrointestinal symptoms* occur from exposure of the intestines. The symptoms include nausea, vomiting, diarrhea, cramps, salivation, and dehydration.

- The *neuromuscular symptoms* occur from brain exposure. The symptoms include fatigue, apathy, sweating, fever, headache, hypotension, and hypotensive shock.

✔ **Recovery stage:** The body is repairing the damage. At high dosages, this stage doesn't occur. If the person survives the first eight weeks, then the chances of recovery are very good.

The symptoms vary depending on the whole body absorbed dose (D_{body}):

✔ D_{body} < 0.25 grays: There appears to be no short-term effects.

✔ 0.25 grays < D_{body} < 1 gray: Hematopoietic symptoms.

✔ 1 gray < D_{body} < 2 grays: Hematopoietic symptoms and mild gastrointestinal symptoms. 0 to 1 out of 20 people will die within eight weeks.

✔ 2 grays < D_{body} < 6 grays: Hematopoietic symptoms, moderate gastrointestinal symptoms, and mild neuromuscular symptoms. 1 to 20 out of 20 people will die within six weeks. (The odds of mortality drop by half with medical treatment.)

✔ 6 grays < D_{body} < 8 grays: Hematopoietic symptoms, severe gastrointestinal symptoms, and moderate neuromuscular symptoms. 19 to 20 out of 20 people will die within four weeks. (10 to 20 out of 20 people will die even with immediate medical treatment.)

✔ 8 grays < D_{body} < 40 grays: Hematopoietic symptoms, severe gastrointestinal symptoms, and severe neuromuscular symptoms. Approximately two weeks to live at 8 grays, and by 40 grays, the person dies in less than two days.

✔ 40 grays < D_{body}: Death occurs within minutes from neurological or cardiovascular failure.

This list doesn't include rates. Quantifying is harder, but it does influence the boundaries in the list. For example, 4 grays of radiation have a 50 percent mortality rate within six weeks if the radiation is spread over four hours. If the person is subjected to the radiation over half an hour, then the mortality rate jumps to approximately 75 percent. If the person is subjected to the radiation over a day, then the mortality rate drops to approximately 25 percent. Note that at very low total dosages and very high total dosages the rate doesn't play that great of a role. Only in the intermediate range from 2 to 6 grays is the rate important.

Living with radiation – long-term effects

Exposure to moderate to high doses of ionizing radiation over a long period of time is known as *chronic radiation syndrome*. The symptoms don't show for a long time, but they're similar to acute radiation syndrome (which I discuss in the previous section) and vary from gastrointestinal to low blood counts to neurological effects and death. Chronic radiation syndrome can also lead to other diseases such as the different types of cancer. In the case of acute radiation syndrome, most people will survive a dose between 1 gray and 4 grays. However, the probability of these people dying from cancer during their lifetime has doubled.

The effects of radiation are cumulative and the doses need to be added, so the longer a person lives, the greater the chance of acquiring cancer. The *cumulative whole body absorbed dose* is the sum of all the contributions to the whole body absorbed dose over the years. The *cumulative effective dose* is the sum of all the contributions to the effective dose over the years. The *cumulative equivalent dose* is the sum of all the contributions to the equivalent dose over the years. These quantities are useful for predicting the long term risk of cancer.

When a large population is involved, then *the collective dose* is more useful for determining statistical averages. The *collective effective dose* is defined as

$$\varepsilon_C = \sum_{i=1}^{N} \varepsilon_i = N\bar{\varepsilon}$$

The N is the number of people and $\bar{\varepsilon}$ is the average effective dose. The units are *person-sieverts* (or *person-rem*). Here are a couple of examples:

✔ A collective effective dose of 100 person-sieverts causes one premature cancer death. The *linear no threshold hypothesis* states that any amount of radiation is harmful. (I discuss this hypothesis in more detail in the next section.) If you assume the linear no threshold hypothesis is correct, then you can calculate the number of premature deaths from the background radiation. The average background global radiation is 0.0024 sieverts per year, so the number of premature cancer deaths each year caused by the background radiation is (0.0024 sieverts) × (7 × 10^9 persons) / (100 person sieverts per death) = $1.68 × 10^5$ deaths. Based on this hypothesis, there are 168,000 premature deaths each year from the background natural radiation.

✔ The collective effective dose of a collective equivalent dose of 10,000 person-sieverts from Iodine-131 causes approximately 60 latent thyroid cancers. Iodine-131 collects within the thyroid and has a very short half-life of 8.04 days. The risk is very low, and Iodine-131 isn't found in nature because of the short half-life. Iodine-131 is a useful radionuclide for nuclear medicine.

Glowing walls — all matter is radioactively decaying

All organic material has carbon, and one part in a trillion of carbon is carbon-14, which means all organic material is radioactive. Of course, the amount of carbon that is carbon-14 is very small, so the radiation level is small. This section is about low-level radiation or the level of radiation you're more than likely to be exposed to. I discuss background radiation and the different models proposed for low dose radiation.

Looking at background radiation around the world

In the United States, the 2006 population was exposed to an average radiation of approximately 6.2 millisieverts per year. Fifty percent was from natural sources, 48 percent was from medical sources, and 2 percent from consumer products, occupational exposure, industrial exposure, and nuclear power stations.

3.1 millisieverts per year was from natural sources. Elements with more protons than bismuth, which has 83 protons, are unstable. The two heaviest elements that are abundant in the soil are uranium and thorium. Uranium-238 is the most abundant isotope of uranium with a half-life of $4.468 × 10^9$ years. This is the same as the age of the planet, so there is only half the uranium left on the planet compared to when the planet formed.

Uranium and thorium and their decay products are in the soil. These elements along with Carbon-14, Calcium-28, and Potassium-40 are ingested with food and water and are producing radiation within everyone's bodies. In the United States the average internal radiation is about 0.38 millisieverts per year.

When uranium decays, one of the daughter products is radon, which is a gas. Radon-220 has a half-life of 55.6 seconds, so it usually decays into Polonium-216 before escaping the ground but some does escape; however, Radon-222 has a half-life of 3.8235 days before decaying into Polonium-218. Radon-222 has plenty of time to escape the soil and enter the air. The amount of Radon-222 in the air at the earth's surface is approximately 0.14 becquerels per cubic foot (5 becquerels per cubic meter), which humans breathe into their lungs. The background radiation (alpha particles) from radon gas in the United States is 2.28 millisieverts per year. This is two-thirds of the natural background radiation in the United States. The natural background radiation in the United States is 3.1 millisieverts per year, whereas the world average is 2.4 millisieverts per year.

The rest of the natural background radiation varies a lot depending on the location. The first source is the cosmogenic radionuclides produced by cosmic rays, which includes Hydrogen-3, Beryllium-7, and Carbon-14, for example. The average in the United States is 0.33 millisieverts per year and a world average of 0.39 millisieverts per year. In Honolulu, Hawaii (sea level) the cosmic radiation is approximately 0.2 millisieverts per year whereas in Colorado Springs, Colorado, it's 0.7 millisieverts per year. Florida as a whole state has the lowest amount of cosmic radiation within the continental United States, whereas at 35,000 feet (10,700 meters) in a passenger airline, the cosmic radiation is about 0.008 millisieverts per hour.

The terrestrial background radiation varies due to the amount of thorium and uranium in the soil. The average US terrestrial background radiation is 0.21 millisieverts per year. The world average is 0.48 millisieverts per year. The last 10 percent of the background radiation comes from the uranium, thorium, potassium, and carbon (plus others) inside your own body through the ingestion of food and water.

The variation in the rock and soil can have a significant effect upon the total natural background radiation. For example, the background radiation in Finland is four times greater than in the United Kingdom (2 millisieverts per year). Also, the thorium and uranium concentration is extremely high in the monazite sand on the coast near Guarapari, Brazil, where the natural background radiation is 175 millisieverts per year. Monazite is a stone containing high concentrations of thorium and uranium. Along ancient coast lines, such as near Guarapari, Brazil, and Kerala, India, (70 millisieverts per year), the ocean has had time to break the monazite down into highly radioactive sand.

Searching for zero radiation – linear no threshold hypothesis

Understanding the short-term and long-term effects of radiation-induced cancer as a function of the dose and the rate of absorption is a nontrivial problem. By the end of the 1950s, governments and people knew high doses of radiation were very harmful, but what level was safe was unknown. Also, it was unknown if different types of radiation were just as harmful. For example, in the 1950s, companies could purchase X-ray machines as a gimmick to attract clients. By 1960, governments were interested in low-dose radiation and the biological effects of radiation. Around this time several models had been proposed:

- ✔ The *linear no threshold model* was proposed, and it states that any amount of radiation is harmful and an increase in dose (both short term and cumulative) will increase the chance of cancer. This model reads as risk(ε) = A ε, where ε is the effective dose and *risk* is the probability of getting cancer. The A is a proportionality constant that depends on the type of cancer. The model works very well at high doses where lots of data exists to support a linear relationship between the dosage and the risk of cancer. However, at low doses it's hard to acquire statistically meaningful data, and even after more than 50 years the jury is still out.

  Supporters of the model argue that biological repair mechanisms aren't perfect; therefore, any damage will leave behind some badly repaired cells. The National Academy of Sciences' Biological Effects of Ionizing Radiation committee (BEIR) has published many reports on the health risks of exposure to low doses of radiation. The 1970 report fully supported the LNT model, but in the 1970s, studies of people living in regions with high levels of background radiation showed no increased levels of cancer. The 1980 BEIR report stated that the LNT was probably an overestimate of the effects of low dose radiation. The 2006 BEIR report has gone back to fully supporting the LNT model.

- ✔ The *nonlinear no threshold model* states that any amount of radiation is harmful; this model is similar to the linear no threshold model. The risk of cancer is risk(ε) = A ε + B ε^2 + . . . , where ε is the effective dose. This modification allows for a better quantitative fit to experimental data at moderate doses. At moderate dosages the risk of cancer appears linear, but in the case of high dosages of Iodine-131, the risk is reversed. Iodine-131 collects in the thyroid, and at moderate dosages it damages the cells and not all the DNA is repaired properly, producing mutagens and leading to cancer. However, at high dosages the radiation kills the cells instead of allowing them to be repaired incorrectly, so the risk of cancer plateaus.

- ✔ The *nonlinear threshold model* states that the radiation is harmful only above a certain level of radiation. This model reads risk(ε) = 0 if ε < $\varepsilon_{cut-off}$ or Risk(ε) = A + B ε + C ε^2 + . . . , if ε > $\varepsilon_{cut-off}$. The *cut-off* is close to the average background radiation level; below this level the body can repair any damage to the cells caused by the radiation. Studies of people living in regions with high amounts of natural background radiation haven't shown an increased rate of cancer. The United Nations Scientific Committee on the Effects of Atomic Radiation has supported the linear no threshold model, but in 2012 submitted a white paper indicating it may not be correct at low doses.

✔ The *radiation hormesis model* (also known as *radiation homeostasis*) is similar to the nonlinear model, but it states that low dosages of radiation near the background radiation level are beneficial, and the dose must increase above a certain level before it becomes harmful. This model reads as risk(ε) = $-A\,\varepsilon + B\,\varepsilon^2 + \ldots$ A recent study has shown very successful DNA repair with no flaws at very low dosages, but the DNA repair mechanisms become saturated at higher dosages.

Looking closer at lung cancer

Lung cancer in the industrial nations can be considered a preventable disease. The No. 1 cause of lung cancer is cigarette smoking, whereas the No. 2 cause is a person's home. Among nonsmokers, the home is actually the No. 1 cause of lung cancer. In fact, smoking and the home cause 95 percent of lung cancer. (Smoking causes 80 to 85 percent of all lung cancer, whereas a person's home causes 10 to 15 percent of all lung cancer.) Air pollution and all other sources contribute only 5 percent toward all the lung cancer cases. In other words, if everyone stopped smoking and people's homes were fixed, then lung cancer cases would drop by at least a factor of 20.

Prior to the 20th century, lung cancer was an uncommon disease. Now, more than 1 million people die each year from lung cancer with the highest rates in the industrial and developing nations. The good news is recent studies have shown that if a person quits smoking before the age of 40, then the lungs are back to essentially those of a nonsmoker within 15 to 20 years.

In a home, the biggest source of natural background radiation is radon gas. It seeps from the soil into the home and gets trapped there. People then breathe in the radon gas where it alpha decays into polonium. The alpha particles travel into the tissue and shred the cells with a *radiation weighting factor* of 20. Until the 1980s, radon wasn't considered a concern. The outside air radon concentration varies between 0.14 becquerels per cubic foot (5 becquerels per cubic meter) and 0.42 becquerels per cubic foot (15 becquerels per cubic meter). The radon concentration above the oceans is approximately 0.0014 becquerels per cubic foot (0.05 becquerels per cubic meter).

In the early 1980s, it was discovered that some homes had radon concentrations as high as 2,800 becquerels per cubic foot (100,000 becquerels per cubic meter). A 1991 study in the United States found an average radon concentration of 1.3 becquerels per cubic foot (45 becquerels per cubic meter) in the living spaces of the homes, which is about ten times larger than the outdoor air radon concentration. Most countries now have regulations that set a maximum radon concentration level of 200 to 400 becquerels per cubic meter for the home. In addition, many places are putting radon prevention into the building codes, especially where the outdoor air radon concentration is high. The new homes in these areas of high radon concentration must have radon preventative measures built into them.

Radon gas appears to produce a linear risk of cancer. There is a 16 percent increase in the risk of lung cancer for every 2.8 becquerels per cubic foot (100 becquerels per cubic meter) concentration of radon gas. Smoking and radon combined together is a very bad combination with the risk for the smoker typically ten times worse than that of the nonsmoker's risk.

Chapter 19

Seeing Good Biophysics in the Medical Field

In This Chapter

▶ Glowing in the doctor's office

▶ Petting cats and non-animal CATs

▶ Looking for non-organic PETs

*B*iophysics has many applications in many different fields, such as in medicine and in two other subfields of physics: medical physics and health physics. *Medical physics* is the application of physics in medicine and includes three primary branches: radiation therapy physics, nuclear medicine, and diagnostic imaging physics. *Health physics* deals with health and safety with radiation, and protection from radiation.

This chapter focuses on how radioactivity and radionuclides are used in medicine. I focus on three applications: the first is radioactivity and nuclear medicine, the second is the use of computerized axial tomography (CAT) scanners (also known as computed tomography (CT) scanners), which allows doctors to image the human body using X-rays, and the third application I discuss is positron emission tomography (PET) scanners, which is a technique for imaging processes within the body.

This chapter doesn't address ultrasound imaging, magnetic resonance imaging (MRI), and functional magnetic resonance imaging (fMRI). Although MRI is based on nuclear magnetic resonance (NMR), it has nothing to do with radioactivity and radionuclides even though it has nuclear in the name. (NMR is based on Faraday's law, which I discuss in Chapter 16.)

Identifying Radiation at Work in Medicine

The application of radioactivity and radionuclides in medicine can be split into two concentrations: nuclear medicine and radiology. Here I discuss radiology and nuclear medicine in greater depth.

Arming dentists and doctors with X-ray machines

Radiologists use the radionuclides and radiation for both therapy and diagnostics. In the case of therapy, it's known as *radiotherapy, radiation therapy,* or *radiation oncology.* The three different methods of radiotherapy are as follows:

- **Radionuclide therapy:** *Radionuclide therapy* involves using radionuclides with pharmaceutical drugs to produce radiopharmaceutical drugs. The drugs are designed to target certain parts of the body thereby delivering a high dose of radioactivity to the target. This therapy is used for both curative and palliative purposes. Depending on the target within the body, some radionuclides will cluster at the target without any drug, and other radionuclides only need to be bound to a molecule.

- **Brachytherapy:** *Brachytherapy* involves the direct implantation of the radionuclides into the tumor. This method guarantees a high dose of radioactivity within the target, which works well for localized tumors.

- **Teletherapy:** *Teletherapy* involves ionizing radiation in the form of a beam being shot at the patient's tumor. In order to maximize the absorbed dose at the tumor while reducing the amount of damage to healthy tissue within the body, the beams of radiation are fired from several different directions with all the beams converging at the tumor. Any form of ionizing radiation can be used, but the most common forms of radiation are X-rays and electrons (β^- particles). The use of gamma radiation isn't as common as the X-rays. The rest of the radiation falls under the category of *hadron therapy,* which includes the particles: protons, neutrons, alpha particles, and heavy nuclei.

You're probably wondering how something that can cause cancer can also be used to cure cancer. Certain cancer cells are very susceptible to radiation whereas others are quite resistant. The more localized and susceptible a cancer is to radiation, the more successful the radiotherapy will be. Usually, radiotherapy needs to be combined with other techniques, such as surgery and or chemotherapy, in order for it to be successful at curing the cancer. Typical doses of radiation will vary between 10 to 80 grays (100 grays will kill living tissue) spread over several cycles. The cycle's duration depends on several factors related to the patient's health. An absorbed dose is a measure of the amount of energy from the radiation that is absorbed by the tissue per unit mass. The units are 1 gray = 1 joule per kilogram = 100 rad = 10.76 foot pound per slug.

The other medical application of radiation is imaging. Imaging can be split into the following two categories.

X-ray imaging

X-ray imaging is one of the most popular methods of imaging the body in a noninvasive manner. X-ray imaging creates X-rays outside the body and passes them through the body. The denser the material (such as bone) the more X-rays that are absorbed by the material and less X-rays will pass through to the detector on the other side.

Both the medical and dental professions use X-rays. In the medical field, X-rays give the medical professionals an immediate view of what's going on inside the body. They also have a low cost and are noninvasive. On the downside, X-rays are ionizing electromagnetic radiation, which kills some cells and damages other cells. If the cell is repaired incorrectly, it can become a mutant and possibly cancerous. Check out the later section, "Focusing Your X-Ray Vision — Computer Tomography (CT) Scans" for more information.

In the dental field, the dose of radiation is small and focused only on the jaw. The dental bitewing exam produces an effective dose of 0.005 millisieverts. (The *effective dose* is a means by which to estimate the cancer danger from different types of radiation irradiating different parts of the body. Check out Chapter 18.) In comparison, the average natural background radiation on the planet is 2.4 millisieverts per year. You would need more than 600 visits to the dentist to equal the background radiation the average person in the United States receives in one year. The panoramic X-ray images are a little more than double the radiation of a bitewing. The average person in the United States receives 3.1 millisieverts of radiation from natural sources and 3.1 millisieverts from medical examination each year. The dental examinations are insignificant in comparison and don't have an impact.

Radionuclide imaging

The *radionuclide imaging* devices place radionuclides, called *tracers,* within the body. The three types of imaging include the following:

✔ **Single photon emission computed tomography (SPECT):** *SPECT* uses a radionuclide with usually a very short half-life. (The *half-life* is the time it takes half the radionuclides to decay and emit radiation.) The radionuclide decays and emits a gamma ray with a certain amount of energy. The detectors, called *gamma cameras* or *Anger cameras*, absorb the gamma rays, allowing the doctor to determine the location within the body from which they were emitted.

✔ **Radioimmunoassay (RIA):** *RIA* is an interesting and highly sensitive technique developed to determine the concentration of molecules within the body using radionuclides. For example, in the case of measuring the concentration of an antigen within the body, a known quantity of an antigen is labeled with a radionuclide and mixed with the unlabeled antigens within a serum or in the body. The antibody protein molecules you're interested in will attach to both the labeled and unlabeled antigens. These complex structures can be separated out from the other material and the radioactivity in the separated complex structure will determine the concentration of unlabeled antigens within the body.

✔ **Positron emission tomography (PET):** *PET* uses a radionuclide or a radiopharmaceutical drug that targets a specific item of the body (a tumor for example). The radionuclide uses decays by emitting positrons, which are anti-electrons. The body has a huge number of electrons and one of them is annihilated by the positron and produces gamma radiation, which is emitted from the body and detected. Check out the later section, "Posing For Pictures — Positron Emission Tomography (PET)" for more information about PET.

Producing radionuclides and radiopharmaceuticals — nuclear medicine

People involved with nuclear medicine combine radionuclides with other elements or with pharmaceutical drugs to form radioactive chemicals or radiopharmaceutical drugs. For example, the nuclear facilities in Chalk River, Ontario, Canada produce one-third of the North American supply of radionuclides and the generators, which produce the radionuclides.

Generators are radionuclides with a longer half-life. Molybdenum-99 is a common generator because it has a half-life of 65.94 hours and emits beta particles when it decays into Technetium-99m, which is the most common radionuclide used in nuclear medicine.

Technetium-99m is used in more than two-dozen different radiopharmaceuticals. Technetium-99m usually emits a 140.5 kiloelectron volt gamma particle with a half-life of 6.01 hours while relaxing into its ground state Technetium-99. Technetium-99 has a half-life of 211,000 years, which is relatively stable. *Gamma cameras*, SPECT machines, and PET machines can detect the gamma radiation and produce an image. When Technetium-99m is combined with molecules to form radiochemicals or with pharmaceuticals to form radiopharmaceuticals, then it will concentrate in specific locations in the body or within specific ailments, such as cancer producing a targeted image and map.

The next most common radionuclides (called *diagnostic radiotracers*) are Iodine-123 (13.2 hours half-life), Iodine-125 (60.14 days half-life), and Iodine-131 (8.04 days). Iodine by itself collects in the thyroid, which makes it well suited for detecting things such as Graves' disease. In addition to being useful as a radiotracer in diagnostics and imaging, iodine can be used in radiotherapy as well. Higher doses of iodine (usually Iodine-131) can be ingested or injected into the body, and the concentrated radiation will kill thyroid disease such as thyroid cancer.

Different elements collect in different spots of the body. For example, Yttrium-90 (half-life of 64.0 hours) is used for synovial disease and Strontium-89 (half-life of 50.5 days) for metastatic bone cancer. Alternatively, any radionuclide can be used in brachytherapy because the element is inside a container, which is placed inside the tumor or in the vicinity of it.

Focusing Your X-Ray Vision — Computer Tomography (CT) Scans

Computer tomography (CT) machines or scanners are one of the most popular ways to perform noninvasive images of the internal human body. CT scans also go by several other names, including *X-ray computed tomography* (X-ray CT) scan, *X-ray computerized axial tomography* (X-ray CAT) scan, *computed tomography* (CT) scan, and *computerized axial tomography* (CAT) scan.

CT machine uses X-rays but they need higher doses of radiation than a regular X-ray image used for broken bones because they're imaging different soft tissues. Because most X-rays pass through a body (about 99.7 percent), a higher dose of radiation is needed to more easily measure the absolute change in the intensity. For example, if 1,000 photons enter the person, then 997 photons will come out the other side. If 1,000,000 photons are used, then

997,000 photons will come out the other side. The missing 3,000 photons are easier to detect than the 3 missing photons in the first case. This higher dosage of radiation increases the resolution of the image.

CT scans are popular even though they use X-rays. The benefits outweigh the hazards and they help save lives. I explain the machine's name to help you understand how it works. I then explain what the machine is used for, and finish with the scanner's benefits and hazards. I hope you can understand why and when the benefits outweigh the hazards.

Zapping the body — how CT works

The CT scanner goes far beyond a regular X-ray image by changing two very important features. First, the resolution is improved by increasing the intensity of the X-rays passing through the patient's body. Second, a two-dimensional image is created by combining multiple shots of X-rays through the body from different angles. A CT scanner can give doctors a clearer picture of the tissue they're examining.

Here is what happens when a patient gets a CT scan. A CT scanner usually consists of a giant ring with the source of X-rays being the X-ray tube inside the ring on one side of the body and an X-ray detector placed inside the ring on the other side.

1. **The patient lies down upon a bed, which is slid into the center of a large ring.**

   The portion of the body to be imaged is aligned with the ring. Inside the ring, X-rays are produced and aimed at the patient. The X-rays pass through the patient and enter the ring on the opposite side. In this part of the ring are detectors that measure the X-ray intensity.

   The medical personnel can't determine where along the path within the patient's body the X-rays were absorbed, only just how much X-ray radiation was absorbed along the path.

2. **The direction of the X-ray source and the detector is rotated about an axis, and a new measurement of the X-rays passing through the patient's body is made.**

   A complete circle of data (rotation of 180 degrees) gives enough information that the amount of X-rays absorbed at each point in the plane can be determined. This produces a two-dimensional image (slice) of the patient. The details of how the image is constructed from this data are very involved and complicated, but computers make it possible.

Examining the evolution of the CT scan

The name *X-ray computerized axial tomography* tells you everything this machine does. Just break it down, word by word:

- **X-ray:** This word says the machine (obviously) uses X-rays, which is ionizing electromagnetic radiation.

- **Computerized:** The X-ray machine is connected to a computer, which makes it computerized. Being computerized helps in detecting small changes in the intensity of the X-ray beam after the beam has passed through the patient. Computerized has been changed to *computed* in CT scan.

- **Axial:** This part of the name comes from the image being slices of the patient perpendicular to an *axis.* (The X-ray generator and detectors usually form a ring to create the two-dimensional slice, and the patient is moved up or down the *axis* of the ring when a new slice is to be made.) New software and computers allow for the combination of the slices so a full three-dimensional image can be formed. Hence the axial has been dropped from the name.

- **Tomography:** This word means imaging an object by sectioning it with the use of penetrating waves such as X-rays from different angles.

So, the X-ray CAT has become simply CT.

---

3. **The patient is then moved further into the ring containing the X-ray source and detector, and another slice is created.**

   This is repeated until the entire volume is covered and a full three-dimensional image is created. Unfortunately, it means each part of the patient is exposed to a relatively high dose of X-rays several times.

Looking inside the body — what CT is used for

A CT scan can be used to see the different soft tissues within the body. It's possible to reduce the X-ray dosages and still obtain high-resolution image by using a contrast agent. The contrast agent readily absorbs X-rays and changes the intensity of the X-rays passing through, which helps with the resolution of the image. Barium and iodine are popular contrast agents in CT scans.

New CT machines can image any part of the body, but the dose varies quite a bit depending on what needs to be imaged. For example, the brain requires a higher dose because it's encased in bone so a head CT scan will produce an effective dose of 2 millisieverts (but the absorbed dose is about four times larger than that for the chest CT scan). A chest CT scan produces an effective dose of 7 millisieverts (300 times greater than a regular chest X-ray). A whole-body CT scan produces an effective dose of 10 millisieverts. (The same as three years of natural background radiation.) Chapter 18 discusses how to compare different types of radiation on different parts of the body.

Staying away — who should avoid CT scans

The effects of a single CT scan usually aren't harmful in adults, but the effects of the radiation are cumulative and multiple scans can be a problem later in life. (The amount of absorbed energy in a scan can reach 0.08 grays.) Each scan does have a small chance of producing cancer or some ill effect, but generally the need and benefits outweigh the risk. Patients should avoid unnecessary scans because the radiation causes some damage to the cells each time.

Pregnant women are an exception. Pregnant women should avoid CT scans or any ionizing radiation while pregnant if at all possible. The fetus is very susceptible to radiation, especially during the first trimester. CT scans that don't directly target the fetus are relatively safe with caution, but an abdomen CT scan has the potential of causing severe harm. In all cases with pregnant women, the potential hazard to the fetus needs to be weighed against the need of the CT scan.

Posing For Pictures — Positron Emission Tomography (PET)

Positron emission tomography (PET) is a very powerful imaging tool used for analyzing functional processes within the body and medical diagnosis. In the 1990s, PET scans were combined with CT scans and MRI scans. The other techniques provided anatomical information, which is lacking from the PET scan. The geometric registration of the PET scan with one of the others gives a very detailed image combined with its anatomical location. These sections describe how the PET machine works and discuss its applications.

Setting up the PET — how does it work

PET scans are very interesting in the way they work. The ability of a PET scanner to target specific functions or objects (such as a tumor) makes it a very powerful tool in noninvasive medical imaging and diagnostics. The process involves using radionuclides, so PET scanners aren't available at all hospitals, clinics, and medical institutes.

Here is what happens when a patient gets a PET scan:

1. **A nuclear medicine technician prepares the radionuclides or radiopharmaceutical drug.**

   PET scans use the following radionuclides: Carbon-11 (20.5-minute half-life), Nitrogen-13 (9.97-minute half-life), Oxygen-15 (2.03-minute half-life), and Fluorine-18 (110-minute half-life). The half-lifes of these radionuclides are very short so they need to be prepared shortly before use.

2. **A technician puts the radionuclides/radiopharmaceutical drug into the patient's body.**

   The radionuclides/radiopharmaceuticals travel within the body and collect at the target.

3. **A positive beta particle is emitted when the radionuclide/radiopharmaceutical drug radioactively decays.**

   A *positive beta particle* is a positron emitted from the nucleus, and a positron is an anti-electron.

4. **The beta particle can't travel far in the body and usually stops within a distance less than 0.02 inches (0.5 millimeters) from where it was emitted.**

   The majority of the positrons used in a PET scan have a mean energy of 250 kiloelectron volts (2.954×10^{-14} foot pound = 4×10^{-14} joules), and they usually stop moving quickly. This length scale sets a lower limit on the resolution of a PET scan to about 0.04 inches (1 millimeter).

5. **The positrons usually stop moving before annihilating with an electron.**

   The positron is an anti-electron, and it will annihilate with a local electron.

6. **The annihilation of the electron-positron pair creates two gamma particles that fly off in opposite directions.**

   The two gamma photons both have an energy of 511 kiloelectron volts (6.037×10^{-14} foot pound = 8.186×10^{-14} joules), which is the rest mass energy of the electron (and the positron as well).

7. **The gamma detectors surround the patient in a ring and record a signal whenever a pair of gamma photons simultaneously hit the detectors.**

A *line of response* connects two detectors that record signals from the two gamma photons. After the measurements are completed, all the lines of response are put together, showing where there is a high concentration of radionuclides within the body.

Picturing the body — what is PET used for

The radionuclides that are used in a PET scan as part of radiopharmaceuticals work best when used for imaging certain physiological properties. Combined with a CT scan or an MRI, a PET scan is good for locating things such as tumors. In fact, a PET scan can light up for the brain, the kidneys, and any cancer in the body be it lung cancer, lymphoma, or something else so doctors can better examine and make a diagnosis.

Besides specific targets, the radionuclides can be attached to molecules, such as, say glucose. These molecules would circulate within the human body and metabolize, providing information about activity within the body.

A PET scan can work with any biologically active molecule. The most common radionuclide is Fluorine-18. One of the radiopharmaceuticals is *2-fluoro-2-deoxy-d-glucose*, more commonly called *fluorodeoxyglucose (FDG)*. These radiopharmaceuticals are also known as *radiotracers* because they bind to specific receptors or drug action sites within the body. FDG, which is a positron-emitting glucose analog, is well suited for detecting cancers and determining the progress of any treatment of the cancer.

Cells, such as brain cells, kidney cells, and cancer cells that require large amounts of glucose, absorb FDG as well. These three types of cells prevent glucose from escaping the cell once absorbed. FDG is missing the 2'hydroxyl, which prevents it from being metabolized within the cell. Therefore, the FDG sits in the cell until it decays and emits a positron. FDG PET scans allow for the imaging of the body where there is a high demand for glucose. It's useful not only in medicine, but also in research, such as within biology and neuroscience.

Part VI
The Part of Tens

In this part . . .

- ✔ Uncover ten important tips that can help you study better and perform better on quizzes and tests, which ultimately can get you a higher grade in your biophysics course.

- ✔ Be exposed to the many exciting careers in biophysics and related fields such as medical physics and health physics. No matter where your talents lie, there is an exciting career in biophysics for you.

- ✔ Discover many benefits of radiation, radioactivity, and nuclear physics in a bonus POT chapter found at www.dummies. com/extras/biophysics.

Chapter 20

Ten (or So) Tips to Help You Master Your Biophysics Course

Many people have a phobia about biophysics and mathematics, and they're very frightened when they take their biophysics courses. This chapter strives to help alleviate some of your fears with some helpful tips to make your biophysics course a tad easier (or less painful). Unfortunately this chapter doesn't give you the answers to the questions on your exams.

Drawing Diagrams and Figures

When working problems in your biophysics course, always draw figures, diagrams, and graphs. Even if you have a figure to look at, draw your own figure. Doing so can help you visualize what's going on with the problem.

Even if you encounter time constraints, don't try and cut corners and not draw your figures and diagrams. Students are tempted to not draw the diagrams because they think they know what is going on and they think they're wasting time by drawing the diagram. However, it's usually the opposite. They end up wasting more time because they don't have the diagram and haven't correctly visualized the problem.

Obeying the Rules

Biophysics is a set of rules, but many people in their first biophysics course think of biophysics as a cookbook with recipes. They think all problems of a certain type should use the same steps and formulas in the same order.

Unfortunately it's not the case. However, whatever approach you take is okay, as long you don't break the laws of the physical universe.

In any problem, spend the time setting up the problem and determining what physical principles you need to use before seeking a formula to use. The steps are as follows:

1. **List everything you know.**

2. **Draw diagrams, figures, and graphs.**

3. **Decide what physical principle you need to apply.**

   Ask yourself which physical rules are relevant.

4. **After you decide what rules are relevant for the problem, you can select the appropriate equations you need to use in order to solve the problem.**

   For example, do you need kinematic equations, Newton's laws of motion, work-energy theorem, Hooke's law, or some other principle?

Creating Your Own Dictionary

You should create your own personalized dictionary, especially if you're planning on continuing on in biophysics, medical physics, or some related field. Your own dictionary gives you a quick source to remind you of ideals and concepts.

In your dictionary, include concepts, definitions, physical laws, mathematical symbols, and mathematical formulas. Having your own dictionary is also important if you're using more than one source. Each source typically has its own set of symbols, so different sources will use different symbols to represent the same thing. For example, within a source, it may use the symbol T, K, or E_K to represent the kinetic energy. Another problem is that sources may use the same symbol to represent different things. For example, the symbol T has been used to represent kinetic energy, temperature, period, and half-life. Keeping your symbols and their meaning straight is very important for making it through your course and your dictionary can help you do so.

Understanding the Concepts

A *concept* is an understanding of something formed by mentally combining all its characteristics or particulars. In a biophysics course, the concepts are more important than the mathematical formulas. (The mathematical formulas are a shorthand representation of the concepts.) If your biophysics textbook has concept problems, then you should work through all of them to get a better understanding of the concepts. If you understand the concepts, then the mathematical formulas become a lot easier to understand and use.

For example, consider Newton's second law of motion, which is the law of angular acceleration. Lots of physical concepts are contained within this law, which aren't evident by just looking at the mathematical formula, $F_{NET} = m\, a$. The mathematical formula makes sense if you understand the concepts, but the concepts aren't evident if you start with the formula.

Not Fearing the Mathematics

In a biophysics course, chances are you'll encounter your fair share of math. To do well, don't fear the math. Make friends with the math; it's meant to help you. The mathematics serves three purposes:

- ✔ Mathematics serves as another language, which can help you understand the concept and describe it.

- ✔ Mathematics is a shorthand writing of physical relationships or physical quantities. For example, instead of writing "gravitational potential energy" every time, I can just write "E_p". I also can write the phrase "gravitational potential energy is equal to the mass of the object times the gravitational constant times the vertical position," or I can write the formula "$E_p = m\, g\, y$". They mean the same thing, but the second is much quicker to write and read if you understand what the mathematical formula means.

- ✔ Mathematics allows you to describe things in a quantitative manner as well.

Work toward feeling comfortable with the mathematics. To do so, you can start by writing things out longhand and then using the mathematics to write it in shorthand. Adding it to your dictionary can also be helpful. (See the earlier section, "Creating Your Own Dictionary" for more information.)

Applying the Knowledge in Your Field

In a typical introductory biophysics course, the students have very different backgrounds and different interests. Because biophysics is an interdisciplinary subject with applications in many different fields, the chances the discussion and or examples are directly related to what you're interested in are slim. Hence, when studying a specific topic in biophysics, try to figure out how you can apply it to your field of interest.

Doing so is a fun exercise and makes the material easier and more interesting to learn. For example, a neuroscience student can think about how electrical circuits can be applied to modeling the neurons in the brain, whereas a bioengineer can think about how to build electrical circuits that can be applied to building a mechanical device.

Networking with Your Classmates

Working with your fellow students is an important method of mastering the material and making it through your class. Each person usually picks up something different from the classes and the textbook, so by combining information with your classmates, you can form a more complete picture. Also, you may have understood something that can help your classmates while they may have understood something that can help you.

If you meet and network with your classmates, you must do some preliminary studying and work prior to the meeting. You need to have a clear picture in your mind of what you know and where you're struggling. If you don't, then you'll meet up with them, listen to what they have to say, and never really learn the material. To understand what I mean, suppose you find an assignment question worked out in detail online. You copy the solution and submit it. You'll probably get a good grade, but you haven't learned anything and you won't be prepared for your exam.

Surfing the Internet

Several excellent biophysics websites exist online that are excellent resources of information. You may discover some helpful information to supplement your course text. Just know that many of these online sites focus on a specific area of biophysics though, which is usually related to the type of biophysics research they're doing.

Don't use the Internet to find the solution of assignment questions. Instead go online for help in understanding the general concepts and ideas. You can also use the Internet to see how biophysical concepts are applied in your area of interest. If you use it to look up solutions, the only person you're hurting is yourself because you aren't learning and understanding the concepts and ideas.

Chatting with Biophysicists

Biophysicists are very passionate about their research and biophysics in general. They love talking about their research. To help you do better in your biophysics class, take time to talk to them about biophysics and their research because they're a great resource of information. They'll usually provide important information that can help you understand the concepts and ideas in your course.

Chapter 21

Ten Careers for People Studying Biophysics

In This Chapter
▷ Considering academia
▷ Eyeing jobs in industry
▷ Working in hospitals and clinics

*T*his chapter mentions a few types of jobs available to people interested in biophysics, medical physics, and health physics. A career in these fields can take place in academia, governments, hospitals and clinics, and in the private sector. This list isn't exhaustive; I keep it general enough to cover a large percentage of the jobs. Consider which ones you might be interested in pursuing if biophysics is your passion.

Experimental Biophysicist in Academia

An experimental biophysicist uses the tools and instruments from physics, biology, chemistry, and mathematics to study biological systems. A *biological system* can be anything from the membrane of a cell to a large environmental system with multiple living organisms.

In addition to knowing how to use instruments, such as a nuclear magnetic resonance machine in experiments, these professionals also need a working knowledge of computers for data collection and analysis. Experimental biophysicists work in a wide variety of university departments, including biochemistry, bioengineering, biology, dentistry, kinesiology, medicine, neuroscience, and physics.

Theoretical Biophysicist in Academia

A theoretical biophysicist uses tools from physics, mathematics, computer science, biology, and chemistry to study biological systems. They work on problems from neural networks to the environmental impact of climate change.

Computers play an important role in modeling, simulation, and visualization of the processes being studied. Theoretical biophysicists work in a wide variety of university departments, such as biochemistry, bioengineering, biology, dentistry, kinesiology, medicine, neuroscience, and physics.

Biophysicists outside Academia

Biophysicists also work in private research labs, industry labs, and government labs. These careers are usually research focused within an interdisciplinary setting, where the person collaborates with people from other backgrounds, such as chemists, biologists, medical doctors, and pharmacists. Depending on where the person is working, the focus of the work can be very different. For example, their research may be focused on molecular biophysics, membrane biophysics, pharmaceuticals, or bioenergetics to mention a few.

Nuclear Power Reactor Health Physicist

The nuclear power reactor health physicist works at a nuclear reactor site and is responsible for all aspects of radiation protection. The professional has multifaceted responsibilities: He or she keeps track of all radiation protection equipment, trains all the plant workers, and monitors and analyzes the radiation data (*radiation dosimetry*).

Governmental Health Physicist

A governmental health physicist focuses on regulatory enforcement and occupational safety. Many government agencies hire health physicists; the list is extensive. A couple of examples include the US Department of Energy and the Consumer Product Safety Commission.

This professional ensures the protection of people and the environment from radiation sources by guaranteeing that all safety regulations and procedures are being followed in the manufacture, use, and disposal of the radioactive material.

Environmental Health Physicist

An environmental health physicist focuses on protecting people and the environment from unnecessary man-made radiation problems. They also work on natural sources of radioactivity. Many agencies' companies hire health physicists, such as the Environmental Protection Agency (EPA). For example, an environmental health physicist would measure for radon gas in homes, high radon and radium concentrations in ground water, and highly radioactive soil and rock.

Medical Health Physicist

A medical health physicist focuses on protecting people and the environment from potential radiation problems from devices that use radiation. This professional works in hospitals, clinics, and other major medical centers that use radiation sources in radiology, nuclear medicine, or radiation therapy departments. The medical health physicist is usually the radiation safety officer and monitors radiation exposure of patients, staff, and visitors to the facility. The person also reviews all scientific research that involves radiation in these facilities.

Radiation Therapy Medical Physicist

This branch of medical physics involves using radiation in the treatment of cancers. Hence this field also goes by the names *radiotherapy* and *radiation oncology*. A radiation therapy medical physicist requires working with radiation oncologists to design treatment plans (*teletherapy, brachytherapy,* or *systemic radioisotope*), and monitors the equipment and procedures. This professional also measures and characterizes the radiation and ensures accurate patient dosimetry.

Diagnostic Imaging Medical Physicist

This branch of medical physics involves developing and improving instruments used in imaging the body. These physicists work in many different surroundings with some in clinics and hospitals, some in research hospitals or institutes, and some in research departments at universities.

Some of the machines these medical physicists work with are used for: ultrasound imaging, magnetic resonance imaging (MRI), positron emission tomography (PET) scans, computed tomography (CT) scans, combined PET/MR imaging, combined PET/CT imaging, fluoroscopy, angiography, and mammography.

Nuclear Medicine Medical Physicist

A nuclear medicine medical physicist closely collaborates with physicians to determine the best treatment for patients. This branch of medical physics involves combining radionuclides with chemical compounds or pharmaceutical drugs for both diagnostic and therapeutic applications. The radiopharmaceuticals produced bind to specific organs or cellular receptors within the patient to allow for diagnosis or radiotherapy. Refer to Chapter 19 for more information on radionuclides.

Index

• *O* •

• *P* •

About the Author

Ken Vos, PhD, graduated from the University of Alberta. He has been at the University of Lethbridge since the mid-1990s teaching everything from general interest physics courses about physics and society to introductory level undergraduate physics courses and graduate level physics courses. He won the distinguished teaching award from the University of Lethbridge in 2008 and has been actively involved in outreach with the local school systems. He is passionate about spreading the joy of biophysics. He enjoys all areas of physics from the abstract to the applied. He has done research in the areas of biophysics, condensed matter physics, and mathematical physics.

Dedication

To everyone who loves to learn.

Author's Acknowledgments

I owe a huge thank you to Chad Sievers (Project Editor), Anam Ahmed (Acquisition Editor), Pauline Ricablanca (Production Editor), and everyone at John Wiley & Sons who helped with this book. I also want to thank those people, such as my students, who helped make this book possible.

Publisher's Acknowledgments

Associate Acquisitions Editor: Anam Ahmed

Project Editor: Chad R. Sievers

Copy Editor: Chad R. Sievers

Technical Editor: David Siminovitch, PhD

Art Coordinator: Alicia B. South

Project Coordinator: Kristie Rees

Cover Photos: ©iStockphoto.com/
browndogstudios